Milestones in Computer Science and Information Technology

Milestones in Computer Science and Information Technology

Edwin D. Reilly

An Oryx Book

GREENWOOD PRESS
Westport, Connecticut • London

Library of Congress Cataloging-in-Publication Data

Reilly, Edwin D. #51258496.
    Milestones in computer science and information technology / Edwin D. Reilly.
        p.   cm.
    Includes bibliographical references and indexes.
    ISBN 1–57356–521–0 (alk. paper)
    1. Computer science—Handbooks, manuals, etc.   2. Information technology—Handbooks,
manuals, etc.   3. Computer science—History.   4. Information technology—History.   I. Title.
QA76.R434   2003
004—dc21   2002044843

British Library Cataloguing in Publication Data is available.

Library of Congress Catalog Card Number: 2002044843
ISBN: 1–57356–521–0

First published in 2003

Greenwood Press, 88 Post Road West, Westport, CT 06881
An imprint of Greenwood Publishing Group, Inc.
www.greenwood.com

Printed in the United States of America

The paper used in this book complies with the
Permanent Paper Standard issued by the National
Information Standards Organization (Z39.48–1984).

10 9 8 7 6 5 4 3 2 1

To my grandchildren—
Katie, Ben, Matt, Patrick, Ryan, Meghan,
Phoebe, Jack, Laura, Andy, Henry,
Zachary, Hannah, and Samantha.

Contents

Preface

Of all my books, this one was the most fun to write. This work is not an encyclopedia—the articles are too short for that—nor is it merely a dictionary of the terms and concepts in current use in computing; there are several good exemplars in both categories, and I am closely associated with one of them. This began as a historical work, and indeed the criterion for inclusion of a topic was that it be either something that advanced the progress of computing or, in the case of the theoretical and technologically oriented articles, something that was an essential prelude to a later development. Included herein are 617 "milestones" of **computer science** and **information technology** and 54 definitions of terms used in their description, for a total of 671 articles. Inevitably, topics were declared to be milestones by the subjective opinion of the author. Though there would be considerable overlap, no other author would make exactly the same choices. The most unusual inclusions may be the **Knuth textbooks** and the digit **zero**, decisions that I defend in those articles.

My career has been spent in mathematics, physics, and computer science, but I have also been both a professional editor and an amateur journalist. In the spirit of the latter, I have tried to use the narrative of each article to give the "who, what, when, and where" of the milestone and, in cases where it would seem to be more of a milestone of mathematics or physics,

"why" it is also a computing milestone. In some cases such as **bit**, **byte**, **computer** (as a person), and **software**, the term is not in and of itself a milestone but is discussed because it was coined by an identifiable person. Article titles that are both milestones and creditable neologisms at the time of origin are, among others, **artificial intelligence**, **compiler**, **computer science**, **data mining**, **emulation**, **firmware**, **laser**, **minicomputer**, **software**, and **software engineering**.

In article narratives, the first instance of the name of another article is set in boldface, as is being done in this preface. These boldface terms are what would be *hyperlinks* in an online **hypertext** version of this book, as they may yet become at the discretion of the publisher. But the reader is encouraged to follow them manually in order to place an otherwise isolated milestone in perspective.

Every effort was made to explain specialized notation in simple terms and to keep the articles accessible to the high school student, even those who may not yet have encountered a mathematical topic such as **logarithm**, a very significant milestone in computing. A good way to explore a topic that is far broader than an individual milestone is to consult the Classification of Articles in the back of the book. For example, there are, among many others, 21 articles that pertain to a particular **calculator**, 45 that describe a type of **computer** or computing,

92 that describe a particular **digital computer**, 22 that describe **algorithms**, 56 that describe computer companies, 44 that discuss a particular **programming language**, 32 that pertain to computer **networks**, 26 on the theory of computation, and 24 on **software**. Each article other than some that are merely definitions ends with a recommendation as to *additional reading*, and an extensive list of the sources cited is given as an appendix. Few readers will want to read the book in order from A (**Abacus**) through X, Y, and Z (**Zero**, **Zilog Z-80**, and **Zuse computers**). If you look for a particular topic and do not find it or a **pointer** to a synonymous term, consult the General Index; it may well be discussed in a related article. The Chronological Index, a year-by-year timeline of milestones, and the Geographical Index may also be of interest. And since milestones are made by people, the Personal Name Index contains all of those cited in articles, their countries of birth and accomplishments, and their birth years or life spans when I have been able to learn them.

Most of the milestones covered occurred within the last 50 years, which is just about the span of my working career. I have personally used many of the historic computers and languages described, and I know or knew or worked with many of those cited in the Personal Name Index. My first program was written in 1955 for the **Univac 1103** in raw **machine language**. I first learned about **high-level languages** in a course taught by Grace Hopper. I was at the **General Electric** Knolls Atomic Power Laboratory near Schenectady, New York, when the first version of **Fortran** was delivered for our IBM 704. I wrote one of the early papers on **time sharing**. I had the pleasure of meeting Seymour Cray in his laboratory in Chippewa Falls, Wisconsin, where he was building the **CDC 6600**. For brief periods, I had my hands on the **LARC** and the **Stretch**. My first **personal computer** was a **Radio Shack TRS-80** in 1980, but my friend and coauthor of several books, the late Francis Federighi, kept taunting me about the alleged superiority of his **Apple II** and later **Apple Macintosh** with its fancy display and powerful **mouse**. And sure enough, my own first **Graphical User Interface**, Windows 3.1, radically changed the way I worked. Finally, writing my books would be immeasurably more difficult without the incomparable resources of the **Internet** and the **World Wide Web**. I have just cited more than ten milestones, but of the 617 covered, which are the most significant? For one person's opinion, check out The Top Ten Consolidated Milestones list at the end of the book.

As stated at the onset, this book began as a historical work, and certainly it is at least that. But what gave me such pleasure in making it was the opportunity it afforded to collect, ponder, describe, interrelate, and nostalgically relive all that I knew about the history of computing and, more important, to reveal to me how very much I didn't really know and had to research by virtually emptying certain sections of two public and three university libraries. Now that the book is done, I find myself wishing that I had something like it over the many years that I taught Introduction to Computer Science at the State University of New York at Albany. Every such course uses a textbook that supports the particular high-level language being taught, but more is needed to make the course a true *introduction* to the exciting topics that majoring students will encounter later in their curriculum. Perhaps some of those still teaching introductory courses will assign this book as a supplementary text. Then instructor and student alike could take pleasure in telling me which topics they enjoyed and, so that I may continue to learn too, which topics I have allegedly slighted or miscredited or omitted entirely.

Finally, my thanks to Brian Randell and Eric Weiss for their constructively critical readings of an early draft of this book and to the editors at Oryx for suggesting that I write it. Doing so has given me a chance to publish my own interpretations of events in the history of computing. Making those value judgments put more of me into the book than just my words.

E. D. Reilly
reilly@cs.albany.edu
Niskayuna, New York, USA
June 2003

Alphabetical Listing
of Articles

A

Abacus
CALCULATORS

The *abacus* is an ancient digital computational aid whose use has continued into modern times, especially in Asia. Because it consists merely of multiple beads that slide along wires, an abacus is more of a counter than it is a **calculator**; the **algorithm** that is essential to its workings resides in the head of its operator, not in the device itself. Archeological evidence places the origin of the abacus to the Near East of about 3500 B.C., but it did not reach China until about A.D. 1200, Korea until about 1400, and Japan until about 1600. The Japanese and Chinese abaci differ somewhat. Each contains anywhere from 11 to 20 vertically parallel dowels fixed in a frame with an intermediate horizontal bar that divides beads on the dowels into two compartments, a top compartment called "heaven" and a bottom one called "earth." But where the Chinese abacus, called a *swan pan* (counting board) has two beads in heaven and five on earth, the Japanese soroban uses only one bead in heaven and four on earth. When either is laid flat, beads can be moved from a lower base position to an upper position that indicates a value. A heavenly bead has the value five when raised, and a lower bead one. Thus any column (dowel) of the swan pan can represent any number from 0 (**zero**) to 15, and any column of the soroban any number from 0 to 9.

In 1542, the Welsh mathematician Robert Recorde wrote the first widely used book on arithmetic in English and was the first to use the symbol "=" to mean "equals." Recorde also coined the word *zenzizenzizenzic*, the word in the *Oxford English Dictionary* (*OED*) with more Zs than any other, to refer to the eighth power of a number. The book, which remained in print until about 1700, was essentially an instruction manual on how to use an abacus. On 12 November 1946, Kiyoshi Matsuzake used a soroban to outperform the American Tom Wood, who used the best electrically driven mechanical calculator of that era in a contest of speed and accuracy.

Additional reading: Aspray, *Computing Before Computers*, pp. 7–15.

ABC
COMPUTERS

The *ABC*, or Atanasoff-Berry Computer, named for its inventor, John V. Atanasoff, and his graduate assistant, Clifford E. Berry, was the world's first electronic **digital computer**. Atanasoff conceived the basic plan of the ABC during the 1937–1938 academic year at Iowa State College. A model of the computer's **central processing unit (CPU)** was built in December 1939. Completed in May 1942, the ABC was a binary **parallel computer** designed to solve systems of up to 29 simultaneous linear equations. In practice, however, results were not reliable beyond five equations because of **input / output (I/O)** and other errors. But the ABC's capacitor **memory** and its **vacuum tube** arith-

Figure 1. J. Presper Eckert, co-inventor of the ENIAC, holding a top-of-the line abacus. Could he be checking results from his new electronic machine? (With permission of the estate of Dr. J. Presper Eckert)

metic unit worked perfectly and established the feasibility of electronic computing once and for all. The **switching theory** embodied in the ABC was used throughout the **ENIAC**, and its principles of both regenerative storage and logical switching **adders** were also used in the later **EDVAC**. This technology and the fact that J. Presper Eckert had visited Atanasoff at Iowa State led to the court judgment in the famous **Honeywell** v. **Sperry Rand** suit of the late 1960s that invalidated the ENIAC patents of Eckert and John W. Mauchly and declared that Atanasoff was the true inventor of the electronic digital computer. (*See* http://www.scl. ameslab.gov/ABC/Trial.html for a blow-by-blow description of the litigation.) But Atanasoff's known penchant for secrecy supports the credibility of Eckert and Mauchly's claim that they learned nothing of significance from him. A replica of the ABC was built at Iowa State University by Delwyn Bluhm and George Strawn, the human analogs of John Atanasoff, and Charles Shorb, the counterpart of Clifford

Berry, and was demonstrated to the public on 8 October 1997. Atanasoff, unfortunately, did not live to enjoy seeing it as he died in 1995, the same year in which Shorb, working on the **Intel** Teraflops **supercomputer**, watched it solve a linear system of 125,000 equations in 125,000 unknowns, 124,995 more than the ABC could handle.

Additional reading: John Gustafson, "Reconstruction of the Atanasoff-Berry Computer," in Rojas and Hashagen, *The First Computers*, pp. 91–106.

Abstract data type DATA STRUCTURES

An *abstract data type* is a programmer-defined **data type** that consists of a specification and at least one implementation. For example, the language **C++** does not have an intrinsic data type that supports complex arithmetic, but the programmer may define a **class**, an abstract data type that will thereafter allow the **declaration** of a **variable** as being of type *complex* (or whatever it is named). The specification gives an abstract description of the behavior of instances of the type, independently of any particular implementation. Users of the type need understand only the specification, rather than the implementation, to know how values of the type will behave. For example, a **linked list** can be implemented with data nodes (**records**) and **pointers**, or with an **array** that is allowed to expand and contract during processing. The user need not know which representation is being used. The practice of *encapsulating* (packaging) the implementation with *methods* (processing procedures) that can be used with the abstract data type is called **information hiding**, the principle that forms the core of **object-oriented programming (OOP)** (*see* **Encapsulation**).

Additional reading: *Encyclopedia of Computer Science*, pp. 1–5.

Accuracy DEFINITIONS

Accuracy is freedom from error. **Precision** is another matter. For example, as an approximation to **pi** (π), 22/7 $\cong$ 3.142857143 is *precise* to nine places after the decimal point, but it is *accurate* to only three figures.

ACE / Pilot ACE COMPUTERS

After Alan Turing completed his cryptographic work at Bletchley Hall in the fall of 1945, he accepted a position reporting to John R. Womersley, superintendent of the mathematics division of the National Physical Laboratory (NPL) near London. Although Turing was noted for the theoretical concept known as a **Turing machine**, he wanted to design and help build a real working **digital computer**, one to be called *ACE* (Automatic Computing Engine). By 1946, Turing had completed five successively more ambitious preliminary designs for the ACE, and two more were made later by his colleagues James Wilkinson and Michael Woodger. But Turing, dissatisfied with the lack of progress in turning any one of the designs into reality, left NPL in 1948, returned very briefly, and then left permanently to join Fred Williams and Tom Kilburn at Manchester University. Earlier that year, Harry Huskey had joined NPL for what turned out to be a brief tenure, but he did succeed in forcing a management decision to begin construction of a scaled-down version of the fifth design. In 1949, after Huskey left, Wilkinson and Donald Davies began work on a machine that was at first called just "the test assembly" but was soon given the more whimsical name *Pilot ACE*. Its **ultrasonic memory** consisted of 300 32-bit **words**. The machine contained only 800 **vacuum tubes** but was several times faster than the **EDSAC** for **matrix** arithmetic. The completed computer ran its first successful **program** on 10 May 1950, a small routine called *successive digits* that turned on console lights one by one at a rate determined by a number set on panel switches. The program, usually referred to by its nickname "suck digs," was used for some years thereafter as a maintenance exercise and continued to be used on the English Electric *Deuce*, the successor to Pilot ACE, of which 32 were made. A full-scale ACE was finally built at NPL in 1959, so relatively late that it was essentially obsolete from the first day it ran. Parts of the Pilot ACE are on display at the Science Museum in London.

Additional reading: Williams, *A History of Computing Technology*, pp. 338–346.

ACM. *See* **Association for Computing Machinery (ACM)**.

Acoustic delay line. *See* **Ultrasonic memory**.

Ada LANGUAGES

The **high-level language** *Ada*, named in honor of computer pioneer Charles Babbage's assistant (Augusta) Ada Byron, Countess of Lovelace, was designed to supersede the hundreds of **programming languages** then in use by the U.S. Department of Defense (DoD). The DoD High Order Language Working Group (HOLWG) with David Fisher as Secretariat was chartered in 1975 to formulate the requirements for a **procedure-oriented language (POL)** suitable for use in DoD **embedded systems**. Competitive bidding resulted in selection of the firm Cii **Honeywell**-Bull (*see also* **Machines Bull**) for design and **compiler** construction. Jean Ichbiah became the principal architect of the first version of Ada, which became an ANSI standard in February 1983 and an ISO standard in 1987. An "Ada 9×" project was launched in 1990 to update Ada, with Tucker Taft of the firm Intermetrics as project leader. The objective was to provide improved facilities for handling **object-oriented programming (OOP)** while maintaining compatibility with the existing 1983 Ada. In February 1995, the International Organization for Standardization (ISO) adopted the revision of Ada, and the American National Standards Institute (ANSI) adopted the new standard, called Ada 95, in April 1995. The original became Ada 83, and "Ada" without qualification now generally refers to Ada 95.

Additional reading: *Encyclopedia of Computer Science*, pp. 12–17.

Adder HARDWARE

An *adder* is a device that accepts two **operands** expressed in a **positional number system** and produces their arithmetic sum. The adders in Babbage's **Analytical Engine** and in early **calculators** were mechanical, and the adders in some early **digital computers** were electromechanical, implemented with relays. The adders in current computers consist of electronic **logic circuits**. The simplest adder is the 1-bit *half-adder*, which accepts two **bits** as input and produces as output their 1-bit sum and 1-bit carry for use in the addition stage to the left. A *full-adder* accepts three inputs, one bit from each of the two numbers being added and the carry from the bit position to the right. Thus a serial adder can be made by using a half-adder to add the bits in the rightmost bit position of its operands and full adders for all of the more significant bit positions to the left. The first 1-bit binary adder may have been the one built on a kitchen table in 1937 by George Stibitz using discarded **telephone** relays, old batteries, and part of a tin can. His device was the forerunner of those used in his complex number calculator (*see* **Bell Labs relay computers**). In 1939, Helmut Schreyer built a 10-bit adder and Atanasoff and Berry built a 16-bit adder, for their **ABC**, laying the foundation for binary adders of any length.

Additional reading: *Encyclopedia of Computer Science*, pp. 17–20.

Adobe Systems, Inc. INDUSTRY

Adobe Systems, Inc. was founded in 1982 by Charles Geschke and John Warnock, who left **Xerox PARC** to do so. The company's early growth stemmed from its *Adobe Photoshop* **desktop publishing software**, first for the **Apple Macintosh** and later for **IBM PC** compatibles and from the **page description language** Adobe **PostScript**. Adobe specializes in software solutions for **network** publishing, including Web, print, video, wireless, and broadband applications, and also markets a popular **electronic book (e-book)** reader. Adobe is the second largest **personal computer** software company in the United States, with annual revenues exceeding $1.2 billion. It employs over 2,800 employees worldwide and has operations in North America, Europe, the Pacific Rim, Japan, and Latin America. Adobe's worldwide headquarters are in San Jose, California.

Additional reading: http://www.adobe.com

Advanced Micro Devices (AMD)

INDUSTRY

Advanced Micro Devices (AMD) is the major competitor to **Intel** in the production of **micro-processors** compatible with the **Intel 80×86** series. AMD was founded in 1968 by eight former **Fairchild Semiconductor** employees led by Jeremiah "Jerry" Sanders who served as AMD chief executive officer until 2002 when succeeded by Hector Ruiz. Until a move to Sunnyvale, California, the company's headquarters was the living room of cofounder John Carey. In 1981 AMD renewed and expanded its cross-licensing pact with Intel and a year later signed a technology-exchange pact regarding 80×86 microprocessor technology. But in 1989 AMD sued Intel over the latter's claim that AMD could not sell 80×86-compatible products based on its own technology. In October 1991 AMD started production of the Am386 chip, the equivalent of an Intel 80386, and, after 11 additional months of litigation with Intel, gained the right to sell them. Since then, the AMD Athlon series of 80×86-compatible chips have proved to be strong competitors to the corresponding Intel chips, but Intel remains the sales leader.

Additional reading: *Computing Encyclopedia*, vol. 1, pp. 17–18, 53.

Algol 60 LANGUAGES

Algol 60 was the most significant **programming language** of the 1960s because its **block-structured language** syntax and readable **control structures** became the norm for so many **structured programming** languages of the next 40 years. Algol was the result of a collaboration of American and European committees whose work began in 1956, concurrent with but independent of the **IBM** work on **Fortran**. The objective was to establish a machine-independent notation for **algorithms**, and hence **programs**, usable on any computer for which a suitable **compiler** was written. In 1957, a working group of GAMM (Gesellschaft für angewandte Mathematik und Mechanik) having this goal joined a similar effort by the **Association for Computing Machinery (ACM)**. In June 1958 a GAMM-ACM working conference in Zurich proposed an algebraic language designed not just for programming but also for describing algorithms in journal publications. Several versions of this variant, Algol 58, were implemented, including one with the colorful name **Jovial** (Jules' Own Version of the International Algebraic Language) written by Jules Schwartz.

In January 1960, a 13-person committee met in Paris to discuss and subsequently approve a new Algol version, Algol 60, designed by Peter Naur. Two influential members of the committee were Friedrich Bauer and Klaus Samelson of Germany, coauthors of the book cited as a reference.

As an important by-product of the techniques adopted for handling programs and **procedures**, Algol 60 became the first block-structured language to support **recursion**. The published description of Algol 60 employed a formal notation later called **Backus-Naur Form (BNF)** after John Backus and Peter Naur who devised it. The first Algol 60 compiler was written for the Dutch *Electrologica* computer by Edsger Dijkstra, and Naur himself created an excellent Algol 60 compiler for the GIER, a Danish computer. A popular variant called *Algol-W* was written by Niklaus Wirth. Algol 60 was used to some degree in Europe but very little in the United States, partially because of the widespread success of Fortran and possibly also because Algol has no standard **input / output (I/O)** syntax; its designers apparently assumed that I/O procedures were too machine dependent. Nonetheless, the cleanliness of Algol's design and the principles it embodied served to set a standard that makes its creation a very significant milestone in **computer science**.

Additional reading: Baumann et al., *Introduction to Algol*.

Algol 68 LANGUAGES

Algol 68, the earliest attempt to extend the **Algol 60** of Peter Naur, was also influenced by the languages *Algol-W* and *Euler* created by Niklaus Wirth. This work came to the attention of a working group of the **International Federation for Information Processing (IFIP)**, which chose to follow a different route proposed by Adriaan van Wijngaarden. The result, Algol 68, bears little or no relation to Algol 60 other than its name. The syntax of the language was well documented by van Wijngaarden through use of a notation that he called a "two-level grammar." Unlike much other **software**,

Algol 68 was indeed released in the year implied in its title, the same year in which a minority report was written by a group led by Edsger Dijkstra and C.A.R. Hoare that disparaged the language as being so elaborate that it would be difficult to develop efficient **compilers** for it. The useful life of Algol 68 itself proved to be very short, but van Wijngaarden's notation has had enduring value.

Additional reading: *Encyclopedia of Computer Science*, pp. 34–36.

Algorithm GENERAL

An *algorithm* is a detailed step-by-step description of how to do something. This informal definition encompasses such commonplace tasks as tying a shoelace or baking a cake all the way to the solution of the complex equations that govern the behavior of a nuclear reactor. A **digital computer** is powerless until given an algorithm expressed in the form of a **program**. The **procedures** we learn in grade school for addition, subtraction, multiplication, and division of integers are examples of algorithms that are built into the circuitry of such computers. Essentially, any program written in a **procedure-oriented language** must be translated, usually by a **compiler**, into sequences of arithmetic operations. Some of the now-classical algorithms that are the subjects of separate articles are the **Euclidean algorithm**, the **Fast Fourier Transform**, **Horner's rule**, the **Kalman filter**, the **least squares method**, **Newton's method**, the **Sieve of Eratosthenes**, the **simplex method**, **Simpson's rule**, and **Quicksort**, one of many **sorting** algorithms. The name *algorithm* and its earlier form *algorism* are derived from Algorismus, the Latinized name of the Arab mathematician al-Khowarizmi, who, although he worked in A.D. 850, centuries after development of the earliest algorithms, was the first to publish a systematic study of their properties. To add to al-Khowarizmi's fame, the word *algebra* stems from the second two words of his book *Kitab al jabr w'al-muqabala* (Rules of restoration and reduction). *See also* **Analysis of algorithms**.

Additional reading: Berlinksi, *The Advent of the Algorithm*.

Figure 2. Ninth-century Egyptian mathematician abu Ja'far Mohammed ibn Musa al-Khowarizmi on a Russian stamp of 1983. His enduring legacy is related to the derivation of the terms "algorithm" and "algebra" from the Latinized version of his last name. (From the collection of the author)

Altair. *See* **MITS Altair 8800.**

Alto   COMPUTERS

On 22 November 1972, the Xerox Corporation began the design of a very early **personal computer (PC)** to be used for research. The resulting *Alto* bore the much-shortened name of the Palo Alto Research Center (**Xerox PARC**), where it was developed through the joint effort of Ed McCreight, Chuck Thacker, Butler Lampson, Bob Sproull, and David Boggs. Their goal was to implement Alan Kay's **memex**-like dream of a *Dynabook*, a desktop device powerful enough to drive a high-quality **operating system** for support of **computer graphics**. The Alto consisted of an 8½" × 11" graphics display **monitor**, a **keyboard**, a **mouse**, and a 6 MHz processor and **hard disk** that was housed in a beautifully formed, textured beige metal cabinet. Xerox valued each Alto at $32,000

but nonetheless donated a total of 50 to Stanford, Carnegie-Mellon, and MIT in 1978. An improved version of the Alto's **graphical user interface (GUI)** was used later for the **Xerox Star** of April 1981. The Star was not a commercial success, but its ideas that stemmed from the Alto were adapted by **Apple**, first in the innovative but failed **Apple Lisa** of 1983 and then in the highly successful **Apple Macintosh** of January 1984.

Additional reading: Hiltzik, *Dealers of Lightning: Xerox PARC and the Dawn of the Computer Age*, pp. 167–177.

Alwac III-E   COMPUTERS

The first **minicomputers** are often reputed to be those of the **DEC PDP series** of the early 1960s, but by any reasonable definition of the term, the *Alwac III-E* of 1954 and it contemporary, the **Librascope LGP-30**, were minicomputers. The machines were small enough to fit in a portion of a room of modest size, and they sold for about $80,000, far less than the **mainframes** of their day. *Alwac* is named for its international financial backer, Axel Wenner-Gren, with the final "ac" meaning *Automatic Computer*. The Alwac was manufactured in the United States by Logistics Research, Inc. of Redondo Beach, California, one of whose founders was Glenn Hagen, who had been project director of **MADDIDA** at Northrop Aircraft. The Alwac was a binary computer with **magnetic drum memory** and a combination of **vacuum tube** and crystal diode logic packaged in removable units that made the machine quite easy to maintain. Because of its 32-bit **word length**, the computer was programmed using the **hexadecimal** rather than the **octal number system**, as were several of the **IAS**-class computers of a few years earlier. Accordingly, the content of its **registers** could be displayed on an oscilloscope in 4-bit groups, 1s being taller blips than 0s, and thus easily mentally translated into hexadecimal equivalents for comparison to expectations (*see* reference). In 1955, while at the U.S. National Security Agency (NSA) in Arlington, Virginia, the author served as project director for an Alwac III-E that was used as a **server** to four remote terminal clients, each of which was granted a 15-minute time slice in round-robin fashion (*see* **Client-server computing**).

Additional reading: http://www.abo.fi/natorn/ History/Page20.html

AMD. *See* **Advanced Micro Devices (AMD)**.

Amdahl Corporation   INDUSTRY

The *Amdahl Corporation* was founded in October 1970 by Gene Myron Amdahl, the lead designer of the **IBM 360 Series**, to build **mainframes** that were plug-to-plug compatible with the IBM 370 Series. The Amdahl series computers were the first to use **VLSI** (Very Large Scale Integration) circuitry. The Amdahl 470/V6 was a quarter the size and four times as powerful as the corresponding IBM 370/165, yet it was priced at the same $3.5 million. As Amdahl revenue increased, **IBM** sales dropped noticeably. Eventually, however, Amdahl moved on to pursue other interests through Trilogy and Elxsi, two companies involved in making high-performance **minicomputers**. The Amdahl Corporation, headquartered in Sunnyvale, California, is now a wholly owned subsidiary of **Fujitsu, Ltd**. and continues to build IBM-compatible mainframes.

Additional reading: Slater, *Portraits in Silicon*, pp. 185–193.

Amdahl's law   GENERAL

Amdahl's law places a theoretical limit on the ability to speed up **program** execution using a **parallel computer**. In 1967, Gene Amdahl of **IBM** and, later, the **Amdahl Corporation** stated that the performance gain that can be achieved from parallelizing a program is limited by that part of the program that runs sequentially. Specifically, if, on an N-processor system, s is the fraction of a program that runs *sequentially* and p is the fraction that runs in parallel (i.e., $s + p = 1$), then the overall speedup is given by $(s + p)/(s + p/N) = [s + (1 - s)/N]^{-1}$, which is bounded from above by $1/s$. For example, if 10% of a program runs sequentially, speedup cannot exceed a factor of 10, regardless of the number of processors assigned to it. But in 1987, the Sandia National Laboratories achieved a 1,000-fold speedup on its massively parallel 1024-processor hypercube system. John Gustafson explained this refutation of Amdahl's law by noting that for many practical problems it is the parallelizable part of the program, p, that scales with problem size. Thus if p scales linearly, while s remains constant, speedup becomes a linear function of N, the number of processors, rather than being bounded by $1/s$. This result, sometimes called *Gustafson's law*, is said to vindicate the use of massively parallel processing.

Additional reading: *Encyclopedia of Computer Science*, pp. 960–963.

America Online (AOL)   INDUSTRY

America Online (AOL), originally called *Quantum Computer Services*, was founded by Steve Case in 1985 and began by offering dial-up **modem** service to grant access to the Q-Link **bulletin board system (BBS)** to owners of **Commodore International** computers. In 1988, service was extended via PC-Link to those with Tandy computers (*see* **Radio Shack TRS-80**) and a year later extended further to owners of the **Apple II** and **Apple Macintosh**. By 1995, by which time service could be offered over the **Internet**, AOL had five million subscribers and has now surpassed 35 million. AOL acquired its rival *CompuServe* in 1997 and **Netscape Communications** in 1999 and then merged with Time Warner in 2001 to form *AOL Time Warner*, the largest provider of Internet access via **fiber optic** cable TV. As of early 2003, the merger was much less successful than had been forecast, and Case had left the company.

Additional reading: *Computing Encyclopedia*, vol. 1, pp. 29–30.

Amiga. *See* **Commodore Amiga**.

Analog computer   COMPUTER TYPES

In contrast to **digital computers**, which obtain their results by enumeration (counting), *analog computers* obtain their results by measurement. More specifically, analog devices measure a physical property that changes in proportion to (or *analog*ously to, hence the name) some other quantity of interest. A mercury thermometer, for example, is based on the fact that the height of a liquid in a tube rises and falls through expansion and contraction in accord with the rise and fall of temperature. A **slide rule**, another analog device, multiplies numbers by adding lengths of wood or plastic ruled in proportion to **logarithms**. Other early analog devices were the sundial, astrolabe, **Antikythera mechanism**, and **planimeter**.

In 1876, James Thomson and his brother William (Lord Kelvin) built the **Kelvin tide machine**, a mechanical analog computer that they called a *Harmonic Synthesizer*. The tide machine was a kind of *simulator* (*see* **simulation**), as was the *wind tunnel* devised by the Wright brothers in the early 1900s. In 1909, an analog *Harmonic Analyzer* was built by Otto Mader of Germany. The first large-scale analog computer was the electromechanical **differential analyzer** invented and constructed by Vannevar Bush in 1930. He and his colleagues developed a **vacuum tube** (and hence electronic) version in 1942, four years after the first special-purpose electronic analog computer, *Polyphemus*, was created by George Philbrick. And although it was not learned until after World War II, Helmut Hoelzer of Germany had also built a special-purpose electronic analog computer called the "Mischgerät" over the period 1936–1941, and this became the basis for the first general-purpose electronic analog computer of about 1945.

Because measurement is faster than counting, an analog computation can be faster than the equivalent digital computation, but the latter is more *precise* (yields more significant digits in an answer—*see* **Precision**) and is much easier to program. It is also the case that an analog computer can be simulated on a digital computer through use of certain **problem-oriented languages** developed for that purpose (*see* **Dynamo**; **Simulation**). For these reasons, few analog computers are still in use, though **hybrid computers** are still occasionally made that combine analog and digital methods.

Additional reading: Thomas Lange, "Helmut Hoelzer—Inventor of the Electronic Analog Computer," in Rojas and Hashagen, *The First Computers*, pp. 323–347.

Analysis of algorithms   THEORY

Although antecedents can be found, the *analysis of algorithms* essentially began in about 1835 when computer pioneer Charles Babbage began to wonder how fast certain problems could be solved on his **Analytical Engine**. With respect to solving problems on a modern **stored program computer**, this concern is now typically expressed as seeking to determine how running time or **memory** space needed increases as n, some measure of problem size, increases. For example, in **sorting**, the problem size is the number of objects to be placed in order. A very good sorting **algorithm** runs in $O(n)$, where the "Big O" means "Order of"; that is, for large, n, actual running time $T(n)$ is bounded by $an + b$ where a and b are constants that depend on the intrinsic speed of the computer and how well the sorting algorithm is programmed. The poorest sorting algorithm have running times $O(n^2)$, and those that depend on comparing one item against another have, at best, running time that is $O(n \log n)$. Since Donald Knuth has contributed so much to the analysis of algorithms, it should not be surprising that so much of the **Knuth textbooks** is devoted to the subject. *See also* **Computational complexity**; **NP-complete problems**.

Additional reading: Dewdney, *The Turing Omnibus*, pp. 89–95.

Analytic geometry   MATHEMATICS

In 1637, René Descartes published *Géométrie*, which unified algebra and geometry by showing that each algebraic equation in two **variables** such as $x^2 + y^2 = 25$ has a unique geometric pattern when graphed as y versus x in Cartesian coordinates (a circle of radius 5 in the example given). The same would be true for three or more variables, but beyond three (using three-dimensional x,y,z coordinates) we cannot visualize the shape of the result. Descartes' discovery is a major milestone in mathematics, but it is also a milestone in **computer science** because it underlies all work in **computer graphics**. Analytic geometry is what allows the graphics programmer to transform discrete numbers in computer **memory** into exactly the right **pixels** to illuminate or the right vectors to draw on a **monitor**.

Additional reading: Struik, *A Concise History of Mathematics*, pp. 102–104.

Analytical Engine

COMPUTERS

The *Analytical Engine* of the English polymath Charles Babbage was the first **digital computer** to be conceived. Although it was not successfully implemented, it is a true milestone in that its design incorporated all but one of the major concepts of the modern computers of 1950 onward: it was not a **stored program computer**. Design of the Engine began in 1833. It was to be a decimal mechanical machine whose numbers were held in a *store* and brought to a *mill* for computation. The store was essentially a **memory** containing 1,000 50-digit numbers. Input was to be via **punched cards** similar to those used with the **Jacquard Loom**. Output was intended to be fed to either a card punch or an online printer. Babbage could even be said to have anticipated **word processing**, although his "words" were multidigit numbers. He designed a printer intended for both his **Difference Engine** #2 and his Analytical Engine that would print justified numbers (flush left and right) in from one to three columns in a mixture of two different typefonts. Although Babbage's failure to complete his dream machine was once attributed to the inability of machinists of his day to meet his stringent design tolerances, historians now believe that the more likely cause was Babbage's procrastination based on an unrelenting drive for perfection of design.

Between the Analytical Engine and the **ENIAC** of a hundred years later, the only known attempt—also an unsuccessful one—to build a digital computer rather than a **calculator** was that of the Irish auditor Percy Ludgate, who proposed an "analytic machine" whose design he worked on from 1903 to 1909. Little of its detail is known other than that it would have used **instructions** that consisted of an **operation code** (command) and four **operands**, an anticipation of the **multiple-address computers** of 40 years later.

Additional reading: Babbage, *Passages from the Life of a Philosopher*, Chap. VIII.

Passages from the Life of a Philosopher

Figure 3. A portrait of Charles Babbage, the father of the Analytical Engine. The image appears on the cover of the 1994 reprint of Babbage's famous autobiographical reminiscences edited by Martin Campbell-Kelly. (Courtesy of Rutgers University Press)

Antikythera mechanism

HISTORICAL DEVICES

In 1901, an intriguing and initially mysterious device was recovered from an ancient shipwreck near the island of Antikythera in the Aegean Sea. Now called the *Antikythera mechanism* (or *Calendar*), it is believed to be the work of the Greeks of 80–50 B.C. It is the earliest example of a mechanical **analog computer**, one that contained an elaborate differential gear mechanism that did not reappear for another 1,700 years. Its purpose was to show the exact phase of the moon in its 29.53-day cycle and to feed that information to dials that show the positions of the sun and the moon relative to the stars over an 18-year period of lunar and solar eclipses. The reference shows a picture of a modern reconstruction of the Antikythera made by Aspray.

Additional reading: Aspray, *Computing Before Computers*, pp. 162–164.

AOL. *See* **America Online (AOL).**

Apache
SOFTWARE

Apache is the **software** used by two-thirds or more of the **servers** used by the *Internet service providers* (ISPs) that provide access to the **Internet**. Apache, along with **Linux**, is one of the two most successful and widely used **open source software** products. Apache, written by the Apache Software Foundation (ASF), was derived from a server program called the NCSA HTTP Daemon written in 1995 at the National Center for Supercomputing Applications at the University of Illinois. As of 2002, the chairman of the ASF was Roy Fielding and its president was Brian Behlendorf. The ASF software began as a series of patches to the **source code** of the NCSA server software, which led to the name "Apache," a pun on "a patchy server." Apache was initially used exclusively with **Unix**, but versions are now available for Linux, BeOS, **Microsoft Windows**, and other **operating systems**.

Additional reading: *Computing Encyclopedia*, vol. 1, p. 38.

APL
LANGUAGES

APL, the most ingenious **high-level language** is both a **procedure-oriented language (POL)** and a **functional language**. APL began as a mathematical notation devised by Kenneth Iverson, one whose title is taken from the initials of his 1962 book *A Programming Language*. The notation was originally intended to influence the prevailing mode of mathematical expression and to document computer **algorithms**, not necessarily to be implemented as a **programming language**. Ultimately, however, an **interpreter** was created for APL by Iverson and Adin Falkoff. Despite the fact that an interpreted program usually runs much more slowly than **object code** produced by a **compiler**, APL concedes very little in this regard because each of its **statements** does so much. For example, if *A* is an **array** of 100,000 numbers, the APL expression to sum them is just $+/A$. What Iverson essentially did was to envision the many situations in which a programmer would write a *do-loop*, *for-loop*, or *while-loop* in a conventional language and then

to devise succinct expressions for those situations that obviate the need to write out an explicit **loop**. In APL, many algorithms—such as the **Sieve of Eratosthenes** for creating a list of **prime numbers**, for example—can be written in just a few consecutive characters. This makes APL a good language for use on **parallel computers**, which can process multiple elements of an array concurrently. In fact, the *CDC Star* used a subset of APL as its **machine language**. APL notation also makes an excellent **hardware description language (HDL)**.

One of the criticisms of the language is that it uses a good many Greek letters, and these were difficult to display or feed to the computers of the era in which the language was created. For example, the expression to create an array of consecutive integers from 1 to the length of an existing array *A* is $\iota\rho A$ (iota rho *A*). Iverson and Roger Hui are working to perfect a successor language called *J* that uses the conventional **ASCII** character set.

Additional reading: Iverson, *A Programming Language*.

Apple
INDUSTRY

Apple, incorporated in January 1977 by Steven Jobs and Stephen Wozniak, was the first mainstream vendor of **personal computers (PCs)**. Wozniak, who had been a **Hewlett-Packard (hp)** employee, was a regular attendee of the **Homebrew Computer Club**. By mid-1976, Wozniak had written a **Basic** language **interpreter** for a new **microprocessor**, the **MOS 6502**, and designed a complete **digital computer** to run it. He passed out photocopies of his design to Homebrew friends and helped them build their own machines from the plans. Wozniak and Jobs had collaborated to create the videogame *Breakout* for **Atari**, where Jobs worked. To raise capital for Apple, they sold Jobs's Volkswagen van and Wozniak's **programmable calculator**. Their business plan was written by friend and investor A. C. "Mike" Markkula, who had retired from **Intel** at age 33 and was later president of Apple for a time. Jobs then landed an order for 50 Apple I computers from one of the first computer retail stores. Manufacture began in the garage of Jobs's parents' home in Cupertino, California.

Wozniak immediately began designing a better computer that incorporated a **keyboard**, power supply, color graphics, and Basic. The **Apple II** was introduced in the fall of 1977 at the first West Coast Computer Faire. Of the 130,000 Apple II computers sold by September 1980, an estimated 25,000 were purchased specifically for their ability to run the VisiCalc **spreadsheet** program. The Apple III was less successful, but in 1983, Apple became a Fortune 500 company and recruited Pepsi-Cola's president John Sculley as its new chief executive. The **IBM PC**, introduced in August 1981, brought formidable competition to Apple products. In early 1983, Apple launched its fourth computer, the *Lisa*. The **Apple Lisa** was slow and expensive and failed in the marketplace, but it paved the way for the smaller **Apple Macintosh**, a breakthrough machine. The Macintosh of January 1984 used a **Graphical User Interface (GUI)**, including **icons**, windows, pull-down menus, and a **mouse**, and set new standards for ease of use. In 1985, Wozniak resigned from the company to start a new video electronics business, and Jobs was forced out by Sculley. From 1987 through 1991, Apple continued to introduce new versions of the Macintosh, but its share of the business market was just slightly more than 10%, with low-priced **IBM** clones proving to be extremely tough competition. In March 1990, Apple lost the visionary Jean-Louis Gassee, who had replaced Jobs as the company's technological leader. In October 1991, Apple reached an agreement with IBM. In an alliance called Taligent, the companies planned to create new technologies, especially cross-platform operating systems that both companies believed crucial to their futures. But the alliance foundered, and Taligent became an IBM subsidiary in December 1995. Other parts of the IBM-Apple partnership survived. They sponsored a new family of **reduced instruction set computers (RISCs)** optimized for PCs and entry-level **workstations**. PowerPC chips are now made by **Motorola** and IBM for both Apple Macintosh and IBM computers and successfully marketed by both. They form the basis for the current Apple product line.

In 1992, Apple introduced its *Newton* **per-sonal digital assistant (PDA)**. The device permitted data entry by handwriting recognition and offered note taking, address book, and calendar functions. But the machine's technology proved faulty, especially its initial handwriting recognizer, and the product was discontinued in 1998. Apple's 1995 announcements included an Apple Internet **server** for the **World Wide Web** and PowerBooks and PowerMacs based on the PowerPC chip. In July 1997, Jobs rejoined the Apple board of directors, and a few months later he agreed to become interim chief executive officer, a position he held until January 2000 when he agreed to drop the "interim." In 1998 and 1999, Apple brought G3 and G4 desktops and new PowerBooks to market, followed later with the lower-priced iMac. More than 800,000 were sold in only five months. The iBook that followed, a G3-powered **laptop computer** with a unique clamshell design, also proved very popular. In 2002, Apple gained attention by announcing a redesigned 800 MHz iMac that uses an adjustable 15-inch flat-panel **monitor** attached to a hemispherical base that contains its **central processing unit (CPU)** and rewritable DVD **optical storage** drive. The machine's 10.6-inch desk footprint is the smallest in the industry.

Additional reading: Malone, *Infinite Loop*.

Apple II   COMPUTERS

The 1977 *Apple II*, a **personal computer (PC)** designed by Steve Wozniak of **Apple**, was the successor to the *Apple I* and the product that launched Apple into the mainstream of the industry. Because of its low price, Wozniak had chosen the 1 MHz **MOS 6502** chip for the Apple I and decided to stick with it for the model II. MOS (Metal Oxide Semiconductor) was founded by the 6502's inventor, Chuck Peddle, who had previously helped design the **Motorola** 6800 (*see* **Motorola 68×××series**). The competing **Commodore PET** also used the 6502, but the **Radio Shack TRS-80** used the **Zilog Z-80**. A cassette recorder for **program** and data storage and a color **monitor** were priced separately, but the Apple II could be hooked up to a TV set in lieu of a monitor. Minimum **random access memory (RAM)** was 4K (K = 1,024 **bytes**), expandable to 48K.

An **interpreter** for **Basic** was stored in separate **read-only memory (ROM)** and immediately available upon powering up the computer. Eight expansion slots were a novelty, the first to be incorporated into a personal computer, but it was many months before Apple or any other company made ancillary products with circuit cards that could use them. Wozniak designed the innards of the Apple II, but it was Steve Jobs who assumed responsibility for its external appearance. He wanted the Apple II to have more widespread appeal than just to his fellow members of the **Homebrew Computer Club**. Jobs modeled the case for the Apple after those that **Hewlett-Packard** used for its **calculators**. The computer looked something like a **typewriter** but yet futuristic enough to be a computer. Jobs also hired **Atari** engineer Rod Holt to design a reliable, lightweight power supply that would stay cool, and Holt also helped design the video interface.

Apple raced to have the Apple II ready for the first West Coast Computer Faire in April 1977. Apple's booth was near the entrance and was visible to everybody entering the convention center. They demonstrated a kaleidoscopic **computer graphics** program on a huge Advent display monitor, catching everybody's attention. But, after the Faire, its organizer Jim Warren, editor of *Dr. Dobb's Journal*, did not think that Apple made a strong showing. Nonetheless, the company received about 300 orders for the Apple II, over a hundred more than the total number of Apple I machines sold, and orders for many thousands more over the next few years. The engineering advances that the Apple II incorporated were officially recognized when, some months later, Wozniak was awarded a patent for a **microcomputer** for use with video display, and Holt was given one for a direct current power supply. The Apple II series was extended to include the Apple II+ in 1979, the 64KB Apple IIe with a 32KB ROM in 1983, and the end of the line, the 4 MHz 12MB (megabyte) Apple IIgs in 1986, two years after the introduction of the **Apple Macintosh**.

Additional reading: *Computing Encyclopedia*, vol. 1, p. 40.

Apple Lisa
COMPUTERS

The *Apple Lisa* of 1983 was **Apple's** first attempt to revolutionize computing through use of a **graphical user interface (GUI)** used in conjunction with an **operating system** that supported **multitasking**. The Lisa, based on the 68000 **microprocessor** of the **Motorola 68×××series**, had 1MB (megabyte) of **random access memory (RAM)**, extraordinary for its time; a 10MB **hard disk**; and an 800K (kilobyte) 3.5-inch **floppy disk** drive that replaced an earlier and troublesome 5.25-inch drive. The **hardware** and **software** project leaders for Lisa were, respectively, Richard Page and William Atkinson. Lisa cost Apple $50 million and 200 person years to develop, but its $10,000 purchase price proved too high for it to be sold as a **personal computer (PC)**, and its performance was not sufficient to interest small business users. The much lower cost **Apple Macintosh** that followed Lisa by only a year proved fatal to Lisa marketing efforts, and only a few thousand were sold. Nevertheless, the Lisa is a historic milestone for having attracted so much attention to the GUI and to multitasking that it laid the groundwork for the very successful Macintosh and, later, for the **Microsoft Windows** operating system that **Microsoft** was compelled to develop to maintain and advance its competitive position.

Additional reading: *Computing Encyclopedia*, vol. 2, p. 215.

Apple Macintosh
COMPUTERS

The *Apple Macintosh* ("Mac") was introduced on 24 January 1984, and it quickly revolutionized the way **personal computers (PCs)** were used. The Macintosh design team was initially headed by Jef Raskin, who laid the groundwork in 1980 on a portable computer based on the Motorola 6800 8-bit **microprocessor** with 64K of RAM. The design eventually evolved into the Macintosh of 1984 with its Motorola 68000 32-bit microprocessor and 128K of RAM (*see* **Motorola 68×××series**). When Steve Jobs was removed from the **Apple Lisa** project in early 1983, he took control of the Macintosh project and eventually forced Jef Raskin out. The Macintosh was the first inexpensive commercially available PC to use a **mouse**-driven **Graphical User Interface (GUI)**; its **key-**

board had no arrow keys to move its screen **cursor**. Although similar to the Lisa interface, the Mac used square **pixels**, making it far easier to map graphics to the screen. Its 3.5-inch **floppy disk** drive stored 400KB (kilobytes), 25% more than the 320KB 5.25-inch disks used with the **IBM PC**. The Mac's 8 MHz Motorola 68000 processor was 60% faster than Lisa's 5 MHz Motorola 6800, so the Mac was quite fast in its time. The original Macintosh shipped with only 128KB of **random access memory (RAM)**, but this was augmented by a 64KB **read-only memory (ROM)** that included routines for sound and graphics that otherwise would have reduced the amount of main memory available to users. When memory prices dropped, **Apple** introduced a "Fat Mac" with 512KB of RAM in September 1984. The original Mac, commonly called the 128K, remained on the market until October 1985, leaving the Fat Mac as the only model until January 1986. Over the next eight years, Apple introduced 40 additional versions of the Macintosh. The first Mac to use a color **monitor** was the Mac II of March 1987.

Additional reading: Levy, *Insanely Great: The Life and Times of the Macintosh, the Computer That Changed Everything.*

Applet
SOFTWARE

An *applet*, almost literally a "little application," is a small **program** that is automatically transferred over the **Internet** to the **memory** of a **personal computer (PC)** that invokes a Web page that contains such a program. When an applet is received, it is executed by a **browser**, an action that makes its parent page dynamically active rather than a static image to be viewed or read. (See the example discussed in **Cosmic Cube**.) Not all Web pages contain applets, and originally none did, so one might validly claim that it was the invention of the applet by **Sun Microsystems**, rather than any syntactical feature, that made its **Java** language a milestone.

Additional reading: *Computing Encyclopedia*, vol. 1, p. 41.

APT
LANGUAGES

APT (Automatically Programmed Tools) is a **problem-oriented language (POL)** for the nu-

merical control of machine tools. APT was developed by Douglas Ross at MIT and released in 1957, just slightly after **IBM** delivered **Fortran**. The language was the first to become an ANSI standard.

Additional reading: http://www.nfrpartners.com/nfraptlang.htm

Arithmetic-Logic Unit (ALU)
HARDWARE

The *arithmetic-logic unit (ALU)* is that part of a computer's **central processing unit (CPU)** that carries out arithmetic and the logical operations of **Boolean algebra** on **operands** stored in **memory** or special **registers**. Historically, the first ALU conceived, but not completely implemented, was the *mill* of Babbage's **Analytical Engine** of the mid-1830s. The ALU of a binary computer may be bit-serial, byte-serial, or parallel, depending on how many **bits** are processed simultaneously by its **adder** circuit. In a serial ALU, the adder adds a pair of bits at a time, propagating any carry to the left. In a byte-serial ALU, it adds successive pairs of **bytes**; in a parallel ALU, it adds two full machine **words**. Machines with a variable **word length** necessarily have byte-serial ALUs. Early computers were called either serial or parallel machines in accord with the mode of operation of their ALU. Eventually, all computers were parallel in this sense, and the term **parallel computer** came to mean one that could perform multiple simultaneous operations, such as a vector computer, **pipelined computer**, or multiprocessor (*see* **Multiprocessing**).

Arithmometer
CALCULATORS

The *arithmometer*, the first successful commercial **calculator**, was made in 1820 by Charles Xavier Thomas of France, also known as Thomas de Colmar, a French insurance executive. The arithmometer may have been based on a similar calculator made by Phillip Mathias Hahn of Germany in 1774, almost 50 years earlier. Thomas's version used the stepped-drum principle of the **Leibniz calculator** but had a fully working carry mechanism. Several thousand arithmometers were sold in Europe from 1820 to about 1900. In 1851, Victor Schilt exhibited a similar but key-driven adding machine during the Crystal Palace exhibition in London.

Additional reading: Kidwell and Ceruzzi, *Landmarks in Digital Computing*, pp. 30–31.

ARPAnet   NETWORKS

The *ARPAnet*, a **packet switching** network established in 1969 with funding from the *Advanced Research Projects Agency* (ARPA), a division of the U.S. Department of Defense (DoD), was the forerunner of the **Internet**. The origins of the ARPAnet (and hence the Internet) can be traced to a memo of 25 April 1963 that J. C. R. Licklider addressed to "the members and affiliates of the intergalactic network" and to a 1966 paper by MIT scientists Lawrence Roberts and Thomas Marill entitled "Toward a Cooperative Network of Time-Shared Computers." The project director of implementation of the ARPAnet was Robert Taylor. Its first link connected the Stanford Research Institute to the University of California at Los Angeles (UCLA), with the University of California at Santa Barbara and the University of Utah added to the **network** soon thereafter. By the end of 1970, the ARPAnet became intercontinental with the addition of Harvard and Carnegie Mellon University (CMU). **TCP / IP** was adopted as the official ARPAnet **protocol** in 1982. In 1989, ARPAnet was replaced by the faster 14-node *NSFnet* (National Science Foundation network), and in 1991, NSF opened the NSFnet to commercial use. The name of the expanded network was changed to *Internet* in 1995, and the rest is history.

Additional reading: Waldrop, *The Dream Machine: J. C. R. Licklider and the Revolution That Made Computing Personal*, pp. 259–332.

Array   DATA STRUCTURES

An *array*, a word brought into English from the French *arai* in about 1350, is a regular arrangement of items. In computing, an array is a homogeneous geometric **data structure** whose contiguous elements in a **random access memory (RAM)** may be quickly accessed by *indexing*, one index for one-dimensional (1D) arrays, two for two-dimensional (2D) arrays, and so on. For example, the element of array A that mathematicians write as A_{ijk} has the **Pascal** des-

ignation A[i,j,k]. The **memory** of a **von Neumann machine** is a natural 1D array in the sense that it is a linear sequence of **bytes** or **words** numbered sequentially from 0 to the highest number that need be assigned, usually one less than a power of 2. A **string** is a 1D array of characters. A **matrix** is a 2D array of numbers with properties widely used in scientific computation. The arrays supported as primitive (intrinsic) **data types** in most **high-level languages** from **Fortran** onward are *Cartesian*. For example, a 2D array has the shape of a rectangle, and a 3D array has the shape of a parallelepiped (or "brick"). Triangular arrays and arrays of other geometric shape must be simulated through use of other data structures, for example, by mapping them to Cartesian arrays or representing them with nodes that carry information and **pointers** that point to neighboring nodes.

Additional reading: Reilly and Federighi, *Pascalgorithms*, pp. 270–315.

Array processor. *See* **Parallel computer**.

Artificial Intelligence (AI)

ARTIFICIAL INTELLIGENCE

Artificial Intelligence (AI) is that branch of **computer science** that deals with the attempt to induce general- or special-purpose **digital computers** to exhibit behavior that, if observed in humans, would be deemed intelligent. The field can be said to have originated with Alan Turing's famous paper "Computing Machinery and Intelligence" in the October 1950 issue of the journal *Mind*. In this article, Turing posed the question "Can Machines Think?" and proposed the *Imitation Game* (**Turing test**) to shed light on the subject. But Samuel Butler was thinking about such things as early as 1872 when he wrote in *Erewhon*: "There is no security against the ultimate development of mechanical consciousness in the fact of machines possessing little consciousness now." Another major antecedent was the late 1940s work of Arturo Rosenbleuth and Norbert Wiener that Wiener called **cybernetics**. The term *artificial intelligence* was coined by John McCarthy, but not until a summer conference on the subject held at Dartmouth College in June 1956.

There are three main branches of artificial intelligence. The first, the algorithmic approach favored by McCarthy and his colleague Marvin Minsky, attempts to **program** general-purpose digital computers to perform such tasks as proving mathematical theorems, playing games at a high level of competence, giving advice through **expert systems**, organizing vast amounts of data in a **relational database**, translation of one **natural language** to another, and when the computer is properly equipped with some means of **computer vision**, **pattern recognition**—that of simple shapes at first, human faces eventually. Advocates of the second branch, the construction of **neural nets** that learn from experience, claim that no brute-force machine that merely follows an **algorithm** can ever be truly intelligent; that if all humans did was to follow algorithms they would have no need to be conscious; and that consciousness is an *emergent* phenomenon that arises from the sheer complexity of biological neural networks (*see* **Artificial Life [AL]**). The third branch, use of **genetic algorithms** that support *evolutionary programming*, is currently the most exciting. Presciently, all three approaches were considered by Turing in the paper cited at the beginning of the article. To date, the algorithmic approach has had the most success, but artificial neural networks and genetic algorithms hold greater prospects for success in the long run.

The first AI program is generally considered to be the *Logic Theorist* of Allen Newell, Herbert Simon, and Cliff Shaw, an **automatic theorem proving** program demonstrated at the Dartmouth conference in 1956, and others followed quite quickly. One lament of AI research people is that once they achieve success in a certain area, the feat is no longer considered an example of artificial intelligence. Two examples are a 1961 **computer algebra** program by James Slagle and, in the same year, an MIT program written by a team led by Bert Green for answering questions about baseball statistics. The latter no longer surprises because of the widespread use of **relational database** programs. Progress in **machine translation** has

been very limited, but the Internet **search engine** *Google* now provides useable translation of Web pages from German, French, Spanish, and Italian into English, its major failing being the amusing transliterations of people's names. Perhaps the major success of algorithmic AI has been in the field of **computer games**. Arthur Samuel of **IBM** wrote a checker playing program in 1955 that could routinely beat its creator and which reigned for 22 years until defeated in 1977 by a program called *Paaslow* written by Eric Jensen and Tom Truscott of Duke University. Humans are no match for Paaslow, but development of grandmaster-level **computer chess** has proved much more difficult. In the heady days of the mid-1950s, it was predicted that a computer would be world chess champion within 20 years, but it actually took twice that time until the **supercomputer** *Deep Blue*, a highly **parallel computer** made by IBM, defeated the human world champion Garry Kasparov in a rematch in May 1997 after having failed a year earlier.

Additional reading: Hogan, *Mind Matters: Exploring the World of Artificial Intelligence*.

Artificial Life (AL)   COMPUTER GRAPHICS

Artificial Life (AL), or Alife, a term coined by Christopher Langton in 1982, is "synthetic biology," the attempt to recreate biological phenomena in alternative media. Its goal is an increased understanding of reproduction in nature and an insight into artificial models with a view to improving their performance. In 1953, John von Neumann wondered whether an artificial machine was capable of self-reproduction. In the reference, Levy gives a good description of von Neumann's (positive) answer to his own question. Von Neumann defined a formal reproductive process in the form of a computer **program** interpreted by a **robotic** device. Other biological phenomena studied are models of the origin of life, evolution, cell construction, animal behavior, and ecology. AL programs make extensive use of **computer graphics**. One of the first examples that displayed results with lifelike behavior was John Horton Conway's **Game of Life**, a simple **cellular automaton**.

An example of the second goal is **software** development through *evolution*, as exemplified by the work of John Koza and his associates published in 1999. Koza's method is based on John Holland's research on **genetic algorithms**. Genetic programming involves a population of programs. Generations of programs are formed by evolution so that in time "survival of the fittest" produces programs that are better able to solve a given problem. Evolution proceeds without human intervention; an initial population is generated at random, and evolution continues until a satisfying result is found.

Another process characteristic of Artificial Life is *emergence*, the fact that phenomena at a certain level arise from interactions at lower levels. In physical systems, temperature and pressure are examples of emergent phenomena; an individual molecule possesses neither. In 1974, F. D. Federighi and E. D. Reilly, computer scientists at the State University of New York at Albany, proposed that intelligence and consciousness are emergent properties of complexity and that, more particularly, consciousness is a being's perception that entropy is being lowering through the creation of something new. Devices that follow **algorithms** cannot invent new things and hence have no need of consciousness. **Artificial intelligence (AI)** has traditionally concentrated on the algorithmic **simulation** of complex human behavior; Artificial Life, in contrast, concentrates on the simplest of natural behaviors, emphasizing learning and survivability in complex environments. The English AL researcher Steve Grand, featured in the *Arts & Ideas* section of the 2 February 2002 *New York Times*, has received several awards for his **computer game** called *Creatures* that embodies an artificial universe populated by fascinating creatures called *Norns*. They are featured in his book *Creation: Life and How to Make It* published by Harvard University Press in 2001. Grand is now working to give his robot *Lucy* artificial life.

Additional reading: Levy, *Artificial Life*.

ASCC. *See* **Harvard Mark I (ASCC)**.

ASCII   CODING THEORY

ASCII (American Standard Code for Information Interchange) is a 7-bit character **code** developed by Robert Bemer in 1967. Because computer **word length**s are usually a multiple of the 8-bit **byte**, characters of the original ASCII were commonly embedded in an 8-bit field in which the high-order (leftmost) **bit** was either used as a **parity** bit or was set to **zero**. Now, however, there are various 8-bit ASCII extensions, among which are the **IBM** extended character set, which includes characters useful in **computer graphics**, and ISO 8859, which includes the Latin and most common Western European alphabets. Some form of extended ASCII is now used on all **digital computers**; the once competitive 8-bit **EBCDIC** developed for the **IBM 360 Series** has been phased out or relegated to a secondary option even on IBM computers. But even 8-bit ASCII is capable of encoding only 256 characters; it cannot encompass the large number of symbols needed to represent the combined character sets of the many common **natural languages** that do not use the Latin alphabet or which, like French and German, have an effective alphabet of more than 26 characters because of the use of umlauts and other diacritical marks. For this reason, the 16-bit **Unicode** character set, promulgated in 1988 by an alliance of several computer companies, may ultimately replace ASCII. This would continue the historic progression from the 5-bit 32-character **Baudot Code** of early computers to the 6-bit 64-character BCD code used on second-generation computers, to 8-bit 256-character extended ASCII, to 16-bit Unicode capable of representing 65,536 characters.

Additional reading: Petzold, *CODE*, pp. 286–300.

Assembler   SOFTWARE

An *assembler* is a program that translates **programs** written in **assembly language** into executable **machine language**. The first assembler, created in 1951 by Maurice Wilkes, David Wheeler, and Stanley Gill for the **EDSAC**, was reverentially called "**initial orders**," but it was very primitive compared to the **IBM 650** SOAP assembler written by Grace Mitchell and Stanley Poley of **IBM** and the SAP assembler developed for the **IBM 700 series** by Roy Nutt in the mid-1950s and to modern assemblers with extensive **macro instruction** capabilities.

Additional reading: *Encyclopedia of Computer Science*, pp. 96–103, 1043–1056.

Assembly language   LANGUAGE TYPES

Assembly language is a symbolic language closely related to **machine language**, one in which numeric commands and addresses are replaced by mnemonic symbols. "Mnemonic" means "of or pertaining to an aid to memory." Thus the symbols most often used for arithmetic operations are ADD, SUB, MPY, and DIV, though there are variations from language to language. Programmers do not get to specify these (although some languages permit the definition of synonyms for them), but they can choose symbols of their choice for **operands**. For example, MPY RATE, TIME, DISTANCE could be used on a three-address **computer** (*see* **Multiple-address computer**) as a virtually self-documenting way to show that DISTANCE is computed as the product of a RATE and a TIME. It would then be up to the programmer to make sure that by the time this instruction is encountered in the flow of execution of the program, sensible values for RATE and TIME have been stored at the locations labeled by those symbols. When the symbolic program is thought to be finished, it is translated into machine language by an **assembler**, and the resulting **object code** is tested by running it. If testing reveals a logical error that requires insertion of a new instruction in midprogram, the correction is no more difficult that editing the **source code**—the prior version of the program—and performing a reassembly. This illustrates the greatest benefit of assembly language programming, namely, that whereas a comparable machine language program with the same error would have had to be either completely rewritten or patched with *unconditional jump* statements, the assembly program's instructions and operands are not permanently bound to specific addresses in the computer.

Additional reading: *Encyclopedia of Computer Science*, pp. 96–103, 1043–1056.

Association for Computing Machinery (ACM)   GENERAL

What was once the *Eastern Association for Computing Machinery*, founded in 1947 with John H. Curtiss as first president, dropped the "Eastern" in 1954, the year of its incorporation, to form the *Association for Computing Machinery (ACM)*. Alston S. Householder was its first president under that name. ACM was formed "to advance the art, science, engineering, and application of information technology, serving both professional and public interests by fostering the open interchange of information and by promoting the highest professional and ethical standards." It promotes these goals by sponsoring conferences, meetings, workshops, and symposia and through an extensive publication program. The leading ACM periodicals are its monthly *Communications of the ACM*; its bimonthly *Journal*, which is a **computer science** research journal; the quarterly *Computing Surveys*; and the monthly *Computing Reviews* (*CR*). Technical activities are carried out through 36 Special Interest Groups (SIGs), most of which have their own specialized publication; the SIG subject areas are given on the ACM Website, which contains links to every facet of an ACM transaction that could conceivably be executed online. ACM, now an organization of some 80,000 members, has 125 professional chapters and 420 student chapters, and many of each are outside the United States. ACM headquarters is at One Astor Place, 1515 Broadway, New York, NY 10036–5701.

Additional reading: http://www.acm.org

Associative memory   MEMORY

An *associative memory*, also called a *content-addressable memory*, is accessed in reverse of a conventional addressable **memory**. Rather than specifying an address and asking for its content, one specifies a search value in the form of a number or character **string** and obtains a list of addresses whose content matches the search value. The first associative memory was patented in France in 1956 by Henri-François Raymond, who used one in his CAB 500 **minicomputer** of 1960. Earlier, Raymond had also designed and built **CUBA**, the first **digital computer** made in that country. In those **operating systems** that implement **multiprogramming** through use of **virtual memory**, associative memories are used to find out which *pages* (fixed-length portions of a running program) are already resident in main memory and, through failure of a search, which ones are not. Another application of increasing frequency is the use of an associative memory to minimize references to **auxiliary storage** by quickly finding out what is and is not currently resident in a **cache**.

Additional reading: *Encyclopedia of Computer Science*, pp. 105–106.

Atanasoff-Berry Computer (ABC). *See* ABC.

Atari   INDUSTRY

The *Atari* corporation, named for the Japanese term in the play of Go that is the equivalent of "check" in chess, was founded by Nolan Bushnell and Allan Alcorn in 1972 to market their arcade game *Pong*. Pong (from "Ping Pong") was not the first electronic arcade game, but it was the first commercially successful one. After making a videogame version of Pong for the home, Atari was sold to Warner Brothers in 1976. In 1984, after marketing a string of unsuccessful products under the Atari name, Warner Brothers resold the company to Jack Tramiel, who had just left **Commodore International**, the company that he founded. A revamped Atari then marketed **personal computers (PCs)** and videogames in the face of such stiff competition from Nintendo and others that it was merged, sold, and resold several more times up through 1996, a small equity in some of its games being all that was left.

Additional reading: *Computer Encyclopedia*, vol. 1, pp. 51–53.

Atlas   COMPUTERS

Atlas, originally called MUSE, was the third in a series of **digital computers** designed by a team led by Tom Kilburn at the University of Manchester, in association with **Ferranti, Ltd**. The previous systems were the Ferranti version of the **Manchester Mark I** and the Mark II (Mercury). Design of Atlas began in 1958. Upon its official dedication at Manchester on 7 December 1962, it was considered to be the most powerful computer in the world. Two more were constructed and installed in 1963— at the University of London and at the Atlas

Laboratory, Chilton. All remained in operation until the early 1970s. Atlas, in many ways a **supercomputer**, was the first major system designed for **multiprogramming**. Its composite **memory** consisted of **magnetic core** and **magnetic drums** linked so as to provide the user with the illusion of a one-level memory (*see* **Virtual memory**). Page switching was controlled by a simple swapping **algorithm**. There was also a ferrite rod memory of 8,000 **words** to hold the *supervisor*, the first full-featured **operating system**. **Word length** was 48 **bits**, equivalent to one single-address **instruction** with two modifiers that provided up to 2^{20} (about a million) addresses. Atlas had 128 **index registers**, an unusually high number for that era. Instructions were normally executed at an average rate of two per millisecond, about a hundred times faster than Mercury. The **magnetic tape** system included both half-inch and 1-inch tape drives. **Hard disks** were not initially provided but were added later to the Manchester and Chilton machines. Multiple **input / output (I/O) channels** provided for **punched paper tape** and **punched card** peripherals and line printers. The supervisor, produced by a team led by David J. Howarth, handled the scheduling and streaming of jobs, automatic control of peripherals, detailed job accounting, and a sophisticated level of operator control.

Additional reading: Lavington, *Early British Computers*, pp. 50–52, 83–85.

ATM. *See* Automatic Teller Machine (ATM).

Augmented reality   DEFINITIONS

Augmented reality, a form of **ubiquitous computing**, is a blend of actual and **virtual reality** in which a *head-mounted display* (HMD) used in conjunction with a **global positioning system (GPS)** allows the user to see supplemental information superimposed on a relevant background. The HMD contains electronics that can follow the user's head and eye motion through use of a **digital signal processor (DSP)** running the **Kalman filter** tracking **algorithm**. For example, if the user focuses on the façade of a movie theater a block or so away, he or she might see floating foreground text that gives the titles and times of the films currently playing or, glancing in another direction, the hours of operation and interest rates of a bank. In other instances, a viewer might see a historic but virtual building sitting on an actual site from which it has long since disappeared. The term *augmented reality* was not coined until the early 1990s, by scientists at Boeing Aircraft in Seattle, Washington, but the field has roots going back to Ivan Sutherland's **Sketchpad** system of the 1960s.

Additional reading: Steven K. Feiner, "Augmented Reality: A New Way of Seeing," *Scientific American*, **286**, *4* (April 2002), pp. 48–55.

Autocode   LANGUAGES

Autocode was a primitive **high-level language** devised by Alick Glennie for the **Manchester Mark I** in 1952, and its corresponding "**compiler**" was called *Autocoder*. There is conflicting evidence as to whether the 1952 Autocoder was truly the first compiler or whether the compiler was Glennie himself who first wrote Autocode and then hand-translated it prior to execution. But by March 1954, a true Autocode compiler had been implemented by Tony Brooker.

Additional reading: http://wombat.doc.ic.ak.uk/foldoc.cgi?/Atlas+Autocode

Automata theory   THEORY

Automata theory, more usually rendered as *theory of automata*, is that branch of theoretical **computer science** that deals with the invention and study of mathematically abstract, idealized machines called *automata* (*see* **Automaton**). The automata of interest are abstractions of information processing devices such as **digital computers** rather than mobile devices such as mechanical toys, automobiles, and the machines of interest in **robotics**. There are many results of the form: The languages *recognized* by a particular class of automata are exactly those formal languages generated by a particular class of grammars (*see* **Chomsky hierarchy**). For example, *pushdown automata* are capable of recognizing the valid syntactic classes of all **Algol**-like languages (*see* **Pushdown stack**). **Automatic theorem proving** is also concerned with language recognition. The language to be recognized is the set of propositions derivable from some set of axioms. Most types of automata are special cases of the **Turing machine**. *See also* **Cellular automaton; Finite State Machine (FSM); Linear Bounded Automaton (LBA); Neural net**.

Additional reading: *Encyclopedia of Computer Science*, pp. 112–117.

Automatic control theory   THEORY

Automatic control theory is a part of **cybernetics** that had a long and distinguished history before Norbert Wiener's influential book of 1948. Despite its third word, automatic control theory is a very practical subject. In 1852 when Elisha Graves Otis invented the safety mechanism that made fast high-rise elevators possible, his company still had to make sure that the elevator car could be brought to a smooth stop with its floor aligned with that of its destination without overshooting, undershooting, or disconcerting oscillation. Initially, this had to be done by sharp-eyed human operators with good reflexes, but their function was ultimately automated by Otis control theorists in the first half of the twentieth century. The automatic pilot developed by Bendix for airplanes and the ship-steering mechanism devised by the Sperry Corporation are more recent examples of automatic control. The basic mechanism by which mechanical, electrical, or biological systems maintain their equilibrium is *feedback control*, more particularly, *negative feedback* in which a tendency for a system to stray from balance automatically generates an opposing force that restores equilibrium. Perhaps the most familiar modern examples are the thermostat on our furnace and automobile cruise control. Cruise control, first used by Chrysler in 1958, was invented in 1945 by Ralph Tector, who had been blinded in an accident at age 5. The earliest known negative feedback system was that of Ktesibios of Alexandria who invented a regulator for a water clock in 270 B.C. This regulator performed the same function as the float in a modern flush toilet. A float regulator was also used by Philon of Byzantium in 250 B.C. to keep a constant level of oil in a lamp.

In about 1624, Cornelis Drebbel of Holland, the inventor of the first successful submarine, developed an automatic temperature control system for a furnace and used it in an incubator for hatching chickens. In 1788 James Watt completed the design of a centrifugal flyball governor for regulating the speed of his rotary steam engine. This device employed two pivoted rotating flyballs that were flung outward by centrifugal force. As the speed of rotation increased, the flyweights swung further out and up, operating a steam flow throttling valve that slowed the engine down. Thus, a constant speed was achieved automatically. Since this flyball governor was clearly visible to public observers already fascinated with steam engines, the governor proved to be a sensation and became the first use of feedback control to achieve popular awareness. In 1840, George Biddell Airy, the British Astronomer Royal at Greenwich, developed a feedback device for pointing a telescope. His speed control system turned the telescope automatically to compensate for the earth's rotation. He was the first to discuss the instability of closed-loop systems and the first to use differential equations in their analysis.

James Clerk Maxwell, who also discovered the fundamental equations of electromagnetics that bear his name, provided the first rigorous mathematical analysis of a feedback control system in 1868. Work of the Russian Aleksandr Lyapunov in 1893 is now central in control theory, but it was generally unknown until approximately 1960, when its importance was finally realized. To reduce distortion in repeater amplifiers, Harold S. Black demonstrated the usefulness of negative feedback in 1927. *Regeneration theory* for the design of stable amplifiers was developed by Harry Nyquist, who devised what is now called the *Nyquist stability criterion*.

Stochastic processes (those subject to chance or probability) in control and communication theory were addressed in 1942 by Norbert Wiener, who developed a statistically optimal filter for stationary continuous-time signals that improved the signal-to-noise ratio in a communication system. The Russian Andrei Kolmogorov had provided a theory for discrete-time stationary stochastic processes in 1941. By another decade, implementation of automatic control devices began to switch from the analog

to the digital realm. In the early 1950s, Irmgard Flügge-Lotz became one of the earliest workers in digital automatic control theory and published the important works *Discontinuous Automatic Control* in 1953 and *Discontinuous and Optimal Control* in 1958. In 1961, she became the first woman appointed professor of engineering at Stanford University. In 1957, Richard Bellman applied the algorithmic method known as *dynamic programming* to the optimal control of discrete-time systems, demonstrating that the natural direction for solving optimal control problems is backward in time. His procedure resulted in closed-loop, generally nonlinear, feedback schemes.

The International Federation of Automatic Control was established on 12 September 1957 with Harold Chestnut of the United States as its first president. By 1958, Lev Pontryagin had developed his maximum principle, which solved optimal control problems relying on the calculus of variations developed by Leonhard Euler. In 1960, Charles Draper invented an inertial navigation system that used gyroscopes to provide accurate information on the position of moving ships, aircraft, or spacecraft. Thus, the sensors appropriate for navigation and controls design were developed. By 1970, with the work of Karl Åström of Sweden and others, the importance of digital controls in process applications was firmly established. Åström is the senior author of *Control of Complex Systems* (Springer-Verlag, 2000). All of these milestones laid the foundation for automatic control theory as a vigorous engineering discipline of continuing interest to the industrial and university research centers of today.

Additional reading: Lewis, *Applied Optimal Control and Estimation*.

Automatic programming

PROGRAMMING METHODOLOGY

In the early to mid-1950s, *automatic programming* was a term used by Grace Hopper and her followers to refer to the expression of **programs** in a **high-level language** such as those that corresponded to her A-0 and A-2 **compilers** rather than in **machine language**. In the USSR, Andrei Ershov used the Russian equivalent of the term to describe similar work he did in conjunction with the **BESM** computer. But in this sense, the term was misleading since there was nothing "automatic" about it; programs still had to be hand crafted. By the early 1960s, programming in **Fortran** and **Cobol** was so common that the term fell into disuse. But it was resurrected in the early 1990s by Michael Lowry to mean the production of sensible high-level language **object code** by a program that was much more (artificially) intelligent than a compiler, one whose input was the specification of *what* was to be computed rather than *how* it should be done. The antecedents of this approach are the 1956 *Logic Theorist* of Allen Newell and Herbert Simon, the advocacy of *application generators* by James Martin in 1962, and Simon's *Heuristic Compiler* of 1963. The intermediate stage of modern automatic programming is considered to be the rise of **expert systems** in the mid-1970s and the 1986 work of Thomas Cheatham and others in developing the Knowledge-Based Software Assistant (KBSA) under U.S. Air Force sponsorship. Current emphasis is on development of powerful **problem-oriented languages** such as *SciNapse*, *Amphion*, *Planware*, and *Mastermind*.

Additional reading: *Encyclopedia of Computer Science*, pp. 119–122.

Automatic Teller Machine (ATM)

APPLICATIONS

An *Automatic Teller Machine (ATM)* is a special-purpose **computer** that dispenses cash to the holder of a credit or debit card who supplies the personal identification number (PIN) associated with it. Cash may also be deposited. In either case, the user's account is properly updated at the main office of the bank that issued the card. A crude ATM was patented in 1939 by Luthor Simjian, who also invented the teleprompter, but Alfred R. Zipf of the Bank of America is credited with inventing the "forerunner" of the modern ATM. The first ATM was installed in September 1962 by the People's National Bank of Gouster, Virginia, but the device did not catch on for another ten years. An ATM of different design was patented in 1973 by Donald Wetzel, Thomas Barnes, and George Chastain of the Docutel Corporation. The concept of a debit card was described by Edward Bellamy in his 1988 novel *Looking Backward*. The first credit cards were issued by hotels in 1900, but only to their best customers.

Additional reading: http://inventors.about.com/library/inventors/blatm.htm

Automatic theorem proving

THEORY

Automatic theorem proving is that branch of theoretical **computer science** that uses a **digital computer** to prove theorems based on a set of given axioms. Automatic theorem proving can also be considered as a kind of **artificial intelligence (AI)**. The earliest theorem proving **program**, the 1957 *Logic Theorist* of Allen Newell, Herbert Simon, and Cliff Shaw, proved theorems in **Boolean algebra**. It performed at approximately the level of a fair-to-good college student on the same theorems. A 1959 program by Herbert Gelernter proved geometry theorems at the level of a good high school student. **Program verification**, proving that a program correctly implements its **algorithm**, is also considered to be a kind of theorem proving. A 1971 program of Robert S. Boyer and J Strother Moore, later called *Nqthm* (New Quantified Theorem Prover), proved the correctness of one of the fastest **pattern matching** algorithms, the correctness of a simple **expression** parser, and the correctness of a **prime number** factorization theorem.

In the early 1960s, Paul Gilmore, Hao Wang, Martin Davis, and Hilary Putnam were the first to program a computer to find proofs in the *first-order predicate calculus* (FOPC), the domain of Russell and Whitehead's *Principia Mathematica*. In 1965, J. Alan Robinson published an inference rule that he called the *resolution principle*, a principle more intuitive and easier to use than are the inference rules used by the early FOPC programs. A general theorem proving program called OTTER (Organized Techniques for Theorem-Proving and Effective Research) written in **C** in 1988 by William McCune of Argonne National Laboratory proved to be much faster than previous programs. Through use of discrimination **trees**, it has proved many difficult theorems, some formerly unknown, and in some cases obtained better proofs for existing theorems. The current language of choice for work in automatic theorem proving, also called **logic programming**, is **Prolog**.

Additional reading: Newborn, *Automated Theorem Proving*.

Automation   GENERAL

According to the eleventh edition of the *Encyclopedia Britannica* of 1911, *automation* is derived from the Greek words for "self" and "to seize" and is characterized by

a self-moving machine, or one in which the principle of motion is contained within the mechanism itself. According to this description, clocks, watches, and all machines of a similar kind are **automata**, but the word is generally applied to contrivances which simulate for a time the motions of animal life.

An early example of automation was the flour mill designed by the American inventor Oliver Evans in 1782. In this mill, grain was transported to the second floor of the building by an elevator, a series of buckets mounted on an endless belt that ran inside a closed shaft. The grain was then dumped into a hopper directly above the millstones that ground the grain into flour, which was then deposited in the mill's basement. From there, the flour was carried by another elevator to a third-floor device that stirred the flour with long, toothed arms. Gravity then took the flour to the boulter, a revolving cylinder that separated the flour from the bran. Very little human intervention was required during the mill's operation. (See Figure 4.)

Despite the above, some sources say that *automation* was coined in 1935 by Delmar S. Harder. But without question, the term was popularized from 1952 onward by John Diebold. To *automate* a process is to replace its intermediate steps that involve human intervention with mechanisms that do not. The result of such automation is a process that is said to run *automatically*, just as the word is used to describe an **Automatic Teller Machine (ATM)**. A **calculator** is not automatic because its human operator must invoke its arithmetic operations one by one, and a **digital computer** is automatic only if it is programmable; early ones were not. The earliest significant clerical operations to be automated involved the **UNIVAC** and **LEO** of 1951 and *ERMA* (Electronic Recording Machine Accounting), built by **General Electric** and placed into service for the Bank of America in 1956. Programmable digital computers are *general-purpose machines*. Automation of factory production is implemented through use of *special-purpose machines* that eliminate or minimize the need for human intervention along an assembly line (*see* **Robotics**). The earliest advocates of factory automation were Frederick W. Taylor, the founder of *systems engineering*, the industrialist Henry Ford in the early twentieth century, and W. Edwards Deming, a tireless advocate of quality control into his tenth decade, the 1990s.

Additional reading: *Encyclopedia of Computer Science*, pp. 122–126.

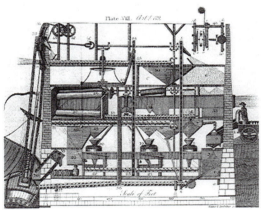

Figure 4. A schematic of a highly automated Evans flour mill of the 1790s. Oliver Evans (1755–1819) believed that a factory should itself be an automatic machine, one to which input—grain in this case—is fed into it at one end and output—flour, say—comes out the other. The image is from Evans's privately printed *The Young Mill-Wright & Miller's Guide*, 1795, plate 8.

Automaton   HISTORICAL DEVICES

An *automaton* is a nonsentient device whose movements simulate a living organism or whose automatic actions produce a result more usually attributed to an intelligent agent. This definition is sufficiently broad to encompass robots (*see* **Robotics**) and **analog** and **digital computers** but also the moving statues of ancient Egypt and China. In 375 B.C. Archytas of

Tarentum (now part of Italy) built a mechanical bird, suspended it from the end of a pivoted bar, and rotated the whole apparatus by means of a jet of steam. All through subsequent history, and especially since the fourteenth century, inventors have sought to create animated machines that reproduce the movements of living creatures. Perhaps the most famous is the 1739 automaton of Jacques de Vaucanson, which, through action of over 400 moving parts, "looked like a duck, walked like a duck, quacked like a duck," and ate and excreted like a duck but, contrary to the adage, was not a duck. Although something of a fraud, Wolfgang von Kempelen's chess-playing automaton was also world famous (*see* **computer chess**). Production of such exquisite mechanisms once reflected the intense desire of philosophic craftsmen to ascertain exactly what it was that made living things "alive." Nowhere is this aspiration better described than in Steven Millhauser's short story "August Eschenburg" from his 1986 collection *In the Penny Arcade*. Even Thomas Edison was fascinated by automata, but although he devoted considerable time and effort in perfecting a talking doll named "Eve," very few were ever sold. The perfection of ever more capable automata is a computing milestone in the sense that the effort is an early attempt at **artificial intelligence (AI)**.

Additional reading: Wood, *Edison's Eve*.

Auxiliary storage

MEMORY

Auxiliary storage is **memory** that supplements the main **random access memory (RAM)** of a **digital computer**. Auxiliary storage generally has more capacity than main memory at the expense of greater time to access its contents. Even the earliest **computers** had some kind of auxiliary storage, most typically **magnetic tape** or a **magnetic drum**. Current auxiliary memory is more likely to be **optical storage** such as CD-ROM and DVD-ROM or magnetic disk memories of one kind or another (*see* **Floppy disk**; **Hard disk**).

Additional reading: *Encyclopedia of Computer Science*, pp. 1137–1144.

AVIDAC

COMPUTERS

AVIDAC (Argonne Version of the Institute's Digital Automatic Computer), a copy of the **IAS computer**, began operation in January 1953 in Argonne, Illinois, at the Argonne National Laboratory operated by the University of Chicago. The machine was named by Donald Flanders and constructed by a team led by Chuan Chu that included Warren Kelleher, Robert Dennis, and James Woody, Jr. AVIDAC was used for problems in reactor engineering and theoretical physics.

Additional reading: http://www.anl.gov/OPA/ frontiers96arch/comphist.html

B

Backus-Naur Form (BNF) GENERAL
Backus-Naur Form (BNF) is a notation for describing the syntax of any context-free language and hence of most **programming languages**. The notation, originally called *Backus Normal Form*, was devised by John Backus in 1959 to describe Algol 58 and modified by Peter Naur in 1960 in order to describe **Algol 60**. To describe **Pascal**, the *Pascal User Manual and Report* of Jensen and Wirth uses a form of pictorial BNF called *railroad diagrams*. In order to allow even more compact syntactical representations, Niklaus Wirth devised an extension to BNF in 1997 called *Extended Backus-Naur Form* (EBNF) that includes the expression syntax of *regular languages* (*see* **Chomsky hierarchy**).

    Additional reading: *Encyclopedia of Computer Science*, pp. 129–131.

Baldwin / Odhner calculator

CALCULATORS
In 1875 Frank Stanley Baldwin invented the first **calculator** that implemented multiplication by repeated addition through a mechanism superior to the stepped wheel of the **Leibniz calculator**. In place of a drum, Baldwin used his newly patented variable-toothed gear in conjunction with a wheel that had nine spring-loaded pins on its edge that eliminated the need to reverse gears to switch from addition or multiplication to subtraction or division. Addition and subtraction required entering two numbers followed by one turn of the handle, either clockwise or counterclockwise, respectively. Multiplication and division was done by turning the handle one way or the other repetitively until the multiplier counted down to **zero** or the dividend became smaller than the divisor. Willgodt T. Odhner, a Swede working in Russia, used the same principles in Europe three years later, and then the German firm *Brunsviga* produced 20,000 Baldwin / Odhner calculators under its label between 1892 and 1912. In 1911, Frank Baldwin joined with Jay Monroe to mass produce his machine as the *Monroe calculator*. As of 1913, Brunsviga calculators were sold under the brand name *Marchant* in the United States.

    Additional reading: Aspray, *Computing Before Computers*, pp. 51–53.

Bandwidth

DEFINITIONS
In the analog realm, "bandwidth" is the range of frequencies that can be sent without appreciable distortion over a transmission line. But in the digital world, *bandwidth* is a measure of the rate of data transfer in units of *baud* or **bits** per second (b/s), which are not necessarily identical. The speed of the fastest **modem** is usually cited as 56Kb/s (56 kilobits per second). Significantly faster devices are rated in megabits per second (Mb/s—millions of bits per second), gigabits per second (Gb/s—billions of

bits per second), and eventually, terabits per second (Tb/s—trillions of bits per second). A single strand of **fiber optic** cable can transmit 400,000 DVD-quality movies simultaneously at a rate of 2.5Gb/s, and a cable can contain as many as 800 strands. In accord with **Moore's Law**, processor speed doubles every two years, but (Scott) McNealy's Law contends that available bandwidth doubles every nine months.

Bar code. *See* **Universal Product Code (UPC)**.

Basic   LANGUAGES

Basic (Beginner's All-purpose Symbolic Instruction Code) is a **procedure-oriented language (POL)** designed and first implemented by John Kemeny and Thomas Kurtz for use on the **Dartmouth time-sharing system**. The goal was a language whose syntax was simpler than that of **Fortran** and thus more suitable for teaching programming to beginning students. The first successful Basic calculation was made at 4 A.M. on 1 May 1964. By 1975, there were at least three implementations of *Tiny Basic*, a version intended to fit in the limited 4K memory of primitive **personal computers (PCs)** such as the **MITS Altair 8800**, one written by the team of Bob Albrecht of the *People's Computer Company* (PCC) and his **computer science** professor Dennis Allison, a second by Li-Chen Wang and the first self-proclaimed *microcomputer consultant* Tom Pittman, and a third by Dick Whipple and John Arnold of Tyler, Texas. Also historically significant were the interpretive versions of Basic written by Paul Allen, Bill Gates, and Martin Davidoff that were stored in the **read-only memory (ROM)** of the **Apple II** and **Radio Shack TRS-80**. Basic has endured to this day because the language has evolved from its original line-numbered **statement** form into versions that fully support **structured programming**. One of the first of these, *True Basic*, was released by Kemeny and Kurtz in 1985 because they had become distraught with the plethora of Basic dialects that they felt were not in keeping with their original philosophy of what a simple programming language should be. *Visual Basic* is now the most popular **visual language** for cre-

ating programs that run under various versions of **Microsoft Windows**.

Additional reading: Lohr, *Go To: The Programmers Who Created the Software Revolution*, pp. 81–98.

Batch processing
                                INFORMATION PROCESSING
In the 1940s and early 1950s when **digital computers** were still something of a novelty, programmers were allowed to sign up for blocks of time to debug their **programs** at the computer's console. But management soon decreed that such practice was a very wasteful use of such a valuable resource and instituted *batch processing* in which a backlog of programs and data for them was "batched" and run sequentially by a professional operator. After that, programmers considered themselves lucky if they received the output of three or four runs per day; an omitted comma in a program could be disastrous. This era lasted until the onset of **Remote Job Entry (RJE)**, which only partially addressed the problem of programmer productivity (or lack thereof), and finally **operating systems** that supported **multiprogramming** and **time sharing**.

Baudot code   CODING THEORY
Morse code is suitable for sending **bits** serially along a wire but less than optimum for encoding data for storage and transmission. The French engineer Jean Maurice Émile Baudot, whose surname later inspired the unit of **data transmission** called a *baud*, devised a telegraphic code in 1874. The *Baudot code* used uniform-length 5-bit codes, each of whose bits could be punched in the form of hole versus no hole across the 5-channel width of a **punched paper tape**. By letting two of the 32 5-bit combinations stand for "letter shift" and "figure shift," the Baudot code can support 62 other **characters** or control codes, enough for both upper- and lowercase letters, all ten digits, and several punctuation marks. Since all 5-bit combinations are legal, no error checking is possible, and the worst error that can occur is one that affects a shift-control character. The 5-bit telegraph code used today is a modified form of the Baudot code devised about 20 years after

the work of Baudot by Donald Murray and called the *International Telegraph Alphabet Number 2* (ITA-2). This is still informally called Baudot code, but it would be more accurate to call it Murray code. Baudot also invented a coding sequence that consisted of a permuted and cyclic sequence of all *n*-bit binary numbers such that each number differed from its predecessor in only one bit position. Such a code is now called a *Gray code* because a device for generating its successive values was patented by Bell Labs scientist Frank Gray in 1953.

Additional reading: Petzold, *CODE*, pp. 288–290.

Bell Labs relay computers   COMPUTERS

The *Bell Labs relay computers* consisted of a series of seven special-purpose **digital computers** built by Bell Telephone Laboratories between 1939 and 1949. The project leader for the initial machine was George Stibitz, and engineering design was done by Samuel Williams and Ernest Andrews. The Model I, which used 400 **telephone** relays, was also called the *Complex Number Calculator* because its **arithmetic-logic unit (ALU)** operated on complex values whose real and imaginary parts were 8-digit **binary-coded decimal (BCD)** numbers. Three teletype terminals allowed multiple but not simultaneous access. The historic first remote access to a computer was dramatically demonstrated at a 1940 meeting of the American Mathematical Society in Hanover, New Hampshire, where a terminal was connected to the Model I at a Bell Labs facility in New York City. A 440-relay Model II was used for development of the M-9 gun director during World War II. The 1,400-relay Models III and IV had a **memory** capacity of only ten complex numbers. Two copies of the Model V, the largest of the series, were built in 1946–1947 and were installed at the U.S. National Advisory Committee for Aeronautics at Langley Field, Virginia, and the Ballistic Research Laboratory at Aberdeen, Maryland. Each 9,000-relay machine had a 30-number memory and hard-wired **floating-point arithmetic**. A rudimentary form of **multiprogramming** allowed the machines to be run unattended throughout the night. The

Model VI, the final machine in the series built in 1949, was a simplified version of the Model V without multiple independent processors, but it did have one of the earliest known **macro** capabilities. None of the Bell machines was a **stored program computer**, although Models V and VI were capable of full conditional branching.

Additional reading: *Encyclopedia of Computer Science*, pp. 135–136.

Bendix G-15   COMPUTERS

Although the Bendix Corporation had earlier produced a **differential analyzer**, the *Bendix G-15* of 1956 was the first **digital computer** marketed by the firm. The computer competed reasonably well with the **IBM 650**, and over 300 were sold. The G-15 was a quite capable computer for its day. It used a **magnetic drum** as **memory** and a combination of **vacuum tubes** and germanium diodes for **logic circuits**. Its magnetic drum contained 2,160 **words** in 20 **channels** of 108 words each; average access time was 4.5 milliseconds. A 108-word buffer channel allowed simultaneous **input / output (I/O)** and computation. **Word size** was 29 bits, allowing single-precision numbers of seven **binary-coded decimal (BCD)** digits plus sign and double-precision numbers of 14 BCD digits plus sign. Each **machine language** instruction specified the address of the **operand** and the address of the next **instruction**. **Software** consisted of an **interpreter** called *Intercom 1000* and a **compiler** called *Algo* that Bendix claimed was the first operational Algol 58 compiler supplied with a commercial machine (*see* **Algol 60**). **Punched paper tape** and, optionally, one to four **magnetic tape** units with 300,000 words per tape reel were used for **auxiliary storage**. The G-15 was superseded in the mid-1960s by the second-generation *Bendix G-20*, which used **transistor** logic and **magnetic core** memory.

Additional reading: http://members.iinet.net.au/~dgreen/g15intro.html

Beowulf   COMPUTER TYPES

In 1993 the U.S. National Aeronautics and Space Association (NASA) supported an R & D project whose goal was to develop a **work-**

station capable of a sustained performance of 1 gigaflops (billions of flops—floating-point operations per second) and that would cost less than $50,000. The intent was to use commercial off-the-shelf (COTS) **software** and **hardware** configured as a **cluster**. As a proof of concept, Thomas Sterling and Don Becker of CESDIS, a NASA subcontractor, assembled a cluster of 16 Intel 80486 **microprocessors** (*see* **Intel 80×86**) in 1994 and called their 70-megaflops system *Beowulf*. The nodes of their cluster were interconnected via a 10Mps (megabits per second) **Ethernet** network. Thereafter, any reasonably similar configuration regardless of the number of nodes came to be known generically as a *Beowulf cluster*. By 1997, NASA personnel met the 1-gigaflops goal by using Pentium microprocessors and a 100 Mbps Ethernet. Beowulf clusters are *scalable*; that is, additional nodes or faster nodes and connectors can be added as financing permits without effect on installed software. Four years later, several thousand such clusters were operating, and in June 2001 a list of the 500 fastest computers in the world included 28 Beowulf clusters. Assembling one is now within the financial ability of a high school, and a 2001 MIT Press publication tells exactly how to do so.

Additional reading: Gordon Bell and Jim Gray, "What's Next in High-Performance Computing?" *Communications of the ACM*, **45**, 2 (February 2002), pp. 91–95.

BESM   COMPUTERS

BESM (Bystrodejstvuyushchaya Ehlektronno-Schetnaya Mashina, meaning "High-Speed Electronic Calculating Machine") was actually a family of computers built in the USSR over the period 1958 to 1995. The chief designer of the series was academician Sergei Lebedev, and the lead designer of **programming languages** for use with it was Andrei Ershov (*see* **Automatic programming**). Only one BESM-1 was made, but several dozen BESM-2 machines were delivered between 1958 and 1962. Single-**precision** numbers were represented in 39-**bit** binary **floating-point arithmetic** using 32 bits for the mantissa, 1 bit for sign, and 6 bits for the signed exponent. A double-precision format was also supported. **Instructions** were 39 bits

long, 6 bits for the **operation code** plus three 11-bit addresses. BESM-2 had 2,048 39-bit **words** of **magnetic core** memory with a 10 μs (microsecond) cycle. **Auxiliary storage** consisted of 5,120-word **magnetic drums** and several **magnetic tape** drives. BESM-2 contained about 4,000 **vacuum tubes**. There was no system **software**. The BESM series concluded with the BESM-6, designed in 1965, delivered in quantity, and used through 1995. This was a highly capable machine with a full complement of system software and **compilers** for all major **high-level languages**.

Additional reading: http://www.icfcst.kiev.ua/museum/Lebedev.html

BINAC   COMPUTERS

The *BINAC* (BINary Automatic Computer), designed and built by J. Presper Eckert and John W. Mauchly in 1948 as a prelude to their **UNIVAC**, was binary in two senses: it used the **binary number system**, and it consisted of dual **digital computers** whose results were compared operation by operation. An error was declared and the machine stopped whenever intermediate results differed. (Had its designers used a third computer and continued calculation if two or more agreed, the computer would have been the first **fault-tolerant computer**, but that honor went to the SAPO six years later.) The BINAC is one of four candidates for having been the first **stored program computer**, the others being the **EDVAC**, the **EDSAC**, and the **Manchester Mark I**.

Additional reading: Lukoff, *From Dits to Bits*, pp. 77–86.

Binary-Coded Decimal (BCD)

CODING THEORY

Binary-Coded Decimal (BCD) is a method of encoding decimal digits on a **binary digital computer**. Since the binary equivalent of decimal 9 is 1001, 4-bit fields (half-**bytes**, or *nibbles*) are needed to stand ready to receive the binary value of any decimal digit from 0 to 9, that is, 0000 to 1001 binary. Thus the three consecutive nibbles that contain 1000 0011 0111 can be said to contain the BCD value of decimal 837. This representation is wasteful in the sense that the largest number that can be encoded in three nibbles is 999, whereas the largest (unsigned) 12-bit binary number (12 consecutive 1 bits) is equivalent to 4,095. Early decimal computers such as the **ENIAC** actually used some form of binary-coded decimal representation in their underlying circuitry, but the representation was not made visible or accessible to the programmer. As the **IBM 360 series** was being designed, **IBM** was concerned that business users of its decimal IBM 702 and 705 computers could not be persuaded to upgrade to its new binary machines unless they had provision for decimal arithmetic as well. Accordingly, the IBM 360 **instruction set** includes commands for the addition, subtraction, multiplication, and division of BCD strings. Historically, there is some ambiguity in the acronym BCD because the **IBM 700 series** and several competitors used a 6-bit "BCD" character code which, since it could encode 64 characters (letters, digits, punctuation, control codes, etc.), should really have been called *binary-code alphanumeric* (BCA) code.

Additional reading: Petzold, *CODE*, pp. 338–339.

Binary number system

NUMBER SYSTEMS

The *binary number system* is a special case of the **positional number system** in which the chosen *radix* (number base) is 2. Only two symbols are needed to express numbers in this system, and the usual choices are 0 and 1. When stored in a computer, the symbols are related to the two possible switching states of the smallest unit of **memory**, the **bit** (a contraction of *binary digit*). Because of its close connection to **Boolean algebra**, the binary system is ideal for use in **digital computers**. For correspondence with that algebra, the symbols 1 and 0 represent *true* and *false*, and suitable circuitry can be devised to operate on any of the four combinations of two bits to implement any desired Boolean function thereof. Since binary numbers are much longer than their decimal equivalents when written out, **hexadecimal** numbers are often used as convenient abbreviations. It takes a minimum of a 4-bit *nibble*, a half-**byte**, to hold the binary equivalent of any one of the 16 hex digits 0 through 15. They are:

Decimal	Hex	Binary	Binary	Hex	Decimal Value
		8 4 2 1	*8 4 2 1*		
0	0	0 0 0 0	1 1 1 1	F	15 or −0
1	1	0 0 0 1	1 1 1 0	E	14 or −1
2	2	0 0 1 0	1 1 0 1	D	13 or −2
3	3	0 0 1 1	1 1 0 0	C	12 or −3
4	4	0 1 0 0	1 0 1 1	B	11 or −4
5	5	0 1 0 1	1 0 1 0	A	10 or −5
6	6	0 1 1 0	1 0 0 1	9	9 or −6
7	7	0 1 1 1	1 0 0 0	8	8 or −7

To inspect the binary values in normal sequence, read down the left column headed *Binary* and then up the column immediately to its right. In accord with the usual rule for positional numbers, the value of any of the 4-bit binary sequences can be ascertained by adding the weight of their 1 bits: 8, 4, 2, or 1 (2^3, 2^2, 2^1, 2^0). Thus 0 1 0 1, for example, does not consist of an 8 or a 2, but it does consist of a 4 and a 1 and hence has the value 5. The second column headed *Binary* is presented in reverse order of value to show that each 4-bit binary number has a corresponding bit-reversed pattern called its *one's complement*. As unsigned numbers, all of the 4-bit strings presented range in value from 0 to 15. But since negative numbers are needed for computational purposes, half of that range could be sacrificed by treating the patterns that begin with 1 (the "sign bit") as representing the negative of their one's complement value. **Complement** arithmetic is normally used only with binary numbers of length 16, 32, or 64 bits.

Any two different symbols can be used to convey binary values. In 1623, the English philosopher Francis Bacon used 5-bit combinations of the letters *a* and *b* to encode the successive letters of the alphabet, the first five being *aaaaa, aaaab, aaaba, aaabb, aabaa*. And the *I Ching* of about 350 B.C. uses all possible 6-bit combinations of broken and unbroken lines to form its 64 hexagrams (see Figure 5). But since neither Bacon nor the Chinese before him *calculated* with binary numbers, credit for discovery of the binary number system goes to Gottfried Wilhelm Leibniz, who described it in

1701, 28 years too late for use with his **Leibniz calculator**.

Additional reading: Ifrah, *The Universal History of Computing*, pp. 86–96.

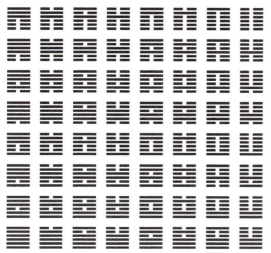

Figure 5. The 64 hexagrams of the *I Ching*. Rotate the page 90 degrees clockwise and consider the little gaps to be 1s embedded in unbroken lines considered as 0s. Then read down the columns from the first through the eighth, column by column, and note the representations of the successive values 0, 1, 2, . . . 63.

Binary tree. *See* **Tree.**

Bit

MEMORY

A *bit* is a digit of the **binary number system**, which is the actual milestone of interest, not one of its constituents. The word is itself a contraction of *binary digit*, and the only sense in which it is worthy of even a short article is the question of its origin. A recent exhaustive search of computer literature found that the earliest published use of *bit* was made in 1946 by John Tukey, a mathematician and statistician who is also credited with coining the word *software* and coinventing the **Fast Fourier Transform (FFT)**. But at least one source claims that Claude Shannon, noted for creation of **information theory**, coined the word as early as 1936.

Additional reading: *Encyclopedia of Computer Science*, p. 306.

Bit map

DEFINITIONS

An image or drawing is said to be *bit-mapped* if it is represented as a two-dimensional **array** of picture elements (**pixels**) each of which is encoded and stored in **memory** as a sequence of **bits**, 1s and 0s. The array itself is a **data structure** called a *bit map*. A black and white image can be encoded in one bit per pixel—0 for white and 1 for black. If four bits (a *nibble*) are used per pixel, an image can be represented in 2^4 or 16 gray levels. If a whole **byte** is used, then its 2^8 combinations can represent 256 different colors, and so on.

Block-structured language

LANGUAGE TYPES

A *block-structured language* is a **high-level language** concept that allows related **declarations** and **statements** to be grouped so as to make **structured programming** possible. The first block-structured language was **Algol 60**, and most **programming languages** thereafter embodied the concept. Block structure allows a sequence of executable statements to be grouped into a single *compound statement*, which may then be used in places where a language normally allows only a single statement (e.g., as the target of either branch of an *if-then-else* statement). Each block introduces a new *scope of variables*, a domain for the definition of symbols. Symbols declared within a block are said to be *local* to that block because they may be accessed only within that block or within a contained subordinate block, thus providing a certain degree of data security. Storage is not allocated for a block's **variables** until the block is entered, and that storage is deallocated when execution of the block is completed. Blocks provide a hierarchical method for controlling access to variables, but the newer concept of a **class** provides even greater data security.

Additional reading: *Encyclopedia of Computer Science*, pp. 151–153.

BNF. *See* Backus-Naur Form (BNF).

Bolt, Beranek, and Newman (BBN)

INDUSTRY

The acoustics consulting firm of *Bolt, Beranek, and Newman (BBN)* was founded in 1948 in

Cambridge, Massachusetts, by MIT professors Richard Bolt and Leo Beranek. They were joined a year later by their former student Robert Newman. In the decade before the company became heavily engaged in the development of **operating systems** and computer **networks**, it consulted in the acoustics design of the United Nations General Assembly Hall and invented ways to measure noise caused by the takeoff and landing of airplanes. The company's entry into computing dates from 1962 when it conducted the first successful demonstration of **time sharing**. A year later, scientists at BBN developed the acoustic coupler, an early **modem**. **LOGO** was developed at BBN in 1966. In 1968, BBN developed **packet switching** and the **protocols** needed to support the **ARPAnet**, the forerunner of the **Internet**. In 1970 BBN programmers wrote the first **virtual memory** operating system in the United States. The first **electronic mail (e-mail)** message was sent from BBN in 1972. In 1973, BBN was hired to analyze the most famous of the Nixon tapes, and Richard Bolt testified as an expert witness that in his opinion its 18-minute erasure could not have been an accident. In 1975, BBN developed the **Telenet** communication network, which later became part of Sprint, and in 1977 developed the TCP protocol for **Unix**-based networks (*see* **TCP / IP**). In 1983, BBN demonstrated IRUS, a **natural language** query **information retrieval** system. Over the next decade, BBN scientists made significant contributions to **speech recognition** and **speech synthesis**. In 1997, BBN was purchased by GTE (General Communications Electronic, Inc.) and became a subsidiary of Verizon when GTE merged with Bell Atlantic. Richard Bolt died in 2002.

Additional reading: http://www.bbn.com

Bombe   HISTORICAL DEVICES

During World War II, the electromechanical *Bombe* was a significant development in the evolution of the Allied cryptanalytic attack on the German *Enigma* rotor machine. From 1928 to 1938, Polish and French cryptanalysts solved both the wiring of the three rotors used at the time and the indicator system and, through the use of a small electrical machine using Enigma

rotors, created a card file that could anticipate the $6 \times 17,576$ possible rotor positions used in setting up the Enigma. In 1938, however, the Germans changed this entire indicator system. To counter this move, Poles under the leadership of Marian Rejewski devised an alternative method of plaintext recovery that could search on command for the correct rotor positions. By November 1938 they had constructed a prototype pseudo-Enigma machine which, because of the loud ticking noises it made as it operated, was nicknamed "The Bombe." Several ersatz "Bombes" of historically accurate appearance can be seen churning away in the 2002 British movie *Enigma*, though its script is almost pure fiction.

In December 1938, the Germans added two more rotors to the three-wheel Enigma. Although the Poles soon solved the wiring, they lacked the resources to proceed further, and the British assumed leadership of the attack on Enigma. Using the Polish and French data, they devised a general method of recovering wheel settings that did not depend on solving the indicator system. Led by Alan Turing, a prototype of an improved Bombe was completed in May 1940. The first fully operational British Bombe was produced in June 1943, and the American Bombe followed in August. That fall, the U.S. Navy sent 600 newly inducted WAVES (Women Accepted for Voluntary Emergency Service) and 200 men to Dayton, Ohio, to help build and train on Bombes manufactured by **National Cash Register (NCR)**. Each Bombe was eight feet wide, seven feet high, two feet deep, weighed 5,000 pounds, and required 64 rotors to be wired to match the rotors actually used in the German Enigma. NCR built a total of 121 Bombes for $6 million and sent them to the Naval Communications Annex where WAVES began operating the machines in shifts, 24 hours a day. A Bombe could test 20,280 rotor settings per second. The Bombes served the Allied cause admirably until the German switch to Lorenz cipher machines caused the British to counter with its **Colossus**, continuing the common historical theme that advances in cryptography often stimulate corresponding advances in computing technology.

Additional reading: Garliński, *The Enigma War.*

Boolean algebra
MATHEMATICS

Boolean algebra is an axiomatic system invented by the English mathematician George Boole in 1847. His system deals with the truth values *true* and *false*, representable in a **digital computer** by the respective **bits** 1 and 0. The unary operator NOT reverses a truth value: NOT *true* is *false* and NOT *false* is *true*. Of the sixteen possible binary operators, the most useful are *A* AND *B*, which is *true* only if both *A* and *B* are *true*; *A* OR *B*, which is *true* if either *A* or *B* or both is *true* (inclusive OR); and *A* XOR *B*, which is *true* if *A* or *B* but not both is *true* (exclusive OR). In the 1860s, the American logician Charles Sanders Peirce showed that Boolean algebra is equivalent to *symbolic algebra* and was just what was needed to analyze the circuits of **switching theory**. The **Jevons logic machine** was built in 1869 by another admirer of Boole's work, and in 1885 Peirce's student Allan Marquand designed, but did not build, an electrical machine to perform simple logical operations. In 1936, Benjamin Burack built a suitcase-sized electrical logic machine that solved *syllogisms*, the kind that begin "All men are mortal. . . ." But it is the Boolean algebra of 1s and 0s that is essential to the construction of a binary **digital computer**. It can be shown that any logic circuit, however complex, can be built with gates that implement the Boolean operations AND, OR, and NOT, or even more simply with only one kind of gate, either NAND (NOT AND, the **complement** of AND) or NOR (NOT OR, the complement of OR).

Additional reading: *Encyclopedia of Computer Science*, pp. 153–159.

Browser
NETWORKS

A *browser* is an interface that enables users to navigate ("surf") the **World Wide Web (WWW)**. The earliest browsers were forms of **text editors**, but current ones take the form of a **Graphical User Interface (GUI)** subordinate to an **operating system** that may itself be a GUI, such as **Microsoft Windows**. A GUI-based browser contains a small rectangular window into which the name of a remote Web page can be inserted. Insertion and activation of a name, called a Uniform Resource Locator (URL), by click of a **mouse** or **trackball** will retrieve and display the desired page. Pages may contain links and activated images that can also be clicked to retrieve other pages. Some browsers use windowing to allow display of portions of multiple pages at the same time.

The concept of a browser is inherent in the invention of the World Wide Web by Timothy Berners-Lee of England in the early 1990s. The *Mosaic* browser written at the National Center for Supercomputing Applications (NCSA) of the University of Illinois and released in February 1993 is often considered the first browser because it is the one that sparked worldwide interest in the Web, but it was no more than the sixth written chronologically, exclusive of some preliminary work at CERN by intern Nicola Pellow and French visitor Jean-François Groff (*see* **Internet**). In April 1992, a team of undergraduates at the University of Helsinki in Finland consisting of Kim Nyberg, Teemu Rantanen, Kati Suominen, and Kari Sydänmaanlakka wrote a browser for the **X window system** that they playfully called *erwise*. (The Finnish abbreviation for their project was OTH, so the full name of the browser thus became *OTHerwise*.) In May 1992, the programmer Pei Wei, who had made a language called *Viola*, wrote another X-based browser called *ViolaWWW*. Then came *Samba* for the **Apple Macintosh**, written by Berners-Lee's colleague Robert Cailliau, and *Arena*, written by Dave Raggett, a CERN visitor from **Hewlett-Packard** of England and Danish staff member Håkon Lie. At about the same time, Tony Jackson of the Stanford Linear Accelerator Center (SLAC) in Palo Alto, California, wrote the X-based *Midas*. Mosaic was next, followed in March 1993 by Lynx 2.0, a text-based browser written at the University of Kansas. The first browser written for Microsoft Windows was *Cello*, written by Tom Bruce. For the story of Marc Andreesen and Eric Bina's *Netscape Navigator* and Microsoft's *Internet Explorer*, the two current Windows browsers that account for all but a small percentage of the "market" (browsers are **freeware**), *see* **Netscape Communications**. A very flexible multiwindow browser called *Opera* has a small but enthusiastic following.

Web pages are plain **ASCII** text files, but to

function as such they must be encoded in a **page description language (PDL)** such as **HTML** or **XML**. Because they are just ASCII files, the encoding and subsequent editing can be done with just a text editor or **word processing** software, but it is now much easier to use a WYSIWYG (What You See Is What You Get) editor to do so. Such an editor is an integral part of the Netscape browser, letting the user toggle between editing and browsing mode. **Microsoft**'s *Internet Explorer* requires a separately sold program called *Front Page* or use of a competitive product like *Adobe Page Mill*. A fine **open source software** browser/editor called *Amaya* is available from W3C, the World Wide Web Consortium.

    Additional reading: *Computing Encyclopedia*, vol. 1, pp. 82–83.

Brunsviga calculator. *See* **Baldwin / Odhner calculator**.

Bubble memory   MEMORY

Bubble memory is a nonvolatile magnetic **memory** invented in 1967 at Bell Labs by a team led by Andrew H. Bobeck. A bubble memory stores **data** in the form of cylindrically shaped magnetic domains in a thin film of magnetic material. The presence of a domain (or "bubble") is interpreted as a binary 1, and absence of a domain as a 0. An external rotating magnetic field propels bubbles through the film. Metallic patterns called *chevrons* deposited on the film steer the domains in the desired directions. Bubble memories are serial high-density storage devices like **hard disk** memories. In disks, however, the stored **bits** are stationary on a moving medium, whereas in the bubble memory the medium is stationary and it is the bits that move. In the absence of power, stored bits ("bubbles") are held intact by the presence of permanent magnets contained within the memory's shielded package. AT&T once used bubble memories to store recorded messages such as, "The number you called is no longer in service." Bubble memory had a short lifetime and has now been almost completely displaced by **flash memory**.

    Additional reading: http://www.xs4all.nl/ ~fjkraan/comp/pc5000/bubble.html

Buffon's needle   ALGORITHMS

In 1733, the amateur French mathematician George-Louis Leclerc, Comte de Buffon, used an analog **Monte Carlo method**, long before that term was coined, to compute a rough value of **pi** (π). What he did was to drop needles of uniform length onto a ruled surface of parallel lines whose spacing was equal to the length of a needle and noting whether the needle did or did not intersect a line (*see* Figure 6). Now, as Buffon knew, the probability of a needle doing so is $2/\pi$. In the diagram, four of the six needles shown, $\frac{2}{3}$ of them, do intersect a line, implying that $\pi = 3$. Of course, six needles is far too few to obtain a good approximation to π; one would have to have the patience to drop the needle a million or so times to obtain a halfway decent value. Buffon's demonstration was little more than a curiosity until the advent of the **stored program computer** made the Monte Carlo method practical (though not for computing π).

    Additional reading: Reilly and Federighi, *Pascalgorithms*, p. 265.

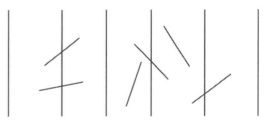

Figure 6. Buffon's needle. A snapshot of a needle-dropping experiment after only six needles have been dropped. Based on the probability that any given needle will intersect a line, the result implies that $\pi = 3$, a crude approximation that would gradually approach a better estimate as additional needles are dropped.

Bug   PROGRAMMING

As used in engineering and computing, a *bug* is an error in a device or **program** that prevents it from functioning properly or from producing a correct answer. Thomas Edison had used "bug" in exactly this context—a glitch in the performance of one of his inventions—in a letter to Theodore Puskas in 1878, and the *Oxford English Dictionary* (*OED*) cites a published 1889 usage that also involved Edison. In September 1945, Grace Murray Hopper, later to be-come a rear admiral in the U.S. Navy, was working on the Mark II at Harvard one night. "Things were going badly," she later wrote. "There was something wrong in one of the cir-cuits of the long glass-enclosed computer. Fi-nally, someone located the trouble spot and, using ordinary tweezers, removed the problem, a two-inch moth. From then on, when anything went wrong with a computer, we said it had 'bugs' in it." Since she and all of Howard Ai-ken's group had been using the term figura-tively throughout their earlier work on the **Harvard Mark I**, Ms. Hopper must have con-sidered the event to be a pun, but the widely reported event caused the use of "bug" in con-nection with computer errors to spread rapidly. Regardless of terminology, finding and remov-ing bugs ("debugging") became a major part of the effort of creating and perfecting a computer program from about that time onward, just as Maurice Wilkes suddenly realized it would be while working on the **EDSAC** in 1947 (*see* Kidwell).

Additional reading: Peggy Aldrich Kidwell, "Stalking the Elusive Computer Bug," *IEEE Annals of the History of Computing*, **20**, *4* (October–Decem-ber 1998), pp. 5–9.

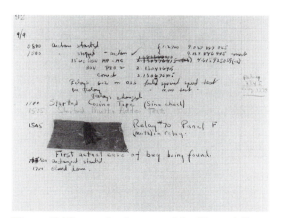

Figure 7. The notebook entry of the Harvard Com-putational Laboratory for 9 September 1945 that doc-uments the first time that a physical rather than a logical bug brought a computer to a screeching halt. The computer was the Harvard Mark II, the bug was a moth, and the (de)mother of computing was Grace Murray Hopper. (Naval Historical Foundation)

Bulletin Board System (BBS)

APPLICATIONS

An electronic bulletin board is a medium for posting announcements and messages of inter-est to a community of online users. The first *bulletin board system (BBS)* was implemented in the 1960s to serve a restricted class of users on the **ARPAnet**, but it was plagued by unre-liable **data transmission**. This problem was solved in 1978 through **software** written by Ward Christensen and Randy Seuss, who coined the contraction "SysOp" for *Systems Operator*. In conjunction with their BBS, Christensen developed the now standard XMO-DEM protocol. XMODEM allows uploading and downloading of **files** without error through unreliable **telephone** lines that are the epitome of the "noisy channels" of concern in Shan-non's **information theory**. Competitive data transmission software called *Kermit* (as in the famous frog) was developed at Columbia Uni-versity in 1981 by Frank da Cruz and Christine Gianone. In some sense, every **World Wide Web** site (*Website*) is now a bulletin board to which its owner can post any information what-soever. Most Web sites are public—anyone knowing their URLs (Uniform Resource Lo-cators) can access them—but some are pro-tected from all but "subscribers" approved by their owners. Most are one-way (read only), but owners can choose to allow others to post mes-sages or data to them by providing them with the relevant *password*.

Additional reading: *Encyclopedia of Computer Science*, pp. 162–164.

Burroughs adding machine

CALCULATORS

In the mid-1880s, the American William S. Burroughs, inspired by the Thomas **Arithmo-meter** and the key-driven **Comptometer** of Dorr E. Felt, invented the first **calculator** that could not only add but also print out a list of the numbers entered and their final total. In 1886 the American Arithmometer Company was formed to market the product. In 1905, the firm was renamed the Burroughs Adding Ma-chine Company. Early mechanical problems were overcome through use of a shock-absorbing dashpot that cushioned the action of the machine's activating lever. The first machines were delivered in December 1892. By 18 January 1908, ten years after Burroughs died, his firm announced that it had sold a total of 63,574 machines of 58 different types. By 1926, the total exceeded a million machines.

Additional reading: *Encyclopedia of Computer Science*, pp. 164, 191.

Burroughs B5000 series   COMPUTERS

The *Burroughs B5000 series* began with the delivery of the B5000, designed by Bob Barton, in December 1962. By a small margin in time, the B5000 was the first large-scale **digital computer** to use zero-address stack-oriented arithmetic (*see* **Pushdown stack**), the other being the **English Electric KDF9** of 1963. This architecture allows fast compilation of algebraic **expressions** because once a **compiler** translates them into reverse **Polish notation**, the result is essentially **machine language**. In fact, with only slight exaggeration, **Burroughs** salespeople claimed that the B5000 had no machine language other than **Algol 60** and that their **Cobol** and **Fortran** compilers were written in that **high-level language**. Another architectural feature of the machine was that the representation it used for **floating-point arithmetic** required no time-consuming conversion from integer to floating-point form, the result of use of a true (unbiased) exponent at the left of a 48-bit word and an integer rather than a fractional *mantissa*. Thus 3 and 3.0, for example, had the same internal representation, and 3.125 could be converted to the integer 3 by simply clearing the exponent field and shifting the mantissa. After reasonable sales success, Burroughs brought out a model 5500 in November 1964, announced a 6500 in February 1967, a scaled-down lower-cost 3500 model in May 1967, and a 6700 in 1969. About 220 B5500s were sold up through 1970. In 1964 an incompatible but still stack-oriented Burroughs B8500 was designed with planned **supercomputer** performance, but this proved too late to compete with the **CDC 6600** or even the Univac 1108 (*see* **Univac 1100 series**). Despite some unfilled orders, the B8500 project died, and the unfortunate result for the industry was that highly efficient zero-address **computer architecture** implemented in **hardware** essentially died with it. But the concept lives on in **software** as the basis for the **byte** code **interpreters** used with UCSD **Pascal** and **Java**.

Additional reading: http://www.cc.gatech.edu/gvu/people/randy.carpenter/folklore/v3n5.html

Burroughs Corporation   INDUSTRY

To produce and market the **Burroughs adding machine**, the American Arithmometer Company (*see* **Arithmometer**) was founded on 20 January 1886 with Thomas Metcalfe as president, William Seward Burroughs as vice president, Richard M. Scruggs as treasurer, and William R. Pye as secretary. The company's initial product line consisted of a single model, a straight adding and listing machine that sold for $475. In 1905, the company was renamed the Burroughs Adding Machine Company. By 1907, the 50,000th Burroughs machine had been manufactured, and over the next two years Burroughs acquired the Universal Adding Machine Company of Missouri and the (William) Pike Adding Machine Company of New Jersey. In 1911, the first model of a Burroughs **calculator** was introduced that bore a striking resemblance to the Felt & Tarrant **comptometer**. By 1926, the one millionth Burroughs machine was produced. In 1951, work began on the Burroughs Electronic Accounting Machine (BEAM), the first **digital computer** development at Burroughs, and a year later Burroughs built an electronic **memory** for the **ENIAC**. In 1953 the company's name was shortened to just the *Burroughs Corporation*. In 1955, the company brought out the first upward-compatible computer series, the B204/205, and sold 112 machines. Sales of the 1955 E101 were even better, with 127 installations. In 1962, Burroughs introduced the very successful **Burroughs B5000**, the first in a line of compatible successors. In 1986, Burroughs bought the Sperry Corporation, the former **Sperry Rand** Univac, for $4.8 billion and named the combined companies **Unisys**. The story continues in that article.

Additional reading: *Computing Encyclopedia*, vol. 1, p. 87.

Bus

A *bus* is a communications pathway in a **digital computer** that allows **data** to flow between the **central processing unit (CPU)** and its **memory** and between and among the CPU and *peripheral* devices. A bus contains one wire for each **bit** needed to specify the address of a device or location in memory, plus additional wires that distinguish among the various data transfer operations to be performed. A bus can transmit data in either direction between any two components of the computing system. The *Unibus*, for example, used on the **Digital Equipment Corporation (DEC)** PDP-11 and **DEC VAX series** computers, had 56 bidirectional lines and a transfer rate of almost 3MB (megabytes) per second. On a **microcomputer**, the bus is usually called an *expansion bus* because its design determines the degree to which the minimum configuration of the system can be expanded with regard to memory, processing speed, sound quality, graphics capability, and peripheral support. *Slots* connected to the bus provide places to plug those cards into, and the bus then provides a mechanism for communicating with them. The **S-100 bus** was a commonly used 8-bit bus for use with early microcomputers that used the **Intel 8080** or **Zilog Z-80** chip. A later 16-bit bus was invented by Mark Dean and Dennis Moeller of **IBM**. To a large degree, the success of the **IBM PC** of 1981 was due to its **open architecture**, use of a flexible expansion bus whose design details were placed in the public domain. The IBM bus became known as the ISA (Industry Standard Architecture) bus. An entire industry flourished in the shadow of IBM, making products that could be placed in the expansion slots of IBM PCs and compatible machines. **Apple** did not recognize the benefits of open architecture until it introduced its NU-Bus in the late 1980s for use with the Mac II. Until then, the **Apple Macintosh** had been a closed machine usable only with Apple peripheral devices. In 1979, Larry Boucher invented the Small Computer Systems Interface (SCSI, pronounced "scuzzy"), an intelligent parallel bus with a standard, device-independent **protocol** that allows multiple peripheral devices to be "daisy chained" (connected in cascade) to the same SCSI port. With the relatively new Universal Serial Bus (USB), devices do not need separate bus interface cards with switches that need to be set correctly; multiple devices can be attached to an external USB socket or *hub*. The USB has a **bandwidth** of 1.5MB/sec, which is shared among all devices connected to it. A still newer technology, **FireWire**, has a **data transmission** rate that is more than 30 times faster.

Additional reading: *Encyclopedia of Computer Science*, pp. 165–167.

Byte

In general, a *byte* is a sequence of **bits** smaller than the length of a computer **word**. According to Fred Brooks, an early **hardware** architect for **IBM** and author of *The Mythical Man-Month*, Werner Buchholz originated the term *byte* in 1956 when working on the IBM **Stretch**. In the reference given below, Buchholz does not claim to have done so, but both he and Brooks were coauthors of the alleged first published use of "byte" in June 1959 in the *IRE Transactions on Electronic Computers*. Usually, the length of a byte is chosen by the computer architect (designer) such that one to eight bytes fit into a word. For example, the 48-bit words of the 1960s vintage **Philco Transac S-2000** and **CDC 1604** could be said to contain eight 6-bit bytes or six 8-bit bytes. For some years, **word lengths** have tended to be a power of two—8, 16, 32, or 64 bits—so that the natural choice for byte length is 8, and this has become a quasi-standard. Also, the 8-bit byte is just the right size to hold a single extended **ASCII** character, or one character from any set of 256 characters. Individual bytes of main **memory** can be accessed in all of the newest computers; on older ones, the programmer had to access the word containing the byte wanted and then use shifting (*see* **Shift register**) or *masking* (*see* **Boolean algebra**) to extract the byte wanted. Some writers call a half-byte, four bits, a *nibble*. A 4-bit nibble is precisely the minimum size needed to hold a **Binary-Coded Decimal (BCD)** digit in those computers such as the **IBM 360 Series** that support decimal arithmetic.

Additional reading: Werner Buchholz, "The Origin of the Word Byte," *IEEE Annals of the History of Computing*, **3**, *1*, p. 72.

C

C
LANGUAGES

C is a general-purpose **block-structured language** developed to implement **Unix**. C was originally designed and implemented by Dennis Ritchie in 1972–1973 for the DEC PDP-11 (*see* **DEC PDP series**), and later enhancements were done in collaboration with Brian Kernighan. C has its roots in an earlier language called BCPL (Basic Combined Programming Language). More specifically, C is the successor to a short-lived BCPL-like language called B that was developed at Bell Labs. C is a traditional **procedure-oriented language (POL)**. Its **data types** and **statements** are straightforward, so C **compilers** are small, and generated **code** is efficient. C relies on library routines for **input / output (I/O)** and other interactions with its **operating system**. The basic data types in C are `char` (usually an 8-**bit byte**); `int`, `short`, and `long`; and `float` and `double` for **floating-point arithmetic**. Integer types come in signed and unsigned variants. An `int` usually corresponds to the natural **word length** of the machine (32 or 64 bits), while `short` is often 16 bits and `long` the longest possible integer. **Strings** are represented as **arrays** of `chars`, usually with a null byte as terminator, and manipulated by library functions. **Pointers** are used to create and access dynamic **data structures**. All C **functions** may be called recursively.

C and its successor **C++** have become the languages of choice for development of systems **software**, and an increasing number of scientific applications are now being written in C in preference to **Fortran**. American and international standards for C and its basic libraries were defined in 1988.

Additional reading: *Encyclopedia of Computer Science*, pp. 171–173.

C++

C++
LANGUAGES

C++ was designed and implemented by Bjarne Stroustrup at the Bell Labs Computer Science Research Center, Murray Hill, New Jersey. Work started in 1979 and culminated with the first commercial release of C++ by AT&T in October 1985. C++ was designed to combine **program** organization as good as that of **Simula**, the efficiency and flexibility of C for systems programming, and increased support for **abstract data types** and **object-oriented programming (OOP)**. After its initial commercial release, Stroustrup continued to refine the language based on user feedback well into the 1990s. An ANSI standard for C++ was promulgated in 1991 and an ISO standard in 1998. C++ itself continues to be widely used for production of systems **software** and as the language of discourse and use in **computer science** courses, and smaller derivatives of C++ such as **Java** and C# (C-sharp) are heavily used to make interactive Web pages for the **Internet**.

Additional reading: Lohr, *Go To: The Programmers Who Created the Software Revolution*, pp. 99–114.

Cache

MEMORY

A *cache*, pronounced *cash*, is a small high-speed buffer **memory** used to hold portions of the contents of some larger memory currently in use. That larger memory is usually main memory, but the principle can also be used for caching **file** lookups in a **directory** or pages on a Web **server**. A cache must be significantly faster (and hence costs more per **byte**) than the memory whose contents it caches and is usually much smaller. Caches are effective because of the *principle of locality*, the fact that information in use is usually either the same that was in use a moment ago or, if not, is information stored quite near to it. Thus a cache operates by retaining copies of segments of the most recently used information and makes use of an **associative memory** to determine whether a needed segment is currently in the cache or needs to be refetched. Cache operation is not noticeable to the user other than through the greater processing speed afforded. The cache principle was first enunciated in 1965 by Sir Maurice Wilkes who, in proposing use of "slave memories," wrote, "By eliminating many data transfers between processor and the main memory, slave memory would allow the system to run within a few percent of the full processor speed at a cost of a few percent of the main memory" (*see* reference below). The slave memory used on the IBM 360/85 was the first to be called a cache, and that is now the prevailing term. Cache memory is now a standard principle of **computer architecture**, and no self-respecting **personal computer (PC)** can be sold without a multiplicity of them.

Additional reading: M. V. Wilkes, "Slave Memories and Dynamic Storage Allocation," *IEEE Transactions on Electronic Computers*, **14** (April 1965), pp. 270–271.

CAD / CAM

APPLICATIONS

CAD / CAM (computer-aided design / computer-aided manufacture) is the use of high-resolution graphics to produce highly detailed engineering drawings that serve as pictorial specifications of a structure or a major part of a structure to be built or manufactured. The concept is a natural outgrowth of the early work in **computer graphics** by Ivan Sutherland and David Evans (*see* **Sketchpad**). The first major commercial CAD / CAM **software**, *CADAM*, was produced by Lockheed in 1965. The leading CAD / CAM software package is *AutoCAD*, produced by *Autodesk*, a company founded in 1982 by John Walker.

Additional reading: *Encyclopedia of Computer Science*, pp. 268–274.

Calculator

DEFINITIONS

A *calculator* is a manually operated computing device. In contrast to a **computer**, it is neither *automatic* nor *programmable*; a human operator must select its arithmetic operations one by one, feeding it appropriate **operands** as a calculation progresses. The minimum calculator can only add and subtract; more capable ones can multiply and divide, and the most advanced ones can do additional operations such as taking a reciprocal or extracting a square root. The earliest mechanical calculator was built in 1623 by Wilhelm Schickard, a colleague of Johannes Kepler. The **Schickard calculator** could add and subtract, but more surprisingly, it could multiply by accessing **logarithms** and antilogarithms through **table lookup**. The next mechanical calculator, the **Pascal calculator** of 1642 called *Pascaline* by its inventor, could only add, but Blaise Pascal invented a gravity-assisted carry mechanism that allowed addition to be performed with far less effort than was the case with the Schickard machine. Then Leibniz went him one better by inventing a *stepped drum* mechanism for his **Leibniz calculator**, which could also multiply. From there on, the story of continually more refined calculators is related in the additional individual articles **Arithmometer** and **Comptometer**, and in the calculator articles beginning with the names **Baldwin / Odhner**, **Burroughs**, **Curta**, **Dalton**, **Friden**, **Grant**, **Stanhope**, and **Steiger**. For descriptions of more obscure calculators, see the translation of Ernst Martin's *Die Rechenmaschinen* as cited in the references.

Additional reading: Aspray, *Computing Before Computers*.

Figure 8. Isaac Newton of England is depicted on more postage stamps—42 at last count—than any other scientist. In his version of calculus, the first derivative with respect to time, t, was called a fluxion and written as $\dot{x}$. Gottfried Wilhelm Leibniz of Germany, in contrast, wrote out the independent variable, t in this example, explicitly, as in dx/dt. Charles Babbage, though himself an Englishman, called Newton's notation dot-age and Leibniz's d-ism, setting up the wordplay in the title of a famous 1811 paper coauthored with his friend John Herschel: "The Principle of Pure Deism, in Opposition to the Dotage of the University," an attack on the alleged uselessness of Newton's fluxions, demanding that his Cambridge colleagues adopt a "real calculus," that of Leibniz, instead. Fans of Newton obtained a small measure of revenge 180 years later with the decision that commercial Internet addresses end in dot com, not dcom. (From the collection of the author)

Calculus

MATHEMATICS

Calculus is the greatest milestone in mathematics, and although it is not a milestone in modern **computer science**, which is so predominantly concerned with **digital computers**, it is nonetheless a milestone with regard to **analog computers**. Calculus, the Latin word for "pebble," was discovered independently by Isaac Newton in 1665 and Gottfried Wilhelm Leibniz in 1673. But Leibniz was the first to publish, in 1684, 20 years before Newton, and it is the Leibniz notation that is the more familiar to us today. Although Newton and Leibniz probably considered it obvious, Isaac Barrow was the first to point out that integration was the inverse of differentiation. The first calculus textbook was published in 1696 by François Antoine Guillaume, Marquis de L'Hospital.

Additional reading: Berlinksi, *A Tour of the Calculus*.

Card Programmed Calculator (CPC)

CALCULATORS

In 1946, **IBM** began to sell its IBM 603 Multiplying Card Punch, an electronic version of a machine that had been offered since 1934. About 100 were sold before an upgrade to the IBM 604, of which 5,600 were sold over the next ten years. But in 1947, an enterprising customer, Northrop Aircraft, connected a multiplying punch to a specially developed storage unit and other **punched card** equipment. The result, the *Card Programmed Calculator (CPC)*, was capable of 1,000 operations per second. It was not a **stored program computer**, but it could add, subtract, multiply, divide, and even extract square roots with a special prewired plugboard. By 1948, IBM had a dozen orders for copies. By 1955, over 700 CPCs had been placed in service over a period during which only 14 **UNIVAC** computers were installed, and IBM's CPC revenue of $21 million in that year was almost double that of its stored program computers. A year later, the ratio would be reversed. The historical significance of the CPC is that although the Watsons always knew that there was a strong market for business **data processing**, it was the surprise success of the machine that made them realize that there was also money to be made by automating scientific and engineering computations.

Additional reading: Bashe et al., *IBM's Early Computers*, pp. 68–72, 333–334.

Cathode Ray Tube (CRT)

I/O DEVICES AND MEDIA

The *Cathode Ray Tube (CRT)* has been the primary image display device for **television** sets for over 70 years and for **digital computers** for over 40. In 1854 Heinrich Geissler of Germany invented a pump that made creation of the first human-made vacuum possible. He then created what became known as a *Geissler tube* for experimentation on conduction in a vacuum. "Cathode rays" were first identified in 1859 by the German physicist Julius Plücker. In 1875, the British chemist William Crookes confirmed their existence by building a tube to display them. English physicist Ambrose Fleming worked with the Crookes tube and found that cathode rays could be deflected and focused. In 1897, German physicist Karl Braun built an oscilloscope to demonstrate how cathode rays could be controlled by a magnetic field. In 1906, German engineer Max Dieckmann tried to create a picture with a Geissler tube. A year later, English inventor A. A. Campbell-Swinton and Russian scientist Boris Rosing independently suggested an electronic scanning system in which a CRT could produce an image on a phosphorus-coated screen. The CRT as we know it is based on the 16 November 1929 patent application of **Radio Corporation of America (RCA)** engineer Vladimir Zworykin, who called it the *Kinescope*. After long and productive use, the CRT is gradually giving way to flat-panel **liquid crystal display (LCD)** screens.

Additional reading: Fisher and Fisher, *TUBE: The Invention of Television*, pp. 114–134.

CDC 1604   COMPUTERS

The *CDC 1604*, designed by Seymour Cray, was the first **digital computer** produced by the **Control Data Corporation (CDC)**, one year after its founding. Both the 1604 and the **Philco Transac S-2000** were 48-bit-word machines introduced in 1958 and thus tied as the first fully transistorized commercially available computers. At a clock cycle of about five microseconds, they were the most powerful computers of their approximate four-year era. The 1604 was one of the first computers used for the support of instruction in **computer science** and was also one of the first to feature a full complement of "standard **software**." **Peripheral** devices included **punched paper tape** and **magnetic tape**, input via teletype or **punched cards**, and line printers. Since software was very primitive and usually loaded from paper tape, a small degree of program control was available through console sense switches.

Additional reading: http://www.cbi.umn.edu/exhibits/cray/cray1604.html

CDC 6600   COMPUTERS

In 1962, Seymour Cray opened a **Control Data Corporation (CDC)** laboratory on property he once owned near Chippewa Falls, Wisconsin, where he designed the *CDC 6600* computer. The machine was announced in 1964 at a price of $7 million. At a clock speed of 100 nanoseconds, the 6600 was the first commercial computer marketed as a **supercomputer**, the first to use Freon refrigerant cooling, and the first commercial computer to use a **cathode ray tube (CRT)** console. CDC checkout engineers created games such as *Lunar Lander* and *Spacewar*, which are thought to be the first **computer games** that used **monitors**. The "first" 6600 was delivered to the U.S. National Center for Atmospheric Research (NCAR) at Boulder, Colorado, in late December 1965 (the 0th version of any powerful new computer, of course, always goes to the National Security Agency [NSA] for cryptographic applications), and two more were delivered the next year to Admiral Rickover's Naval Reactors laboratories, the Westinghouse Bettis Atomic Power Laboratory near Pittsburgh, Pennsylvania, and the GE (**General Electric**) Knolls Atomic Power Laboratory near Schenectady, New York. The CDC 6600 was a large-scale, solid-state, general-purpose **digital computer**. Its **computer architecture** consisted of a **central processing unit (CPU)** with multiple **Arithmetic-Logic Units (ALUs)** and a *peripheral processor* that was time-shared to give the effect of ten separate functional units that handled **input / output (I/O)** operations. Main **memory** consisted of 65,536 60-**bit words** that afforded high-precision arithmetic. The 6600 was a **reduced instruction set computer (RISC)** many years before the term was coined. Input to the computer was by **punched cards** or seven-channel **magnetic tape**. A large **hard disk** and six high-speed **magnetic drums** served as **auxiliary storage**. Several 6600s were sold until the line was phased out in 1977 in favor of its successors, the CDC 7600 and CDC Cyber Series.

Additional reading: *Encyclopedia of Computer Science*, pp. 563–564.

CD-ROM. *See* **Optical storage**.

Cellular automaton COMPUTER TYPES

A *cellular automaton*, or *tessellation structure*, is an *n*-dimensional **array** of processing elements (PEs) that compute in synchronization with their nearest neighbors. An example of one way in which the PEs might operate is described in the two-dimensional **Game of Life**, an example of **artificial life (AL)**. Cellular automata are studied by computer scientists interested more generally in **automata theory**. As so many topics in computing originated, John von Neumann wondered in 1953 whether a cellular automaton could be a *self-reproducing automaton*, then resolved his own question in the affirmative. MIT scientist Edward Fredkin has speculated that the universe may actually be a cellular automaton, and that idea has now been forcefully promulgated by Stephen Wolfram, author of the *Mathematica* **computer algebra** program, in his 2002 book *A New Kind of Science*. Among the distinguished scientists who have written on cellular automata are John Holland, Stanislaw Ulam, John Myhill, Edward F. Moore, and J. W. Thatcher. Others who have done significant work in the field are Charles Bennett, Tommaso Toffoli, and Norman Margolus. The **parallel computers** capable of being programmed as cellular automata include the **ILLIAC IV**, the **Cosmic Cube**, and the **Connection Machine**.

Additional reading: Burks, *Essays on Cellular Automata.*

Cellular telephone COMMUNICATIONS

The first wireless **telephone** of any kind was invented in 1912 by Torigata Uichi of Japan. A *cellular telephone* (cell phone) is a wireless telephone whose voice signals are received and transmitted from tower to tower (cell to cell) between communicating parties. The first cellular telephone system began operation in Tokyo in 1979, and the first system in the United States began operation in 1983 in Chicago. Verizon is a leading producer of cell phones, as is Nokia in Finland, the most wired country in the world. Cell phones containing **digital signal processors (DSPs)** are gradually merging with **handheld computers** that will soon make **ubiquitous computing** worthy of its adjective.

Additional reading: Golambos and Ambrahamson, *Anytime, Anywhere: Entrepreneurship and the Creation of a Wireless World.*

Central Processing Unit (CPU)
HARDWARE

The *central processing unit (CPU)* of a **stored program computer** executes its essential data manipulation, arithmetic, and control tasks. The principal element of the CPU, its **arithmetic-logic unit (ALU)**, performs the arithmetic operations of addition, subtraction, multiplication, and division, as well as the operations of **Boolean algebra** on individual **bits** of binary numbers. It can compare numbers and determine whether one is greater than the other, or whether they are equal. The CPU's *control unit* supervises the functioning of the machine as a whole. It accesses **words** or sequences of **bytes** from **memory** and decides whether to send them to the ALU for processing as data or to an *instruction decoding unit* for interpretation as **instructions**. An important part of the control unit is the *program counter*, whose content normally advances sequentially from instruction to instruction, but the result of a comparison or direct instructional request can reset the counter to effectuate a "jump" to a new address in memory. Memory is not an intrinsic part of the CPU, but the two form the *processor subsystem*. With the advent of **microprocessors**, an entire CPU can now be contained on a single **integrated circuit (IC)** chip. These CPUs are faster and more powerful than those that once required entire cabinets of **hardware**. No names or dates are associated with CPU because its development was necessarily coincident with that of the stored program computer; it is not an independent milestone.

CGI. *See* **Common Gateway Interface (CGI)**.

Channel DEFINITIONS

Channel has a related meaning in both telecommunications and **computer science**, one consistent with the usual English-language definition of a channel as a narrow passageway.

In the former, it is a segment of the electro-magnetic spectrum that has a defined **band-width** allocated to a particular broadcaster, a **television** station perhaps. Within a certain geographic range, one and only one station may use its allocated channel. In the configuration of a **digital computer**, however, a channel is a conduit that may be used by several high-speed **input / output (I/O)** devices at the same time through **multiplexing**, in contrast to a *port*, which is used by only one I/O device at a time.

Chaos theory
THEORY

In 1960, meteorologist Edward Lorenz was using an early **digital computer**, a Royal McBee, to solve equations relating to long-term weather conditions. Because his **program** was subject to interruptions due to the unreliability of that machine, he paused at certain times to print out computed values supposedly accurate to several decimal digits. But when he reentered the data to resume the program at an earlier point in simulated time, he did so to only three digits and was astounded to see what happened. Fairly quickly, the results diverged from what he had obtained before, but they did cluster in a pattern called a *strange attractor*, a term coined by the French scientist David Ruelle in 1971. (There is a beautiful picture of the *Lorenz Attractor* on page 117 of the reference.) Thus began what has now become a thriving field of endeavor called *chaos theory*, a term coined in 1975 by James Yorke and Tien-Yien Li. In the mid-1970s, Mitchell Feigenbaum discovered that there is a universal constant, now called the Feigenbaum number, that is associated with all chaotic systems. This irrational number is approximately 4.6692016090. Initially, there was more art than theory in chaos "theory," but gradually scientists have begun to understand the onset of chaos. French scientist Henri Poincaré had anticipated this in his 1903 essay "Science and Method," when he wrote that "it may happen that small differences in initial conditions produce very great ones in the final phenomena. A small error in the former will produce an enormous error in the latter, and prediction becomes impossible." The effects of "chaos" are sometimes described as "sensitive dependence on initial conditions," or more col-orfully as the *butterfly effect*. The image raised is that the flapping of a butterfly's wings in, say, California can affect the weather days later in Florida. It is now known, of course, that butterflies in Florida can cause chaos in the whole country.

Additional reading: *Understanding Computers: The Puzzle Master*, pp. 108–121.

Character
DEFINITIONS

A *character* is one of the finite set of printable or displayable symbols that comprise a *character set* (*see*, for example, **ASCII**, **EBCDIC**, and **Unicode**). Character sets comprise the uppercase and lowercase alphabets; the digits 0 to 9; punctuation symbols such as the period, comma, colon, semicolon, and hyphen; special characters such as ~, @, #, $, %, ^, &, and *; and in some extended sets, mathematical symbols and accented letters used in languages other than English. A sequence of characters stored in computer **memory** is called a **string**.

Charge-Coupled Device (CCD)
HARDWARE

A *charge-coupled device (CCD)* is the image-sensing **semiconductor** mechanism that is the basis for the **digital camera** and the **optical scanner**. Light rays that strike an **array** of CCDs produce analog signals that, once converted by an *analog-to-digital converter*, can be stored in **memory**, in the **hard disk** of a **personal computer**, perhaps, or on small magnetic disks ancillary to a digital camera. The concept of a CCD was developed by George Smith and Willard Boyle of Bell Labs in 1969, and CCDs were first used in actual products in 1970 by Bell Labs and **Fairchild Semiconductor** in 1974.

Additional reading: *Computing Encyclopedia*, vol. 1, p. 118.

Chess. *See* Computer chess.

Chip. *See* Integrated Circuit (IC); Microprocessor.

Chomsky hierarchy
THEORY

In 1956, the logician Noam Chomsky showed that there are only four basically different forms of grammar, which, in decreasing order of sophistication, he called grammars of Type 0, 1, 2, and 3. Type 0 grammars are called *unrestricted* (or *phrase structure*) grammars and are recognizable only by a **Turing machine**. Type 1 grammars are *context sensitive* and are recognizable by a **Linear Bounded Automaton (LBA)**. Type 2 grammars are *context free*, are recognizable by **pushdown stack** automata, and are of great interest to those who write **compilers** because **high-level languages** come very close to being context free. In **Pascal**, for example, a "**set**" is a "set," regardless of what comes before or after it, whereas in a context-dependent **natural language** like English, "straight set" has a very different meaning than "set straight." Type 3 grammars are called *regular* (or *right linear*) grammars and are recognizable by **Finite State Machines (FSMs)**. Subsequently, results in **automata theory** showed that all computational models (and hence **computers** built in accord with those models), no matter how complex, were equivalent to one or another of the four basic machines that correspond to the levels of the *Chomsky hierarchy*.

An **automaton** is said to be able to *recognize* a grammar if, when given an arbitrary **string** of characters taken from the *alphabet* of the grammar, it is able to determine whether the string is or is not a *well-formed formula*, that is, whether it conforms to the syntactical rules of the grammar. There are both deterministic (ordinary) and hypothetical *nondeterministic* versions of each type of computer in the hierarchy. A nondeterministic automaton acts like an oracle that can instinctively guess the correct answer to a problem (in this context, the precise sequence of steps needed to determine whether a string is well formed) and then proceed to prove that the sequence actually does the job. Nondeterministic automata, if only they existed, would obviously run faster than their deterministic counterparts, but they cannot necessarily recognize languages that are any more complex. For example, it has been proved that nondeterminism adds nothing to the capability of a Turing machine or to a finite-state machine, but it does increase the capability of a pushdown-stack automaton. For example,

a nondeterministic pushdown stack automaton can tell (by guessing when to stop pushing and start popping) that OTTO, RADAR, and ABLE_WAS_I_ERE_I_SAW_ELBA are *palindromes*, but an ordinary pushdown-stack automaton cannot. But, so far at least, no theorist has been able to determine whether a nondeterministic LBA is any more powerful than a deterministic LBA.

Additional reading: Dewdney, *The Turing Omnibus*, pp. 42–48.

Church-Turing thesis   THEORY

When Alan Turing created his **Turing machine** model of computation in 1936, he speculated that anything that could conceivably be computed could be computed by his machine—no more complicated computing device need be invented, no matter how complex the **algorithm** under consideration. This thesis is more of an axiom than a theorem susceptible to proof, but it is almost universally believed. The thesis is often called the *Church-Turing thesis* because Alonzo Church created an equivalent computational model at about the same time. (So did the logician Emil Post, but his name has not been appended *ex Post facto*.) Since a **stored program computer** to which an attendant is willing to add additional units of **memory** as needed during program execution is equivalent to a Turing machine, and hence can execute any algorithm, the Church-Turing thesis and the concept of *algorithm* itself are the two foremost milestones of **computer science**.

Additional reading: Dewdney, *The Turing Omnibus*, pp. 389–396.

Cisco   INDUSTRY

The husband and wife team of Leonard Bosack and Sandra Lerner founded [San Fran]Cisco in 1984 in response to frustrations they experienced as graduate students at Stanford University. University **mainframes** were often unable to handle simple messages passed between incompatible **hardware** systems. Bosack and Lerner developed an electronic box called a **router** that was capable of simultaneous translation. Their router enabled **IBM** computers to talk to **Digital Equipment Corporation (DEC)** computers and identical computers run-

ST. VINCENT & THE GRENADINES 20¢

1937: Alan Turing's theory of digital computing

Figure 9. St. Vincent & The Grenadines honors the (Church-)Turing thesis, one of the two most significant milestones of computer science. (From the collection of the author)

ning different **operating systems** to intercommunicate. A router can cost upwards of $50,000, but they are indispensable to **Internet** infrastructure. In 1990, the very year in which Cisco went public, Bosack and Lerner were forced out by the professional managers that venture capitalists had installed. But Cisco sold 5,000 routers that year, and by 1998 sales were running close to a million units. Cisco is the third most valuable NASDAQ company after **Microsoft** and **Intel** with 2001 revenues of $18 billion.

Additional reading: Chandler, *Inventing the Electronic Century*, pp. 172–173.

Class   PROGRAMMING

The concept of *class* originated in 1967 with the programming language **Simula 67** as an alternative to the **procedure** mechanisms of **block-structured languages** such as **Algol 60**. The class concept is now used in **C++** and **Java** as a mechanism for **information hiding** and the implementation of **abstract data types**. A Simula 67 class resembles an **Algol 60** pro-

cedure and, like a procedure, may also have formal **parameters**. A class is defined through **declaration** of the variables, functions, and procedures that are local to the class, and is followed by the body of the class which is usually a block of **high-level language** program **code**.

Additional reading: *Encyclopedia of Computer Science*, pp. 212–214.

Client-server computing   NETWORKS

The term *client-server computing* was first used in the 1980s in reference to **time-sharing** networks in which the remote outposts, the *clients*, are intelligent **personal computers (PCs)** rather than merely "dumb terminals" with no processing capability of their own. The host computer of the **network** is called the **server** since it responds to inquiries or other requests for service from its clients. Thus client-server computing is a form of **distributed system**, that is, one in which the workload and relevant data are shared among multiple interconnected computers, most of which are clients and one or more of which are servers. But since clients also have processing capability, a server and a client can exchange roles during certain transactions. **Local Area Networks (LANs)**, **intranets**, and the **Internet** all qualify as client-server computing networks.

Additional reading: *Encyclopedia of Computer Science*, pp. 215–218.

CLU   LANGUAGES

CLU, created by an MIT group led by Barbara Liskov, was the first **programming language** to provide direct linguistic support for *data abstraction* (*see* **Abstract data type**). CLU is an **extensible language** whose base syntax looks like **Algol 60** but whose semantics is closer to that of **Lisp**. The design of CLU began in 1973, but its first production **compiler** was not released until 1980.

Additional reading: Bergin and Gibson, *The History of Programming Languages*, pp. 471–510.

Cluster   COMPUTER TYPES

A *cluster* is a collection of interconnected **digital computers** used as a single computing resource whose nodes share a common **memory**.

Clusters have been used since the 1970s; Tandem, Inc. marketed a 16-node uniprocessor in 1975. But the term *cluster* was not coined until 1983, when the **Digital Equipment Corporation (DEC)** used it to refer to a collection of **software** and **hardware** that made several **DEC VAX series** computers appear to be a single **time-sharing** system called the VAX-cluster. That same year, Teradata made a 1,024-node **database** machine. In the early 1990s, many scientific and engineering users began replacing expensive **supercomputers** with clusters of **workstations**. **IBM's** *Deep Blue*, which defeated the reigning human chess champion Garry Kasparov on 11 May 1997, was a cluster with special-purpose added hardware (*see* **Computer chess**). Clusters are usually formed by grouping a specific number of computers in close proximity, but clusters of a variable number of remotely separated computers are often formed as described in **cooperative computing**. A Network of Workstations (NOW) or Cluster of Workstations (COW) is a type of cluster formed from individual workstations already present in an office, connected to form a **local area network (LAN)**. A *Massively Parallel Processor* (MPP) contains a very large number of computing elements, perhaps a thousand or more. By the late 1990s, the processing elements (PEs) in most such systems were complete **microprocessors**, effectively making them clusters. Clusters are now the predominant form of supercomputer, although the fastest, Japan's *Earth Simulator*, is a cluster of *vector computers* (*see* **Parallel computer**) rather than a cluster of **microcomputers** (*see* **Beowulf**; **Supercomputer**).

Additional reading: Gordon Bell and Jim Gray, "What's Next in High-Performance Computing?" *Communications of the ACM*, **45**, 2 (February 2002), pp. 91–95.

Cobol LANGUAGES

Cobol (Common Business-Oriented Language) was created by a committee that was formed as the result of a 28–29 May 1959 meeting called to consider the need for a common business-oriented **programming language**. **Fortran**, the only language in significant use at the time, was not well suited to **data processing**. The meeting was chaired by Charles A. Phillips, then director of the Data Systems Research Staff, Office of the Assistant Secretary of Defense of the U.S. Department of Defense (DoD). Representatives of seven U.S. government organizations, 11 users, and 15 manufacturers attended. The meeting led to the creation of CODASYL (Committee on Data Systems Languages) and its Executive Committee. A short-range committee was formed consisting of representatives from six computer manufacturers— **Burroughs**, **Honeywell**, **IBM**, **Radio Corporation of America (RCA)**, **Sperry Rand**, and Sylvania—and three government organizations: the U.S. Air Force, the David Taylor Model Basin, and the National Bureau of Standards (NBS). In October a six-person subcommittee consisting of Gertrude Tierney and William Selden of IBM, Howard Bromberg and Norman Discount of RCA, and Vernon Reeves and Jean Sammet of Sylvania was formed to create the specifications for what became Cobol, and in April 1960 the resulting specifications were released as a U.S. government report. As a by-product, the committee developed a *metalanguage* comparable to extended **Backus-Naur Form** (EBNF), which continues to be widely used to define programming language syntax. The first version of Cobol to be widely implemented was *Cobol-61*, followed closely in time by *Cobol-61 Extended* of 1962 and *Cobol-65*. The first ANSI Cobol standard was issued in 1968, based on *Cobol-65*. Later standards were issued in 1974, 1985, and 2000. Cobol is one of only two **high-level languages**, the other being Fortran, to continue in widespread use over more than 40 years.

Additional reading: Lohr, *Go To*.

Code DEFINITIONS

Code can be a synonym for all or part of a computer **program**, in which case the intended meaning is either **source code**, or **object code**, depending on context. In cryptography (*see* **Data encryption**), a distinction is made between a *cipher*, which is the encryption of the letters or letter sequences of the words in a message, and a *code*, which is the replacement of whole words or phrases of *plaintext* with symbols that are usually numeric. Mathematics,

computer science, and science in general do not make that distinction, so that **Baudot code**, *bar codes*, **Morse code**, and the *genetic code* of biology would be considered ciphers in the terminology of the cryptologist even though they are nonsecret. *See also* **ASCII**; **EBCDIC**; **Unicode**; **Universal Product Code (UPC)**.

Coding theory THEORY

Coding theory is that branch of mathematics and **computer science** that is concerned with the development and use of **codes**, alternative representations of symbolic information. This may be done for purposes of concealment, as is done in cryptography (*see* **Data encryption**), minimization of transmission time and **memory** storage as is done in **data compression**, or to add redundancy as is done with **error-correcting codes**. Coding theory is closely related to **information theory**.

Additional reading: *Encyclopedia of Computer Science*, pp. 223–227.

Colossus COMPUTERS

Colossus was the first large programmable electronic **digital computer**, though not a **stored program computer**. Proposed by T. H. Flowers, it was developed in secrecy under the direction of Allen Coombs and Donald Michie at the British Post Office Research laboratories at Dollis Hill in North London during World War II to break German messages enciphered by the Lorenz SZ40 and SZ42 cipher machines. The basic operation of the machines was deduced by cryptanalysts led by William Tutte. The 1,500–**vacuum tube** Mark I Colossus was working by December 1943. When installed at Bletchley Park, the Colossus filled a large room. The machine itself was 7.5 feet tall by 15 feet wide by 8 feet deep. Its decimal **logic circuits** operated in parallel at 5,000 pulses per second. The 2,500-tube Mark II, designed by Irving "Jack" Good, followed on 1 June 1944, five days before D-Day. The basic clock rate and reading speed were the same, but five-stage **shift registers** and additional logic increased the processing speed by accessing five characters at a time. Eight more Colossi followed. A fully working rebuilt Colossus is installed at Bletchley Park.

Additional reading: Brian Randell, "Colossus," in Metropolis, Hewlett, and Rota, *A History of Computing in the Twentieth Century*, pp. 47–92.

Comit LANGUAGES

Comit, created by Victor Yngve in 1957, is considered to be the first **string processing** language. Actually, Comit was designed to facilitate **machine translation**, so that many of its primitive operations related to **natural language** words rather than to **strings**. But Comit did have a string-oriented **pattern matching** facility, and the language influenced the design of the later and more widely used **Snobol** language.

Additional reading: Wexelblat, *History of Programming Languages*, pp. 616–618.

Command. *See* Operation code (opcode).

Commodore 64 COMPUTERS

The *Commodore 64* (C-64), the 1982 successor to the low-cost *VIC-20*, was **Commodore International**'s most successful **personal computer (PC)**. Its price-performance ratio proved so attractive that over 20 million units were sold over its ten-year lifetime, more than any other **digital computer** ever sold (unless one were to lump all **Advanced Micro Devices (AMD)** and **Intel 80×86**–based machines sold by many tens of manufacturers). The C-64 had 64 kilobytes (65,536 **bytes**) of **memory**, a 25 line by 40 cpl (characters per line) **monitor**, good digital sound, and a large repertoire of **computer games** that made the computer a highly desirable holiday or birthday gift for children. An upward-compatible successor, the *Commodore 128*, was introduced in 1985. The C-128 had twice the memory of the C-64, which enabled it to run **C / PM**, the most popular **operating system** of that time.

Additional reading: *Computing Encyclopedia*, vol. 1, pp. 141–142.

Commodore Amiga COMPUTERS

The *Commodore Amiga* of 1985 was the first **personal computer (PC)** capable of true **multitasking**. Its users had a choice of operating it under its own excellent **operating sys-**

tem or running it in emulation of an **IBM PC**, an **Apple Macintosh**, or a **Unix** machine. Because of its photographic-quality **computer graphics** and advanced digital sound, its adherents considered it the finest PC of its era. But its adoptive parent, **Commodore International**, was never able to market the machine effectively, and the company was sold in 1995.

Additional reading: *Computing Encyclopedia*, vol. 1, pp. 141–142.

Commodore International INDUSTRY

Commodore International was founded in 1958 by **typewriter** repairman Jack Tramiel to manufacture typewriters, adding machines, digital watches, and handheld **calculators**. The company expanded into **personal computers** in 1976 when it bought *MOS Technologies*. Up through 1984, when Tramiel left Commodore and purchased **Atari**, the company's products included the KIM PC kit, the **Commodore PET**, the CBM (Commodore Business Machines) 8000 series, the VIC-20, and the **Commodore 64** (C-64). In 1985, Commodore bought the Amiga corporation in order to market a **Commodore Amiga** and brought out the Commodore 128, a successor to the C-64. The company's fortunes waned over the next decade, and it was sold to the German company ESCOM in 1995.

Additional reading: *Computing Encyclopedia*, vol. 1, pp. 141–142.

Commodore PET COMPUTERS

The *Commodore PET* (Personal Electronic Transactor), along with the **Radio Shack TRS-80** and **Apple II**, was one of the three **digital computers** that dominated the era of fully assembled **personal computers (PCs)**, 1977–1982, that immediately followed the period of do-it-yourself kits. The PET had a 96KB (kilobyte) **memory** (98,304 **bytes**), an 80 cpl (characters per line) **monitor**, and a **character** set much more extensive than **ASCII**, all three characteristics being superior to its competitors.

Additional reading: *Computing Encyclopedia*, vol. 1, p. 141.

Common Gateway Interface (CGI)

The *Common Gateway Interface (CGI)* is a **protocol** in the form of a language in which *scripts* can be written for execution at the **server** side of a **client-server computing** network. CGI, invented by the developers of **Unix**, is what makes the **World Wide Web (WWW)** interactive. Beyond the simple reading or downloading of static **files**, a CGI **interpreter** running on the server enables users to search for information, maintain a page-hit counter, fill out and submit forms, and so on. There are two parts to a CGI **program**: the executable script itself and an **HTML** page that drives it. The executable element can be anything compatible with the server **platform**, including **operating system** calls, Unix shell scripts, and programs written in a **high-level language** that generates CGI **code**. Since CGI is highly succinct, almost **APL**-like, the most popular high-level languages that produce new CGI scripts as if they were **object code** are **Perl** and variants of **Java**. To promote **software** reuse, servers typically store a library of commonly used CGI scripts that may be accessed by HTML programs.

Additional reading: *Computing Encyclopedia*, vol. 1, p. 143.

Compact disc. *See* Optical storage.

Compaq INDUSTRY

Compaq was founded in 1982 by Rod Canion, Jim Harris, and Bill Murto, who left **Texas Instruments (TI)** to do so. Compaq was the first company to produce a clone for the **IBM PC** and the first to produce a portable version of that **personal computer (PC)**. Everything about an IBM PC was in the public domain except for the proprietary **code** embedded in the machine's Basic **Input / Output** System (BIOS). To produce a clone, the BIOS had to be reverse-engineered into alternative code that performed identically to the original. Compaq's initial PC weighed 30 pounds and hence was more "luggable" than it was a **portable computer**, but the company went on to mass-produce desktop machines and captured the major share of the PC market in the 1980s and

1990s. In April 1986, only four years after its founding, Compaq became a Fortune 500 company, and by 1995 it was the world's largest manufacturer of **IBM**-compatible PCs. From then on, however, loss of sales to competitors **Dell**, **Gateway**, and **Hewlett-Packard**—especially with regard to **laptop computers**—led to management difficulties and loss of valuation, and on 19 March 2002 Compaq stockholders voted overwhelmingly to sell the company to Hewlett-Packard.

Additional reading: *Computing Encyclopedia*, vol. 1, p. 145.

Compiler
SOFTWARE

A *compiler*, a term coined by Grace Hopper, translates a **program** written in a **high-level language** into executable **object code** expressed in **machine language**. So does an **interpreter**, but the difference is that an interpreter follows a **statement**'s translation with immediate execution of its object code equivalent and proceeds this way statement by statement, as a human interpreter at the United Nations does sentence by sentence. A compiler, on the other hand, is comparable to a person who translates a whole book from, say, French to English before anyone else "executes" the result by reading it. The development of compilers and interpreters correlates with the development of high-level languages because one or the other must be written to implement one. The candidates for the first true compiler are *Autocoder*, written in 1952 by Alick E. Glennie for the **Manchester Mark I** (*see* **Autocode**); one by J. Halcombe Laning and Neal Zierler for an unnamed mathematical language for the **Whirlwind** in 1954; and Grace Hopper's A-0, A-2, and B-0 compilers of the early 1950s. The latter was released commercially in 1957 under the name *Flow-Matic*, a business-oriented language that antedated **Cobol**. Early compilers were written with *ad hoc* methods, but theoretical computer scientists soon showed that a compiler's performance was closely related to the **algorithm** used for the *lexical scan* and *parsing* of the statements of the language being compiled in accord with the grammar of the language. The work of Lewis, Rosenkrantz, and

Stearns was particularly influential in this regard.

Additional reading: Lewis, Rosenkrantz, and Stearns, *Compiler Design Theory*.

Complement
NUMBER SYSTEMS

The *complement* of a number is the result of subtracting it from a reference value. For decimal numbers, one of the two useful reference values for an *n*-digit number is the number consisting of *n* consecutive 9s. Thus the so-called *9's complement* of 37416 is 62583, the result of subtracting it from 99999. When 37416 is subtracted from, say, 89257 in the usual way, the result is 51841. But the same result can be obtained by adding the complement of that number to 89257 if it is done in accord with the correct **algorithm**, which is to add any carry out of the leftmost position back into the units position of the result. Thus $89257 + 62583 = 51840 +$ the carried 1, giving the correct result of 51841. Alternatively, one can use 10's complement numbers that anticipate the carry. The 10's complement of a number can be formed either by adding 1 to its 9's complement or by subtracting the number from 1 more than "all 9s," which, in either case, yields 62584. Now the rule for subtraction through addition of the complement is modified; any carry out of the leftmost position is discarded. So $89257 + 62584$ is 51841, the correct answer.

In the **binary number system**, the complements comparable to the decimal 9's and 10's complement are called the *1's* and *2's complement*. Obtaining the 1's complement of a binary number is trivial; just reverse every **bit**. Thus the 1's complement of 1001101011101 is 0110010100010, and the 2's complement would be 1 higher: 0110010100011. Artificial subtraction would then proceed in accord with the appropriate algorithm for handling a carry out of the leftmost position of the minuend. In a **digital computer**, one would ordinarily need (among others) two different kinds of **logic circuit**, an **adder** and a *subtractor*. But some designers prefer to implement a *complementer*, simpler to build than a subtractor, in which case only an adder is needed, and the programmer need not be concerned with or even know which method is used to effect subtraction.

Additional reading: Petzold, *CODE*, pp. 143–154.

Complex Instruction Set Computer (CISC) COMPUTER TYPES

The term *CISC*, for *Complex Instruction Set Computer*, did not exist until the need arose to distinguish *CISC* from **reduced instruction set computer (RISC)** architecture. All **von Neumann machines** from **EDVAC** through most of those still in use are CISCs, that is, they have an extensive **instruction set** and do not embrace the attributes described in the RISC article. The **Stretch** and the **DEC VAX series**, perhaps, had the richest instruction sets ever devised, but many of those **instructions** would seldom, if ever, be generated by a **compiler**. The **CDC 6600**, in contrast, had a relatively sparse instruction set that did not even include an integer multiply command. Thus its designer, Seymour Cray, believed in at least part of the RISC ethos 20 years before the term *RISC* was coined by David Patterson and embodied in several of the latest computer designs. A RISC generates far more instructions per **high-level language** statement compiled, but, in theory at least, its circuitry can execute the longer but simpler sequences faster. With **memory** now so inexpensive, the extra space consumed by RISC **object code** no longer matters very much. And given the declining use of **machine language** programming that can make the most of CISC complexity, RISC has a lot going for it. Nonetheless, it is likely that there will continue to be advocates of both schools of design. *See also* **Very Long Instruction Word (VLIW)**.

Additional reading: *Encyclopedia of Computer Science*, pp. 314–315.

Comptometer CALCULATORS

In 1886 the American Dorr E. Felt, while working for the Pullman Company, invented what he later named the *comptometer*, a key-driven adding machine that embodied the only significant advance over the variable-toothed gear of the **Baldwin / Odhner calculator**. Comptometers were affectionately known as "macaroni

boxes" for many years because Felt's prototype was built on such a frame. Comptometer addition was significantly faster than on earlier calculators because one number could be added to another in a mechanical **register** as it was being entered as input. Felt formed a partnership with Robert Tarrant, and the firm of Felt & Tarrant sold comptometers for over 50 years.

Additional reading: Kidwell and Ceruzzi, *Landmarks in Digital Computing*, pp. 37–38.

Computational complexity THEORY

Computational complexity is concerned with trying to determine, for a given class of problems, limits on the performance of **algorithms** that might conceivably be found in the future, in contrast to the **analysis of algorithms**, which is concerned with analyzing particular known algorithms. Perhaps the simplest complexity result is that it can be proved that no **sorting** algorithm based on comparison of n items can possibly have running time better than $O(n \log n)$. (For an explanation of "Big O" notation, *see* **Analysis of algorithms**.) Then, those who analyze algorithms can take over and prove that the bubble sort and insertion sort algorithms do not attain such performance and that **Quicksort**, among others, does. There are many problems, the traveling salesman problem for one, for which the best known algorithm is, horrendously, $O(a^n)$, that is, its running time grows exponentially with increasing problem size. Despite the fact that $O(n^2)$ is considered poor for sorting, theorists consider that finding that an algorithm is $O(n^a)$—so-called *polynomial running time*—is a good result, whereas exponential running time or worse is considered very disappointing because, for even moderate problem sizes, running time can be so long that the problem is essentially *intractable*, a fancy word for very impractical.

The important difference between an exponential-time and a polynomial-time algorithm was first called to the attention of computer scientists in 1965 by mathematicians Jack Edmonds and Alan Cobham. In the late 1960s, Manuel Blum developed an axiomatic theory of complexity. At about the same time, Juris Hartmanis and Richard Stearns showed that there may be as many as eight different levels (classes) of complexity, although to this day, certain of the classes cannot be shown to be distinct. The classes are like layers of an onion, where the outermost layer contains the undecidable problems—the **halting problem**, for example—for which no algorithm can be devised at all. Then come the decidable problems whose best performance is worse than exponential, and so on down through exponential-time problems, **NP-complete problems**, polynomial-time problems, and simpler problems whose running times grow more slowly than linearly proportional to problem size n. Even when a problem is known to have an $O(n^a)$ polynomial-time algorithm for its solution, the question can often be asked as to whether the exponent a is as small as possible. For example, it would first appear that it should take $O(n^2)$ time to multiply two n-bit **operands**, but a reasonably straightforward algorithm is known that runs in $O(n^{1.585})$ time, and Arnold Schönhage and Volker Strassen found an interesting but impractical algorithm to further reduce this to $O[n(\log n)(\log \log n)]$. Computational complexity is now the most active area of theoretical **computer science**.

Additional reading: *Encyclopedia of Computer Science*, pp. 260–265.

Computer DEFINITIONS

Until about 1946, still the age of the **calculator**, a term was needed to describe those who used them occupationally. Since *calculator* would have been ambiguous, the name chosen as early as 1646 was *computer*. This was done by Sir Thomas Browne, the same scientist who coined the term *electricity*. Thus the first "computers" were human. In the age of the computer as machine, the corresponding occupational titles are *programmer* and *operator*. Notwithstanding the oxymorons **programmable calculator** and **card programmed calculator (CPC)**, it is programmability that distinguished a computer from a mere calculator. Thus the programmable **BINAC** (BINary Automatic Computer) of 1949 and **UNIVAC** (UNIVersal Automatic Computer) of 1951 were aptly named, but certain machines of the same era such as **IBM**'s Automatic Sequence Controlled Calculator of 1944 (*see* **Harvard Mark I [ASCC]**) and Selective

Sequence Controlled Calculator (**SSEC**) of 1948 (which were computers) were not. As late as 1954, Thomas Watson was still resisting the difference in nomenclature by calling the **IBM 650** a "Magnetic Drum Calculator," but as of the **IBM 700 series** of just a year later, IBM finally admitted they were marketing "computers," and there has been no turning back.

Additional reading: Bashe et al., *IBM's Early Computers*.

Computer algebra SOFTWARE

The origin of *computer algebra*, which might more appropriately be called *symbolic computation*, can be traced to the vision of Babbage's assistant Augusta Ada Byron King, countess of Lovelace. With regard to the **Analytical Engine**, she wrote in her famous 1843 addendum to a paper of Luigi Menabrea (*see* **Program**): "Many persons . . . imagine that because the business of the Engine is to give its results in numerical notation, the nature of its processes must consequently be arithmetical and numerical, rather than algebraical and analytical. This is an error. The Engine can arrange and combine its numerical quantities exactly as if they were letters or any other general symbols. . . . The Analytical Engine weaves algebraic patterns just as the **Jacquard loom** weaves flowers and leaves." The realization of this insight did not come until 1954 when Grace Hopper wrote a program that she called an *Analytic Differentiator* for the UNIVAC. Given a reasonably well-behaved mathematical function, her program could compute and print out the symbolic representation of its derivatives to any order. In 1964, Jean Sammet of **IBM** wrote *FORMAC* (FORmula MAnipulation Compiler), the first reasonably full featured computer algebra program. All current programs of this variety can handle far more than algebra; they "know" a good deal of **calculus**, vector analysis, and **discrete mathematics**, and they support extensive **computer graphics**. Classic early systems were *Reduce*, written in the late 1960s by Anthony Hearn, and *MACSYMA* (Project MAC SYmbol MAnipulation), written by Joel Moses and William Martin in 1972. The latter has gone through many upgrades. Perhaps the most comprehensive package is *Mathematica*, whose first version was written by Stephen Wolfram and released in 1988. Other packages of interest are *Maple*, begun in 1980 at the University of Waterloo in Montreal by Gaston Gonnet and Keith Geddes, and *Derive*, written in **Lisp** in 1988 by Albert Rich and David Stoutmeyer.

Additional reading: *Encyclopedia of Computer Science*, pp. 282–301.

Figure 10. Augusta Ada Byron King, Countess of Lovelace, in her day (presumably not a casual Friday). (Image Select/Art Resource, New York)

Computer animation

COMPUTER GRAPHICS

Computer animation is a branch of **computer graphics** that is concerned with the automated production of image *frames* which, when displayed in rapid succession, convey to the viewer the illusion of motion. Except for the addition of the important word "automated," the definition would be identical to the production of motion picture films in which individual frames are either photographs or *cels* drawn by humans. Computer animation is the use of a program that can modify the positions of elements of one frame ever so slightly so as to produce the next frame in succession. **Iteration** of the process then produces enough frames to constitute an entire scene. Computer animation is most often used to produce cartoons and full-length features with the complexity of, say, animated Disney productions of the mid-twentieth century. But it can also be used to superimpose animated special effects on a photographed background, as was done in the *Spider-Man* movie of 2002.

Computer animation dates to about the mid-1970s. George Lucas considered using an animated sequence that showed five X-Wing fighters flying in formation for possible use in the second *Star Wars* movie, *The Empire Strikes Back*. He decided not to but did assemble a group of experts who later formed the leading firm in computer animation, *Industrial Light and Magic*. In 1976, Systems Simulation, Ltd. of London was commissioned to provide computer graphics for a landing sequence in *Alien*, arguably the first example of pure computer animation in a movie. In the 1980s, computer **hardware** increased greatly in performance to the point where it allowed computer animators the freedom to create new graphics tools and techniques. **Fractals** began to be used to create realistic-looking terrain, and a technique called *ray tracing* could render objects that were lit by beams of light just as they would be in real life. Another new graphics technology was pioneered in 1988 when LucasFilm used "morphing" in *Willow*, directed by Ron Howard. In the film, characters seamlessly changed ("morphed") from one image to another, a trick now so popular that it is seen again and again in movies and on **television**.

In 1996, a large part of Industrial Light and Magic split off to form *Pixar*, a studio now owned by Steve Jobs of **Apple**. At Pixar, John Lasseter worked on improving computer animation as a storytelling medium, creating short films such as *Tin Toy* in 1988, which became the first computer animated film to win an Academy Award for best animated short feature. In 1991, Dennis Muren convinced Steven Spielberg to use computer animation for several scenes involving the dinosaurs of *Jurassic Park*, and in 1995, Pixar produced the big hit *Toy Story*. The movie *Final Fantasy* of June 2001 was the first completely 3D animated movie with realistic characters that looked and moved so much like real people that viewers began to wonder whether live actors would ever be needed again. Probably because of its weak story, the film was not a box office hit, but later in the year, the movie *Shrek* was immensely popular and won the Academy Award for best animated full-length film.

Additional reading: *Encyclopedia of Computer Science*, pp. 301–304.

Computer architecture   GENERAL

Computer architecture, a term coined by Fred Brooks of **IBM** in 1963, encompasses the design of a **stored program computer** as perceived by a **machine language** programmer. It embraces the computer's **word length**, the number and function of its **registers**, its *addressing structure*, its **instruction set**, and the layout of its **instructions** with regard to the arrangement of commands and **operand** addresses. Variously different architectures use instructions having 4, 3, 2, 1, or 0 operands. The arithmetic commands of a zero-address **digital computer** such as the **Burroughs B5000** operate on the top two operands "popped" from its **pushdown stack** and then push the result back into the stack. In **multiple-address computers**, instructions typically have prefix form; that is, their commands are to the left of their operand addresses. One-address computers are of the form `command-address` in which, for example, ADD B means "Add the contents of address B to (an implied) register usually called the *accumulator*. The **IAS computer** was one of the very earliest one-address computers.

Additional reading: Blaau and Brooks, *Computer Architecture: Concepts and Evolution*.

Computer-Assisted Instruction (CAI)

APPLICATIONS

Computer-assisted instruction (CAI) is the use of specially written computer **programs** called *courseware* to support teaching and self-instruction at any educational level. The alternative term *computer-assisted learning* (CAL) is beginning to be used somewhat more often. Computers were used as teaching machines as early as 1958 at such places as the **IBM** Watson Research Center, the System Development Corporation, and the University of Illinois. In the fall of 1962, Patrick Suppes and Richard Atkinson established a laboratory at Stanford University dedicated to the investigation of learning and teaching. Suppes, who considered the laboratory as an experiment in educational psychology, is usually cited as the foremost pioneer in the field of CAI. In the mid-1960s, Donald Bitzer, Robert Wilson, and Gene Slottow at the University of Illinois designed a system called PLATO (Programmed Logic for Automatic Teaching Operation) and an associated language called TUTOR that provided instructional computing to about a thousand simultaneous users throughout the university and also to other colleges and schools in Illinois. For about 15 years starting in 1975, PLATO was marketed commercially by **Control Data Corporation (CDC)** for use on **mainframes** that supported **time sharing**. At about the same time Stanford University operated a CAI system to distribute instructional computing material to a number of centers throughout the country, and Mitre Corporation developed TICCIT (Time-shared, Interactive, Computer-controlled Informational Television) that was marketed by the Hazeltine Corporation. But CAI did not have a significant impact on either education in general or on the development of **personal computers (PCs)** and their proliferation in elementary and secondary schools. Computers can be enlisted as an aid to learning any subject, so it is fitting that the language **LOGO** was developed to help young students learn how to program. The combination of a PC and multimedia **software** now provides enormous opportunities for self-study and so much augments and reinforces what the human teacher can do that CAI is having a profound effect on education at all levels.

Additional reading: *Encyclopedia of Computer Science*, pp. 328–336.

Computer-assisted proof   MATHEMATICS

In **automatic theorem proving**, a **digital computer** is used to construct an entire proof. A *computer-assisted proof* is one constructed through the joint partnership of a **computer** and a human. On rare occasions, a mathematician is able to prove a theorem for all but a finite number of special cases, a number too large to examine one by one. By writing a computer **program** that confirms that the conjecture holds for each and every potentially troublesome case, the computer-assisted proof is completed (provided, of course, that the program is correct). In 1852, Francis Guthrie, while coloring a map of English counties, was surprised that he needed only four colors. He wondered whether it was true that *any* map can be colored using at most four colors in such a way that adjacent regions (i.e., those sharing a common boundary segment, not just a point) are colored differently. The four-color map conjecture can also be expressed as a vertex coloring of the map's equivalent **graph** (*see also* **Graph theory**). In 1977, Kenneth Appel and Wolfgang Haken were able to show that if no more than four colors were needed for any of 1,936 basic cases, more than four would never be needed. They completed their proof of what is now the famous Four-Color theorem through use of a computer program that eliminated these cases. Other mathematicians were initially skeptical, but the proof has now been independently verified.

For as long as soap bubbles have been studied, it has been recognized that the sphere has the minimum surface of all those that might be used to enclose a given volume. The proof of this 2,000-year-old theorem is remarkably recent, however, dating back to Hermann Schwarz in 1882. But until 1995, the most efficient shape for enclosing *two* equal volumes remained uncertain. Mathematicians believed, but had not proved, that use of two overlapping spherical bubbles of equal radii would be optimum. Preliminary analysis narrowed the problem down to two possible families of bubbles, the standard double-bubble and the donut-shaped "torus" bubble, a donut with a dumbbell through the hole in the middle. In 1995, Joel Hass and Roger Schlafly found a computer-assisted proof that the double-bubble is indeed optimum. The final solution describes two identical bubbles that meet at 120-degree angles and share a disk-shaped wall whose radius is $\sqrt{3}/2$, about 87% of the radius of the overlapped spheres.

A third example of a computer-assisted proof relates to a 400-year-old conjecture of Johannes Kepler regarding the packing density of identically sized spheres, oranges perhaps. He proposed that letting each sphere fall into the depression formed by the group of four below it—the "grocer's" or "cannonball" arrangement—was optimum. In 1998, Thomas Hales completed an elaborate 250-page proof that showed how to reduce the interstitial spaces in any conceivable packing to one of 5,094 "stars," each having a computable score. The lower the score, the greater the packing density. He then transformed each star into a problem in *linear programming* (*see* **Simplex method**) and wrote a program that found that none of the 5,094 stars had a score lower than that of the Kepler packing.

Since so much of **computer science** is based on a mathematical foundation, it is satisfying that computer science can occasionally advance the progress of mathematics.

Additional reading: *Encyclopedia of Computer Science*, pp. 1096–1103.

Computer Associates   INDUSTRY

Computer Associates (CA) was founded in 1976 by Charles Wang, Judith Cedeno, and Russell Artzt. The initial strategy was to provide **IBM**'s **hardware** customers with performance-enhancing **software**. In 1987 CA acquired UCCEL and along with it the current Computer Associates' president and CEO, Sanjay Kumar. The acquisition effectively doubled the size of CA and greatly augmented its line of systems software solutions. In 1989, CA acquired Cullinet to strengthen its line of **database** products and became the first $1 billion software company. In 1993, CA executed a bundling agreement with **Hewlett-Packard** that comprised the initial offering of Unicenter, the first comprehensive **distributed system** software running under **Unix**. In 1994 CA and **Electronic Data Systems (EDS)** signed a 12-year global software licensing agreement, the largest and longest of its kind to that date. The next seven years saw a continual string of additional acquisitions and software agreements with companies such as **Sun Microsystems** and **Microsoft**. By 2001, CA was heavily engaged in the support of **electronic commerce** and remained the world's largest **mainframe** software company with annual sales of $2.5 billion.

Additional reading: http://www.cai.com/about/history.htm

Computer chess

ARTIFICIAL INTELLIGENCE

Intelligent machines that could play chess better than all or most humans were envisioned as early as the eighteenth century. The first such machine, "The Turk" of 1769, allegedly an **automaton**, was created by Baron Wolfgang von Kempelen and eventually sold to Johann Nepomuk Maelzel. Maelzel exhibited it widely for several decades, and it is known to have played Napoleon Bonaparte, Benjamin Franklin, and Edgar Allan Poe. The machine was a fraud; as was deduced by Poe in his April 1836 story "Maelzel's Chess Player" in the *Southern Literary Messenger*, the Turk's imitation machinery actually concealed a small but skillful human player. In the 1896 Ambrose Bierce story "Moxon's Master," a chess-playing automaton chokes its creator after being checkmated. About 1900, the Spaniard Leonardo Torres Quevedo built an electromechanical machine that could correctly play king and pawn endgames; the machine still functions in Madrid's Polytechnic Museum. In the early 1940s, but unknown outside Germany, Konrad Zuse encoded an elaborate chess-playing **program** in a language of his own invention called **Plankalkül**. But the language was never implemented, so neither could the program be.

The foundation of *computer chess* is considered to be Claude Shannon's 1950 paper "Programming a Computer for Playing Chess" in *Philosophical Magazine*. In England a year later, Dietrich Prinz wrote a program for the **Manchester Mark I** for solving mate-in-two problems, and Alan Turing and Donald Michie of the National Physical Laboratory (NPL) wrote a chess program called TurboChamp with advice from David Champernowne. Turing, us-

ing himself as a **central processing unit** for lack of a sufficiently capable computer, played a complete game that he lost to his colleague Alick Glennie (*see* **Autocode**). In 1958, the team of James Kister, Paul Stein, Stanislaw Ulam, William Walden, and Mark Wells of the U.S. Los Alamos National Laboratory wrote a program for the **MANIAC** that played "anti-clerical chess," a bishopless game restricted to a 6×6 board. The first complete chess program was written in **assembly language** for the IBM 704 (*see* **IBM 700 series**) in 1959 by Alex Bernstein, Michael de V. Roberts, Thomas Arbuckle, and Martin Belsky, but its level of play was unimpressive. Also in 1959, Allen Newell, Herbert Simon, and Cliff Shaw of Carnegie Tech used their IPL language to write a program called *Nss* for the full 8×8 board (*see* **IPL-V**). It was the first operational chess program written in a **high-level language** and the first to use von Neumann's depth-first *minimax theorem* (*see* **Game theory**) to prune the game **tree**, but its play was mediocre. Writing ever-better chess programs became an important endeavor in the field of **artificial intelligence (AI)**. The first programs pitted against one another occurred in 1966 when the Russian program *ITEP* (for Institute of Theoretical and Experimental Physics), written for the M-20 computer by a team led by Vladimir Arlazarov, played a four-game **telegraph** match with a program written for the Stanford University IBM 7090 by Alan Kotok and John McCarthy. The match took nine months, with ITEP the victor by a score of 3–1. *MacHack*, the first program that played respectably enough for tournament play against humans, was written for the PDP-6 (*see* **DEC PDP series**) in 1967 by MIT students Richard Greenblatt, Donald Eastlake, and Stephen Crocker.

The first international computer chess championship, organized by the **International Federation for Information Processing (IFIP)** in 1974, was won by the Russian program *Kaissa* running on an ICL 4/70. In 1976, the program *Chess 4.5* written by Lawrence Atkin, David Slate, and Keith Gorlen of Northwestern University was the first to earn an expert rating. An improved version of the program won the IFIP tournament of the following year. A table of the winners of the many tournaments held over the next 20 years is given in the reference. Some of the most successful programs of that period were *Belle*, written by Ken Thompson and Joe Condon of Bell Labs; *Cray Blitz* of Bob Hyatt, Al Gower, and Harry Nelson; *Hitech*, developed by Hans Berliner; and *Deep Thought*, written by a group of Carnegie-Mellon graduate students led by Feng-hsiung Hsu, Joseph Hoane, and Murray Campbell. IBM then hired Hsu and Campbell and three others to produce *Deep Thought II*, a program that tied the German program *Mephisto* in 1990 and defeated grand master David Levy 4–0. In 1996, the successor to *Deep Thought II*, *Deep Blue* running on an IBM **RISC** System/6000, lost a match 4–2 to world champion Garry Kasparov. But in a rematch the following year, *Deep Blue* defeated Kasparov 5–4 to become the new, though nonhuman, world champion.

By 2000, Kasparov had lost his own world championship to his former pupil, Vladimir Kramnik, and IBM had dismantled Deep Blue. In October 2002, Sheikh Hamad bin Isa al-Khalifa, ruler of Bahrain, posted a $1 million purse and arranged an eight-game match in his country between Kramnik and the program *Deep Fritz*. Deep Fritz was developed by a team of German scientists led by Mathias Feist. The program runs on a mere **laptop computer** of intrinsic speed less than 2% of Deep Blue, but it plays at about the same level because of superior **pattern recognition** software. Kramnik was relieved to escape with a 4–4 tie. Then, in February 2003, Kasparov agreed to a six-game match with Deep Junior, a program developed by Israeli scientists Amir Ban and Shay Bushinsky. Kasparov met the same fate as Kramnik, settling for a draw in the sixth game that ended the match at a score of 3–3.

The first **microcomputer** chess program was written in 1976 by Sidney Samole, owner of Fidelity Electronics.

In 1979, Daniel and Kathe Spracklen wrote a program embodied in the base of an actual chessboard and Samole marketed it as the *Fidelity Chess Challenger*, which gave the casual player the first opportunity to play computer chess. Successively more powerful versions of the *Challenger* were sold over the next ten

years until the popularity of **software** for **personal computers (PCs)** dried up the demand. By the late 1990s, the most powerful inexpensive PC program was a Mattel product called *Chessmaster 8000* that embodied the *King v3.11* "chess engine" written by Johan de Koning (whose last name means "king"). The same engine is used by the Dutch company Tasc. Only grandmasters can hold their own against the King program when it is allowed to play at its highest level on a 2 GHz PC or better. And *Deep Fritz*, of course, is even more powerful.

A good computer chess program, even a PC version, is intimidating to all but a handful of human players. No advantage can be gained early in a game because of the program's huge repertoire of well-analyzed opening moves. No advantage not already earned can be exploited in the end game because the program knows the standard **algorithms** for perfect play when there are six or fewer pieces left. That leaves only the middle game, where the computer's brute force search deep into the game tree tends to prevail over humans whose ability to look ahead seldom matches that of the computer. The only weakness, and a very small one, is that a human's evaluation of a given position may be superior to that of a computer that must rely on a possibly fallible arithmetic score. But the computer is relentless, has no feeling, does not tire, and never makes an outright blunder such as failing to protect a major piece or overlooking a fork. Attention can now turn to the more complex game of *Go*, which should keep AI researchers busy for at least another 50 years.

Additional reading: Newborn, *Kasparov versus Deep Blue: Computer Chess Comes of Age.*

Computer chip. *See* Integrated Circuit (IC).

Computer cluster. *See* Cluster.

Computer engineering GENERAL

The design of **digital computers** and associated components in industry is generally done in a department with "engineering" in its title, and their theoretical and experimental study in uni-

Figure 11. "The Turk," aka Maelzel's Chess Player, allegedly showing that he has no weapon of mass obstruction to hide. (Mary Evans Picture Library)

versities is usually done in an electrical engineering department or, increasingly more often now, in a department of *computer engineering*. Since the field parallels **computer science**, there are also academic departments called *computer science and engineering*, *electrical and computer engineering*, and *computer science and electrical engineering* (or the reverse). The origin of the term *computer engineering* is obscure and likely to be of multiple heritage, but it is clear from the context that it deals much more with computer **hardware** than **software**. But attempts to develop software to rigorous engineering standards have now given rise to the field of **software engineering**, which might be done in any one of the kinds of departments cited.

Additional reading: *Encyclopedia of Computer Science*, pp. 353–354, 615–616.

Computer game GENERAL

Computer game is a broad term that includes both videogames and traditional adult games such as **computer chess** and **programs** that play checkers, backgammon, bridge, Go, Scrabble, and Monopoly (*see* **Artificial Intelligence [AI]**). The analysis of strategies for games such as Go, checkers, and chess that do not have a chance element is usually supported by use of the **data structure** called a **tree**. The vertices of the *game tree* represent positions, and a given vertex has as its successors all vertices that can be reached in one move from the given position. The complete game tree for checkers has about 10^{40} vertices, while the complete game tree for chess has about 10^{120}. Complete game trees for most nontrivial games are too large to be exhaustively searched or even stored in computer **memory**. Strategies for computerized games are effectively strategies for deciding how the complete game tree should be pruned and for choosing a move on the basis of information in the pruned game tree. The more formal analysis of economic, military, political, business, and social interactions that can be cast in the form of a "game" is the subject of **game theory**.

To many, *computer game* connotes *videogames*, action-oriented games whose principal appeal is to children and college students. The first videogame, *Spacewar*, was conceived in 1961 by Martin Graetz, Stephen Russell, and Wayne Wiitanen. It was first realized on the PDP-1 (*see* **DEC PDP series**) in 1962 by Russell, Peter Samson, and other students at the MIT Computing Center. One of the students, Nolan Bushnell, revised it and sold it to Nutting Associates, who marketed it as the arcade game *Computer Wars*. After graduation, Bushnell founded the firm **Atari** in 1972. The first Atari arcade game, *Pong*, was introduced on 29 November 1972. Possibly the most famous computer game of all time is *Pac Man*, created by Namco game designer Tohru Iwatani in 1979. His inspiration for the Pac Man **icon** was a round pizza with one slice missing. Pac Man began as an arcade game but quickly spread to **personal computers (PCs)** and remains popular to this day.

Additional reading: *Encyclopedia of Computer Science*, pp. 357–368.

Computer graphics GENERAL

Computer graphics is that branch of **computer science** that deals with the creation, display, and printing of drawings, graphs, and computer-generated images. Before the development of graphics-oriented **monitors**, crude graphs of very low resolution were constructed on line printers by carefully printing spaces and symbols on successive lines. Computer-driven **cathode ray tube (CRT)** displays were a part of the output of the MIT **Whirlwind** computer of 1950 and the **SAGE** Air Defense System of the mid-1950s. Ivan Sutherland, the foremost pioneer of interactive computer graphics, developed his **Sketchpad** drawing system in 1963. This system included input and interactive techniques for creating line drawings, including pointing, dragging, and **icons** that are still in use today. This system also used a hierarchical method for building objects from simpler components that has had a strong influence on **software engineering**.

Computer graphics was first applied to computer-aided design and manufacturing (*see* **CAD / CAM**) in the mid-1960s. In that era, displays were monochrome and vector-based, with all images drawn as a set of lines that had to be continually and quickly "refreshed" (redrawn) to avoid flicker. In the late 1960s, *Tektronix* developed an alternative display device called the *direct-view storage tube* (DVST). In this technology, there was a special mesh near the phosphor that could store the image being displayed. This mesh would continually attract electrons to the phosphor needed to display the image, so it was constantly refreshed. Display devices now use a technology called *raster graphics*. As with **television**, an image is created by lighting discrete spots of phosphor, each called a *pixel* (picture element), and letting the eye merge them into a picture. As part of a raster graphics system, there is a *frame buffer* that contains enough memory to hold the value to display for each pixel, and a control unit that translates these values into the voltages needed to produce the correct amount and color of light. Changing the image displayed on these systems is now as simple as changing the values stored in this frame buffer.

The creation of a realistic computer graphics image involves two major stages. The first stage is the description of the world and is called *modeling*. The second stage is the drawing of the world and is called *rendering*. In the reference, Jeffrey McConnell states:

As objects become more complex, we have more parameters that can be specified. For three-dimensional objects, we typically specify the following types of properties:

- *Reflectivity* What portion of light that strikes an object reflects off it? Is the reflection sharp like a mirror (specular) or dull like metal? Does the reflection stay focused, or does it diffuse and spread?
- *Refractivity* What portion of light that strikes an object refracts through it? How much does the refraction shift or bend the light (as a pencil in a glass of water appears bent)? Is the object transparent like clear glass or translucent like tracing paper?
- *Absorptivity* How much of the light energy striking an object is absorbed by the object and neither reflected nor refracted?
- *Texture* Does the object have a smooth or rough surface? If rough, is there a pattern to the roughness or is it random?

The speed and **memory** sizes of current computers and the high resolution of available color monitors have made it possible to construct images that have near-photographic realism. This has allowed the production of animated movies in which the "actors" are so lifelike that humans will find increasing difficulty in filling their roles unless they work for wages substantially below current rates (*see* **Computer animation**).

Additional reading: *Encyclopedia of Computer Science*, pp. 368–382.

Computer music APPLICATIONS

Recorded music can be played through the speakers of a **personal computer (PC)**, but all that does is to turn the PC into an expensive compact disc player. What is meant by *computer music* is a composition that is created to take advantage of the fact that a **computer** can be programmed to use a combination of **heuristic** and algorithmic **procedures** to produce original music that is esthetically pleasing to the human ear. An intriguing way to "compose" such music is to used *constrained randomness*, that is, establish limits that conform to an overall pattern but then use pseudo-random numbers to create variations that lie within those limits (*see* **Random number generation**). Using this technique, composer David Cope's computer program *Emmy* composes music that is difficult to distinguish from real music composed by the immortals such as Bach, Mozart, and Beethoven. Thus computer music is a branch of **Artificial Intelligence (AI)**.

Historically, the first notable computer composition was the 1957 *Illiac Suite for string quartet* by Lajaren Hiller and Leonard Isaacson. In this work, the **ILLIAC** was used to emulate musical stylistic rules and derive some of its own. In the late 1950s and early 1960s, Max Mathews and his colleagues at Bell Laboratories introduced the first programs for digital sound synthesis. Computers are now used in all aspects of music, including composition, live performances, musicology, music notation and score printing, sound production, the editing of digitized audio signals, and sound reproduction.

Additional reading: *Encyclopedia of Computer Science*, pp. 396–404.

Computer science   GENERAL

The term *computer science* was coined in the 1950s by George Forsythe. The need for a name for a new discipline related to digital computing arose in conjunction with the burst of activity surrounding the development of **stored program computers** in the mid-1940s. Like most disciplines with "science" in their name, computer science is not a *science* at all since it does not deal with naturally occurring entities. Its theoretical aspects are highly mathematical, and its **hardware** and **software** aspects could more accurately be called *engineering* and *linguistic*, respectively. Before *computer science* became the established term, Louis Fein had argued in 1959 that the field would command maximum respect if it had a one-word name as does mathematics, physics, chemistry, and so on. He suggested "synnoetics." **Cybernetics**, favored by the author of this book, might have caught on but, except for its prefix (*see* **Cyberspace**), that term has preserved its original and narrower connotation of dealing principally with **analog computers**. The first academic department of computer science was established at Purdue University on 1 October 1962 by its founder, first chairman Samuel Conte, and the first Ph.D. in computer science was awarded to Richard Wexelblat by the University of Pennsylvania in 1965, the same year that a graduate department of computer science was established by Alan Perlis at Carnegie Tech in Pittsburgh. Perlis then went on to chair a prestigious computer science department at Yale in 1971.

A 1964 report of the **Association for Com-**puting **Machinery (ACM)** states, "Computer science is concerned with information in much the same sense that physics is concerned with energy." For that reason, some countries call computer science *informatics* or *informatique*, the term suggested by Philippe Dreyfus. But by the end of the 1960s most computing theorists had adopted the position that computer science is the study of **algorithms**, which they call *algorithmics*. In the summary article cited as a reference, Peter Denning lists 11 principal subfields of computer science as being algorithms and **data structures**; programming languages (*see* **Procedure-** and **Problem-Oriented Language [POL]**); **computer architecture**; **operating systems** and **networks**; **software engineering**; **databases** and **information retrieval**; **artificial intelligence** and **robotics**; **computer graphics**; human-computer interaction (*see* **CAD / CAM**; **Computer-Assisted Instruction [CAI]**; **Computerized tomography**; **Data visualization**; **Global Positioning System [GPS]**; **Graphical User Interface [GUI]**); computational science (*see* **Computer algebra**; **Numerical analysis**); and organizational informatics (*see* **Information Technology [IT]**; Management Information Systems [MIS]). Interestingly, that leaves no place to classify the design and construction of **digital computers**, which he considers to be the parallel but separate field of **computer engineering**. But analysis of the performance of computers and their hardware components is firmly within the purview of computer science.

Additional reading: Peter Denning, "Computer Science," in *Encyclopedia of Computer Science*, pp. 405–419.

Computer virus   GENERAL

A *computer virus* is a **program** that copies itself into other programs, causing the modified programs to perform differently, perhaps dangerously so. Corrupted programs seek others to "infect." Successful viruses reveal their presence only when they cause damage. A virus can spread when users download programs from a **Bulletin Board System (BBS)** or the **Internet**, or if they share diskettes (**floppy disks**) or other exchangeable media. Most commonly, viri are transmitted as attachments to **electronic mail** (e-mail) messages; many users will not open the attachment to a message from an unknown person. The idea of a maliciously self-propagating computer program is believed to have originated in David Gerrold's 1972 novel *When Harlie Was One*. The probability of a serious computer virus attack was predicted by Fred Cohen of the University of Southern California in 1983, before one actually occurred, but now they pose a continuing threat.

Although not technically accurate, the term *virus* is often used to refer to other kinds of malicious **software**. A *logic bomb* or *time bomb* is an infection that lies low until activated by a certain event or arrival of a certain date, as one did at a university in Jerusalem on 13 May 1987. A *Trojan horse* is a section of **code** surreptitiously planted into a trusted computer program; it does not spread like a true virus, but it does cause the modified program to perform unintended actions. A *worm* is a widely distributed program that invades networked computers. Worms spread by replication, as viruses do, but they may run as independent processes rather than insinuating themselves into other programs. Recent viruses are designed to "mutate" each time they copy themselves to other **files**, making them difficult to detect. The best protection comes from installation of a software **firewall** and subscription to a service from which new virus "patterns" are downloaded periodically and compared to stored files by a program running in the background as part of normal **multitasking**.

At 9 P.M. on 2 November 1988, Cornell graduate student Robert T. Morris launched a worm on the Internet by exploiting certain security flaws in systems running **Unix**. Fortunately, it was quickly detected, and in 1990, Morris was convicted and fined. On 1 March 2002, Timothy Lloyd of Wilmington, Delaware, began serving a 41-month sentence for activating a virus that wiped out millions of dollars worth of his company's software three weeks after his dismissal by Omega Engineering of New Jersey in 1996. And on 1 May 2002, David Lee Smith was sentenced to 20 months plus three years of probation for unleashing the infamous *Melissa* virus of 1999.

Additional reading: *Encyclopedia of Computer Science*, pp. 1839–1841.

Computer vision   APPLICATIONS

Computer vision, a subfield of **artificial intelligence (AI)**, is the process whereby a **digital computer** is used to extract useful information about physical objects in its "field of view," where the latter is an image captured by a camera. The camera may be online and local, as it would be in a mobile computer, remote but still online through a wireless connection, or remote and offline, in which case computer processing begins with use of an **optical scanner**. Computer vision has many applications, including **robotics**, industrial **automation**, document processing, remote sensing, navigation, microscopy, medical imaging, and the development of visual prostheses for the blind. Computer vision is closely related to **image processing**, which transforms one image into another; **computer graphics**, which converts descriptions into images—the inverse of computer vision; and **pattern recognition**, the classification of identified objects. Computer vision is difficult because it is underconstrained. For example, an image is a two-dimensional (2D) projection of a three-dimensional (3D) scene, but there can be infinitely many 3D scenes that project the same 2D image. For example, the Big Dipper is not a planar configuration of stars. One of the primary tasks in computer vision is to find a set of constraints that allow a computer to interpret images unambiguously. An engineering approach to doing so relies on the intuition and prior knowledge of the system designer as to what the important image features should be and how they should be interpreted. Alternatively, one might try to design a system that can "learn" what the constraints are by observing the world through sensory input. This is the approach used in statistical pattern recognition, and more recently in artificial **neural net**works (*see also* **Perceptron**).

Additional reading: *Encyclopedia of Computer Science*, pp. 431–435.

Computerized Tomography (CT)
                                          APPLICATIONS
Tomography is a diagnostic technique using X-ray photographs in which the shadows of

structures before and behind the section under scrutiny do not show. Tomography was called *sectional radiography* by its inventor, Gustave Grossmann, a German who patented the process in France in 1934, long before a **computer** could play any role. British radiologist Edward Wing Twining was the first to coin *tomography*, from the Greek *tomos*, meaning "section." *Computerized tomography (CT)* is a relatively recent development in which only the section under scrutiny is irradiated, and a **digital computer** is used to construct an image of the section directly from the irradiation data without having to scan X-ray film. The result of applying the technology is called a CAT-Scan, where CAT is the acronym formed from the term *computerized axial tomography*. CT produces images of cross sections of the human body from measured attenuation of X-rays passed through the cross section at multiple angles. The first commercial CT scanner was built in 1972 by EMI, Ltd. of England. **General Electric (GE)** began producing X-ray equipment within a year of Roentgen's discovery of X-rays on 8 November 1895. A very successful full-body CT scanner was developed in 1977 by J. M Houston and N. Rey Whetten of the GE Research Laboratory. The 1979 Nobel prize in medicine was awarded to EMI scientists Allan M. Cormack of South Africa and Godfrey N. Hounsfield of England for their pioneering contributions to CT.

Additional reading: Gabor Herman, "Tomography, Computerized," in *Encyclopedia of Computer Science*, pp. 1783–1786.

Concurrent programming

PROGRAMMING METHODOLOGY
Concurrent programming relates to the theory and practice of *communicating sequential processes*, a term that took hold with the publication of a 1978 paper and an influential 1985 book of that title by C. A. R Hoare (*see* reference). A concurrent program is composed of multiple processes or *threads*, each of which is individually a sequential computation. **Parallel computers**, **time sharing** operating systems, **real-time systems**, and **distributed systems** all must necessarily run concurrent programs, and **multiplexing, multiprogramming, multitask-**

ing, and **multithreading** all must coordinate communicating sequential processes. Concurrent programming is much more difficult than sequential programming. A sequential program is correct if it produces correct results for all possible inputs, but a concurrent program must also produce correct results for all possible timings (interleaved sequences) of concurrent statements. To prevent interference from other processes, certain sequences of instructions must be *atomic*—that is, indivisible. A block of **code** that must be atomic (uninterruptible) to guarantee correct execution is called a *critical section* and is implemented by providing *mutual exclusion* through an interlock mechanism.

It is difficult to pinpoint a date that marks the beginning of concurrent programming. *Ad hoc* techniques for dealing with the concept arose as soon as computers were built that allowed simultaneous **input / output (I/O)** and computation and when the **interrupt** became common. In the reference, Hoare credits Robin Milner as having made "the major breakthrough in the mathematical modeling of concurrency" in 1980, but Hoare himself made important contributions, as did Edsger Dijkstra, Per Brinch Hansen, and Leslie Lamport.

Perhaps the first language specifically designed for concurrent programming was *Concurrent Pascal* in 1975. Other languages that support concurrency include **Modula-2**, based on a synchronization concept called a *monitor*; *Occam*, which uses *message passing*; **Ada**; **Java**; and *SR*, a language that supports several synchronization methods. There also exist concurrent variants of the languages **C**, **C++**, and the **functional language** *ML*.

Additional reading: Hoare, *Communicating Sequential Processes*.

Connection Machine
COMPUTER TYPES
In the mid- to late 1980s, the *Thinking Machines Corporation* (TMC), founded by Danny Hillis in 1983, produced a family of high-performance **parallel computers** marketed as *Connection Machines*. The largest member of the family was the 65,536 processor CM-2 with performance in excess of 2,500 MIPS (millions of instructions per second), and floating-point performance above 2,500 Mflops (millions of

floating-point operations per second). Connection Machines were used in a range of applications including **information retrieval** from **databases**, **image processing**, computer-aided design (*see* **CAD / CAM**), and a wide range of floating-point–intensive scientific calculations. The CM-2 had 512MB (megabytes) of **memory** at a **bandwidth** of 300GB (billion **bits**) per second and eight **input / output (I/O)** data **channels** rated at 40MB per second that collectively provided a maximum transfer rate of 320MB per second. At the height of their popularity, there were over 80 Connection Machines installed, including sites such as the U.S. Los Alamos National Laboratory, the National Center for Atmospheric Research (NCAR), and dozens of universities. TMC underwent Chapter 11 reorganization in the early 1990s, splitting into several smaller companies with the one still called *Thinking Machines* being primarily a **software** company.

Additional reading: *Understanding Computers: Alternative Computers*, pp. 56–59.

Constant
PROGRAMMING

A *constant* is a number or character **string** defined in **assembly language** or in a **high-level language** that cannot or should not be changed during execution of a **program**. The only "constants" that literally cannot be changed by a malicious program reside in **read-only memory (ROM)**, but modern high-level languages are quite good at generating **object code** that protects items declared to be constant by monitoring their **memory** addresses. For examples of how constants are defined and used, *see* **Declaration**.

Control Data Corporation (CDC)
INDUSTRY

Control Data Corporation (CDC), founded in 1957 in St. Paul, Minnesota, by William C. Norris, was the first computer company to be publicly financed. Early success was based primarily on the **CDC 1604** and its successor, the CDC 3600. By the mid-1960s, the **CDC 6600**, designed by Seymour Cray, was recognized as the world's foremost **supercomputer**. By 1965, **IBM** and CDC were the only two profitable computer companies. An early CDC acquisition, Cedar Engineering, manufactured **peripheral** equipment for computers. Cedar Engineering eventually grew into Imprimis Technology, Inc., the largest supplier of high-performance data storage products for the OEM (original equipment manufacturer) market. CDC's computer-based services operation expanded in 1967 when it acquired the Arbitron Company as part of CEIR, a **software** company that is no longer in business. In 1968, when IBM announced its intent to add an allegedly powerful model 360/80 to its **IBM 360 Series** line, sales of the 6600 dropped precipitously. Calling the model 80 a "paper tiger," Norris filed an antitrust suit against IBM and won. CDC realized more than $100 million in the suit and, as part of the settlement, acquired the Service Bureau Company from IBM.

In 1973, CDC's business was severely affected by the loss of Seymour Cray, who left to form his own company, **Cray Research**. CDC sales slowly declined over the next two decades, and in 1992, the company's board of directors separated the corporation into two independent businesses—Control Data Systems Inc., an open systems integrator, and Ceridian Corporation, an information services company. By 1997, CDC had tripled its depressed 1992 value. On 23 September 1997, Control Data Systems, Inc. completed a transition from a public company to a private one, the new owners being the investment firm of Welsh, Carson, Anderson & Stowe (WCAS). The new owners intend to provide **electronic commerce** solutions to large organizations worldwide by focusing on two industries: information services and health care. Thus former **mainframe** maker Control Data Systems evolved into a systems integrator. For the fiscal year ending in December 1996, CDS earned $16 million on sales of $306 million, a one-year growth of 32.8%. But by 1999, its annual revenue had fallen to $180 million. On 1 September 1999 the company was sold to British Telecommunications (BT), which added CDS to its Syntegra Systems Integration division.

Additional reading: *Computing Encyclopedia*, vol. 1, p. 158.

Control structure

A *control structure* is a **high-level language** statement pattern that captures a unit of logic without using explicit transfers of control (e.g., `goto` statements). In a 1964 paper, Corrado Böhm and Giuseppi Jacopini proved that every "**flowchart**" (**program**), however complicated, could be rewritten to use only repeated or nested control structures of no more than three different kinds—a *sequence* of executable **statements**, a *decision* clause of the form `if-then-else`, and an *iteration* construct that repeats a sequence of statements `while` some condition is satisfied. The **Fortran** *do-loop*, the **Algol 60** and **Pascal** *for-loop*, and the Pascal *repeat-until* are forms of **iteration**. Pascal and some other languages also support an additional control structure invented by C. A. R. Hoare called a *case statement*. This structure allows the programmer to list a sequence of statements of which only one will be selected and executed depending on the value of a case **parameter**. Each control structure has a single entry point and a single exit, which allows their interconnection in accord with the precepts of **structured programming**.

Additional reading: *Encyclopedia of Computer Science*, pp. 454–460.

Cooperative computing

Cooperative computing (sometimes called "farming") is a form of highly *distributed computing* done by several thousand or even, potentially, several million **digital computers** (*see* **Distributed system**). The "cooperators" are the owners of the widely dispersed **personal computers (PCs)** who voluntarily devote machine cycles that would otherwise be wasted to computations "farmed out" to them by some central authority. This sponsoring authority then harvests and combines the partial results. Cooperative computing can be thought of as inverse **time sharing**; instead of multiple remote terminals competing for the attention of a single host computer, the host computer asks for a share of time on thousands or potentially millions of remote personal computers. The magnitude of the world's unused machine cycles is prodigious. Current high-performance PCs are capable of executing 2,000 million **instructions** per second (2,000 MIPS, or 2 GIPS, where the *G* stands for "giga"), 7.2 trillion per hour, or about 50 trillion instructions over a typical "third shift" during which its owner typically sleeps. At least another 50 trillion instructions can be appropriated during the day by running the cooperative **software** in the background while the primary user is running other programs (or none at all) in the foreground (*see* **Multitasking**). So either the PC wastes those 100 trillion instructions, or its owner can enroll in a consortium of altruistic users, very few of whom are known to one another. When organized this way, a loosely coupled but coordinated **network** of some thousands of PCs is more powerful than the world's fastest **supercomputer**, but only for a restricted class of problems. Typical are the attempts to find previously unknown **prime numbers** or to find the factors, if any, of a number thought likely to be composite. Large composite numbers are a staple of **public-key cryptosystems (PKCs)** where the supposed difficulty of factoring 100-digit or larger numbers (in order to recover the decryption key) is the basis for the security of the system.

The search for ever-larger prime numbers has centered on finding more so-called *Mersenne primes*, primes that have the form $2^n - 1$ (and which therefore consist of a sequence of n consecutive 1s in binary) where n itself is prime. Since not all numbers of this form are prime, number theorists value the discovery of new Mersenne primes. On 14 November 2001, 20-year-old Michael Cameron, a member of a consortium called GIMPS (Great Internet Mersenne Prime Search), used spare computer time on his 800 MHz PC to find the thirty-ninth Mersenne prime, $2^{13,466,917} - 1$, which has more than 4 million digits. The result was the culmination of a two-and-a-half-year search by tens of thousands of GIMPS volunteers (*see* www.mersenne.org/prime.htm). It is projected that the first billion-digit Mersenne prime will be discovered in 2009. The largest currently running cooperative computing project is the Search for Extra-Terrestrial Intelligence (SETI) run by the University of California at Berkeley (*see* www.seti.org). Almost 2 million users, about half active at any one time, use a special Web **browser** that processes radio astronomy data from SETI in the background.

Additional reading: *Encyclopedia of Computer Science*, pp. 460–462.

Coroutine   PROGRAMMING

A *coroutine* is one of a pair of subprograms that remains active throughout the duration of a running **program** and which furthers the program's goal by passing and receiving control to and from the "coroutine" with which it is paired. The term was coined in 1958 by Melvin Conway, who invented the concept and used it in an **assembler** that he wrote. Independently and at about the same time, Joel Erdwinn and Jack Merner also developed the same idea, which they called "bilateral linkage." Each time control reenters a coroutine, it resumes execution where it left off with its data state retained. Use of coroutine pairs simplifies the implementation of some **algorithms**. For example, a program that needs to determine whether two **trees** of differing structure contain the same terminal (leaf) nodes in the same order might use a separate coroutine to traverse each tree. Other applications of coroutines include **operating systems**, **compilers**, and discrete event **simulation** programs. The language **Simula** supports discrete event simulation with flexible coroutine mechanisms, and **Modula-2** provides a *processes* library module with **procedures** that support synchronization, scheduling, and mutual exclusion for coroutine-based threads. Coroutines are also used in **text editors**, **artificial intelligence**, and **sorting** algorithms. The abstraction data languages Alphard and **CLU** support coroutines (*see* **Abstract data type**), **Java** supports coroutine threads, and **C++** has a coroutine library. **Functional languages** that use *lazy evaluation* (compute values only as needed) also employ coroutine-like control.

Additional reading: *Encyclopedia of Computer Science*, pp. 465–466.

Cosmic Cube   COMPUTER TYPES

The *Cosmic Cube* was an experimental **super-computer** developed at the California Institute of Technology in the early 1980s. The largest machine had 64 processors that were coupled in the form of a hypercube. A *hypercube* consists of p nodes, where $p = 2n$ for some positive integer n. Additionally, each node had a direct connection to n other nodes. One advantage of this **computer architecture** is that each processor has several direct lines of communication to other nodes in the system. This was particularly important for the early generation of **parallel computers** in which message-passing was quite expensive between nodes that were not directly connected. The hypercube architecture was used in the iPSC/1, iPSC/2, and iPSC/860, the first three generations of **Intel** parallel machines. A disadvantage of this design is the cost of increasing the machine size; it does not scale well. To add processors to a machine with 64, or 2^6, nodes and remain consistent with hypercube architecture, the number of nodes must be doubled to 128, or 2^7, nodes. But then the number of interconnections quadruples, since there would be 128 nodes with direct connections to seven other processors instead of the original 64 nodes with direct connections to six other processors. This contrasts with the mesh architecture of the 20×20 **Intel Paragon** of 1993, which was something like a large **ILLIAC IV**. In the Paragon, each node had a direct connection to only four other processors, no matter how large the machine. This **cellular automaton** design proved more flexible because it allowed a machine to grow incrementally rather than by the factor of four needed for each successively larger Cosmic Cube.

Additional reading: http://www.sdsc.edu/Gather Scatter/gsmar94/GS_Mar9.html

CPC. *See* Card Programmed Calculator (CPC).

CP / M   OPERATING SYSTEMS

CP / M (Control Program / **Microprocessor**) was a command-line **operating system** written by Gary Kildall in 1973 for use with the many **S-100 bus** computers of that era that used **Intel 8080** or **Zilog Z-80** microprocessors. Kildall created CP / M to support the language PL / M (Programming Language / Microprocessor) that he had previously implemented for the **Intel**

4004 while working as a consultant to **Intel**. CP / M used a file system in which **files** were given (up to) eight-character names followed by a *file extension* consisting of a period ("dot") and three additional characters such as .txt and .exe. This addressing structure, along with the use of file *volumes* labeled A:, B:, C:, and so on, carried over into the later **MS-DOS** and endured for over 20 years. Kildall formed the company *Digital Research* to market CP / M, and after Intel declined interest, IMSAI became the first vendor to license CP / M for use on its **IMSAI 8080**. Other contracts followed, but the biggest opportunity was missed. When Digital Research reacted tentatively to a proposal from **IBM** to upgrade CP / M for use on 16-**bit** rather than just 8-bit microprocessors, **Microsoft** moved aggressively to license its MS-DOS for use on the **IBM PC** as PC-DOS. Kildall's company did make a 16-bit operating system called CP / M 86 (the "86" standing for both 1986 and the last two digits of **Intel 80×86**). Computer professionals considered an upgraded version called DR-DOS to be superior to MS-DOS, but the combined marketing prowess of IBM and Microsoft proved too great to overcome.

Additional reading: *Computing Encyclopedia*, vol. 1, p. 159.

Cray-1   COMPUTERS

The *Cray-1*, designed by Seymour Cray, was announced by **Cray Research** in 1976 with first delivery made in March to the U.S. Los Alamos National Laboratory. Ten times faster than the **CDC 6600**, the Cray-1 had 12 functional units and extensive buffering between the instruction stream and the **central processing unit (CPU)**. **Memory** options ranged from 265,000 to a million 64-bit **words**. The Cray-1 was a *vector computer* capable of a high degree of parallelism (*see* **Parallel computer**). The machine used only four types of chips, each containing only a few densely packed **logic circuits**. Circuit modules, arranged in a three-quarter circle to reduce interconnection lengths, were cooled by liquid Freon, as was the CDC 6600.

Additional reading: *Encyclopedia of Computer Science*, pp. 564–565.

Cray Research   INDUSTRY

Cray Research, founded in 1972 by Seymour Cray after leaving **Control Data Corporation (CDC)**, was the first of three computer companies to bear his name. Up until 1989 when Cray left to form the *Cray Computer Corporation* (CCC), Cray Research produced the Cray-1 and successively more powerful **supercomputers** such as the multiple-processor X-MP in 1982, the Cray-2 in 1985, and the Y-MP in 1988. After Cray left, Cray Research built the first computer capable of a sustained processing rate of 1 Gflops (billions of floating-point operations per second), the C90 of 1991, and the 2 Gflops T90 of 1995. In 1996, Cray Research was sold to **Silicon Graphics, Inc.**, which produced the first Tflops (teraflops, 1 trillion flops) computer, the Te/900. In March 2000, SGI sold its Cray division to the Tera Computer Company, which has continued operations as Cray, Inc. In the meantime, the Cray Computer Corporation had not fared well, never completed its planned Cray-3, and the company closed in 1995, the year before Seymour Cray's tragic death in an automobile accident.

Additional reading: *Computing Encyclopedia*, vol. 1, p. 163.

CRT. *See* **Cathode Ray Tube (CRT)**.

Cryptography. *See* **Data encryption**.

CTSS   OPERATING SYSTEMS

CTSS (Compatible Time-Sharing System) was a historic early **time sharing** system implemented in 1963 for the IBM 7094 (*see* **IBM 700 Series**) at the MIT Computation Laboratory by a team led by Fernando Corbató and Robert Fano. To allow CTSS to run on the 7094, **IBM** added several new **instructions** and a **memory** boundary **register** to the processor. Additionally, the machine had two **magnetic core** memory banks of 32,768 36-bit **words** instead of one. Core bank A held the CTSS supervisor, and core bank B was used for user **programs**. When the machine was running a program in B-core, **input / output (I/O)** and certain other instructions were forbidden; executing them caused an **interrupt** that transferred control to the supervisor in the A-core. CTSS, which ran its first four-terminal demonstration in November 1961, was the first time-shared **operating system** administered under **Project MAC**, and experience with it led to its successor, **Multics**.

Additional reading: Waldrop, *The Dream Machine*, pp. 190–192.

CUBA   COMPUTERS

CUBA, the *Calculateur Universal Binaire de l'Armement*, the first **digital computer** built in France, was designed by François-Henri Raymond, the founder of the company Société d'Electronique appliquée et d'Automatisme (SEA), and delivered to the unit of the French government from which it derived its name in 1952. CUBA was a binary three-address **stored program computer** (*see* **Multiple-address computer**). SEA went on to develop many other computers during the period 1950–1955, including the CAB 2000 and 3000.

Additional reading: Moreau, *The Computer Comes of Age*, p. 63.

Cursor   DEFINITIONS

A *cursor* is a small movable blip of light that marks the **monitor** location where the next **character** typed at the **keyboard** of a **personal computer** or **workstation** will be displayed or indicates the choice of an action to be invoked (*see* **Event-driven programming**). Cursors usually have a distinctive shape. In the first instance discussed, the cursor may be a simple flashing vertical bar of the size of the character just typed; in the second, it is typically a small arrow that can be moved with a **mouse** or **trackball** to indicate a desired choice—selection from a menu, perhaps, or invocation of a **program** by clicking on its **icon**.

Curta calculator   CALCULATORS

The *Curta*, designed by the Austrian Curt Herzstark in 1943 while a prisoner in Buchenwald, was a pocket-sized handheld **calculator**. Despite its small size, it was based on the stepped-drum principle of the **Leibniz calculator**. Over 14,000 Curtas were sold from 1948 to 1972. The Curta was the last stepped-drum calculator and, given the advent of the **electronic calcu-**

lator, will undoubtedly remain the smallest mechanical calculator ever built.

Additional reading: Kidwell and Ceruzzi, *Landmarks in Digital Computing*, pp. 31–32.

Figure 12. The Curta of the late 1940s, the world's first and last mechanical handheld calculator. (Courtesy of Hewlett-Packard)

Cybernetics   COMMUNICATIONS

Cybernetics was coined by Arturo Rosenblueth, Norbert Wiener, and Julian Bigelow in a 1943 paper and popularized through its use as the title of Wiener's classic 1948 book on the subject. Derived from the Greek *kybernetes*, or "steersman," Wiener defined the field as "the study of control and communication in the animal and machine." Inspired by work done before and during World War II on mechanical control systems such as servomechanisms and artillery targeting mechanisms, Wiener sought to develop a general theory of organizational and control relations in systems. **Automatic control theory** has since developed into a full discipline in its own right, but what distinguishes cybernetics is its emphasis on control and communication not only in engineered artificial systems but also in natural systems that evolve, such as organisms and societies. Cybernetics is a multidisciplinary subject that can be viewed from many perspectives. Philosopher Warren McCulloch considered it to be an experimental epistemology concerned with the interaction between an observer and his or her environment. Management consultant Stafford Beer defines cybernetics as the science of effective organization. Anthropologist Gregory Bateson noted that whereas physical science deals with matter and energy, cybernetics focuses on form and pattern. For educational theorist Gordon Pask, cybernetics is the art of manipulating defensible metaphors, showing how they may be constructed and what can be inferred as a result of their existence. The Belgian scientist Ilya Prigonine won the Nobel Prize in Chemistry for applying cybernetic principles to show how life can arise and evolve in a closed system containing a power source (such as a sun) in apparent contradiction of the Second Law of Thermodynamics. Cybernetics has been a crucial influence on the development of **computer science**, in particular in **information theory**; **game theory**; **automata theory** including **cellular automaton**, **artificial intelligence (AI)**, and **neural nets**; computer modeling and **simulation**; **robotics**; and **artificial life (AL)**. Although there are European universities that offer degree programs in cybernetics, the term is no longer widely used. But its prefix remains popular and is continually being used to coin neologisms such as **cyberspace**, *cybernaut*, and *cybercrime*.

Additional reading: Wiener, *Cybernetics*.

Cyberspace   GENERAL

As described in William Gibson's 1984 novel *Neuromancer*, "cyberspace" was an artificial environment created by and maintained by **computers**. Gibson's cyberspace conveyed realistic detail to all five senses, a technologically simulated experience that has come to be known as **virtual reality**. *Cyberspace* is now associated primarily with interaction over computer **networks**, most particularly the combination of the **Internet** and the **World Wide Web** called "the Net." In this sense, cyberspace has become a synonym for *information superhighway*, a term popularized in 1978 by then-Congressman Albert Gore to refer to a unified, interactive system of electronic communication analogous to the U.S. Interstate Highway System. Cyberspace encourages the formation of *virtual communities* that share a common interest. Such communities operate without regard to national boundaries, spreading the concept of free speech to many areas of the world to which that idea had been utterly foreign. And to this author, among many others, the concept of a technological *cyberspace* is reminiscent of the more ethereal noösphere of the French Jesuit philosopher Teilhard de Chardin.

Additional reading: *Encyclopedia of Computer Science*, pp. 474–475.

Cyclone   COMPUTERS

In 1956, Iowa State University started construction of *Cyclone*, a large-scale IAS-class **digital computer** similar to the **ILLIAC** at the University of Illinois (*see* **IAS computer**). Cyclone was named for Iowa State sports teams, and the university now has a *Cyclone Computer Laboratory*. Upon completion, the machine's **memory** was increased to 16,384 40-bit **words**. Cyclone underwent many modifications until the mid-1960s when it was retired and replaced by a computer called *Symbol*.

Additional reading: http://www3.ee.iastate.edu/pop/History%20of%20ECPE/ 1950s/Cyclone%20Lab.html

D

Dalton adding machine CALCULATORS

The *Dalton adding machine* of 1902, designed by Hubert Hopkins for James L. Dalton, was the first ten-key adding machine and soon became one of the most popular **calculators** of this kind. Earlier models had glass inserts to allow customers to watch the gears move during calculations. Over 150 models of the Dalton were designed in the 20 years after its introduction. Hopkins later teamed with his brother William, a prolific inventor in his own right, and John C. Moon to form the Moon-Hopkins Company and market the Moon Hopkins Adding Typewriter, a combination adding machine and **typewriter**. The company sold 3,226 machines by June 1921, but because of poor profits and patent disputes with Dalton, Moon-Hopkins was absorbed by **Burroughs** later that year. Dalton, in turn, was one of several companies that merged to form **Remington Rand** in 1927.

Additional reading: Martin, *The Calculating Machines*, pp. 133–137, 256–263.

Dartmouth Time-Sharing System (DTSS) OPERATING SYSTEMS

The *Dartmouth Time-Sharing System (DTSS)* was created in 1964 by John Kemeny and Thomas Kurtz of Dartmouth College in Hanover, New Hampshire, in order to promote *computer literacy* by making their **Basic** language available to virtually all students on campus.

DTSS is one of the four historic early **time sharing** systems, the others being **CTSS** and **Multics** of **Project MAC** at MIT and a system for the PDP-1 at **Bolt, Beranek, and Newman (BBN)** (*see* **DEC PDP series**). DTSS was initially implemented on a GE 235 computer (*see* **GE 200 series**) and later moved to the **GE 600 series** and to the Honeywell 6600.

Additional reading: Campbell-Kelly and Aspray, *Computer: A History of the Information Machine*, pp. 209–210.

Data DEFINITIONS

Literally, *data* is the plural of *datum*, so that you might find a sociologist who writes, "The data are inconclusive." But information scientists are more likely to consider *data* to be a collective noun and write something like, "The data is available on CD-ROM." To them, *data* is information collected by humans or automatically recorded in **machine-readable form**. Three decades ago, the qualification "machine-readable form" essentially meant data on **punched cards**, **punched paper tape**, or **magnetic tape**, but the term is seldom heard now because a **digital computer** can be equipped with **peripheral** devices that can record and digitize sound, radiation, infrared and **radio** signals, and visual images (*see* **Digital camera**; **Optical scanner**). A glance at the article titles that follow indicates that data can be based, compressed, encrypted, caused to flow, mined, processed, recognized, structured, transmitted, typed, visualized, warehoused, and reduced to **bits**.

Data compression

INFORMATION PROCESSING

Data compression reduces the **memory** space needed to store **data** by changing its representation. It is widely used to pack more **files** onto **hard disks**, to decrease the time taken to send files over the **Internet**, to transmit fax messages quickly over **telephone** lines (*see* **Facsimile transmission [fax]**), and to increase the apparent speed of **modems**. A compression **algorithm** is either *lossless* or *lossy*. Lossless methods enable compressed data to be restored to their precise original form. Lossy methods make small changes to the data to make it more compressible—for example, they might reduce the amount of detail in an image or decrease the quality of an audio recording. The popular **MP3** music compression algorithm is lossy, but almost undetectably so to the human ear. An early compression method was **Huffman encoding** of 1952. This works on a principle similar to **Morse code**—the more common **characters** are represented by shorter **codes**. An important advance of the 1970s was the discovery of *Ziv-Lempel coding*, currently the most widely used approach for commercial compression systems. The method is named for its inventors, Jacob Ziv and Abraham Lempel. Variants of the Ziv-Lempel method are used by archivers such as Phil Katz's *PKZIP* of 1986 and later variants such as *WinZip*, and for *graphics interchange format* (gif) files. A similar **program** called *Stuffit* was written for the **Apple Macintosh** by Raymond Lau in 1987. These systems are very fast and give good compression. The effectiveness of different methods can vary greatly depending on the type of file being compressed. The best lossless methods compress English text to about half the original size. Less dense text such as program **source code** can be compressed to a greater degree. A black and white **bit map** image can usually be compressed to a tenth of its original size, whereas a full color image resists lossless compression.

Additional reading: *Encyclopedia of Computer Science*, pp. 492–496.

Data encryption

INFORMATION PROCESSING

Data encryption consists of the encipherment of messages and stored **files** using cryptographic techniques so that their information cannot be read by unauthorized persons. *Cryptography* is an ancient art and science whose mechanized use during World War II was, along with the need to solve complex partial differential equations related to nuclear weapons, one of the two driving forces that led to the development of electronic **digital computers** (*see* **Colossus**; **Hagelin machine**). The current leading methods of data encryption are the **Data Encryption Standard (DES)**, **Pretty Good Privacy (PGP)**, and **public-key cryptosystems**, based on the **RSA algorithm**.

Additional reading: *Computing Encyclopedia*, vol. 1, p. 175.

Data Encryption Standard (DES)

APPLICATIONS

The *Data Encryption Standard (DES)* was designed in 1971 by Horst Feistel of **IBM** and approved as a standard by the U.S. National Bureau of Standards (NBS) in 1976. The DES, which Feistel said was inspired by and based on Claude Shannon's **information theory**, enciphers a 64-bit message block under control of a 56-bit key to produce a 64-bit ciphertext. The enciphering operations consist of 16 **iterations** of an exchange of left and right halves of the 64-bit message, followed by replacement of the right half with the bit-wise *exclusive OR* of the right half and a 32-bit word that is a complicated function of the left half, the key, and the iteration number (*see* **Boolean algebra**). DES has been implemented by a large number of manufacturers on special-purpose **VLSI** chips that can encipher at Mb/s (megabit per second) rates. DES was widely used for over two decades, but in the face of criticism that its key length is too short to withstand discovery by brute-force search on ever more powerful computers, the standard is gradually yielding to newer techniques such as **Pretty Good Privacy**

(PGP) and **public-key cryptosystems** based on the **RSA algorithm**.

Additional reading: Levy, *Crypto*, pp. 37–65.

Data mining   INFORMATION PROCESSING

Data mining, sometimes called *information archeology* or, more formally, *knowledge discovery in databases* (KDD), is the process of automating the discovery of patterns in information contained in a large **database**. Statisticians have been analyzing masses of **data** for centuries, but their usual practice was to start with a hypothesis about the relationships among data attributes and then use statistical tools to validate or disprove it. With data that exhibits tens or even hundreds of attributes, this methodology becomes an impossibly time-consuming process. The origin of data mining can be traced to an influential 1962 paper by the American statistician John Tukey, who suggested that the search for relationships could be automated through the use of **artificial intelligence (AI)** and subsequent **data visualization**. But even though a conference on the subject was organized by Leo Breiman and held in Dallas, Texas, in 1977, it was not called *data mining* until the mid-1990s when **IBM** did so in conjunction with several patents that they filed. Breiman, emeritus professor of statistics at the University of California at Berkeley, was elected to the U.S. National Academy of Sciences in 2001 for his fundamental work in data mining, decision **trees**, and **pattern recognition**. Examples of data mining applications include fraud detection in banking and telecommunications; marketing; analysis of scientific data involving the cataloging of objects; and problem diagnosis in manufacturing, medicine, and **networks**. Data mining techniques are particularly relevant to massive data collections for which the processes that generated the data are not well understood.

Additional reading: *Computing Encyclopedia*, vol. 1, pp. 179–180.

Data path   DEFINITIONS

The *data path* of a binary **digital computer** is the number of **bits** that can be fetched from main **memory** in one memory-access **instruction**. For traditional **mainframe** and minicomputers, the computer's data path tended to be equal to its single-**precision** integer length, which was usually the same as its **word length**—at least 32 bits, but perhaps 36 or 48, as was the case on the scientific members of the **IBM 700 Series** and on the **Philco Transac S-200**, respectively, or even 60 bits, as it was on the **CDC 6600**. But since **microprocessor** development could only approach those lengths incrementally, the term *data path* was coined to distinguish word length from the portion of a **word** that could be quickly accessed. The **Intel 4004**, the first microprocessor, had a path length of only 4 bits, and its successor, the **Intel 8008**, only 8. But the Intel 8086 and the **Motorola 68×××series** reached 16 bits, and the latest members of the **Intel 80×86** family have finally reached the 32-bit data path needed to fetch a full integer in one memory access.

Data processing

INFORMATION PROCESSING

Data processing is a 1950s term of uncertain origin that initially connoted use of **digital computers** to process business **data**, although there is certainly such a thing as scientific data processing as well. The more recent and obviously generalized term is *information processing*.

Additional reading: *Encyclopedia of Computer Science*, pp. 502–504.

Data recognition   DEFINITIONS

Data recognition is the use of a **digital computer** and its associated **peripheral** equipment to analyze and identify patterns in input **data**. Since speech, typed **characters**, and photographs are all obvious examples of patterned information, **speech recognition**, **optical character recognition (OCR)**, and **pattern recognition** are all types of data recognition. But *see also* **data mining**, which seeks to extract nonobvious patterns from data thought to be devoid of them.

Data reduction   APPLICATIONS

Data reduction is a statistical term that connotes the replacement of an extensive sequence of measured values with a small number of **parameters** that can be used to characterize or reconstruct the trend of the original **data**. Perhaps the earliest example of data reduction is the invention of the **least squares method** by Adrien Marie Legendre in 1805. The method can be used to compute the slope m and intercept b of the straight line $y = mx + b$ that best characterizes data with a linear trend in the sense that the sum of the squares of the deviations between measured values and their corresponding points on the approximating line is minimized.

Additional reading: *Encyclopedia of Computer Science*, pp. 963–964.

Data structure   GENERAL

A *data structure* is a collection of data values, the stated interconnections (if any) among those values, and the one or more operations to be performed on them. If any one of these attributes is missing, the entity under consideration is not a data structure. Since the simplest form of interconnection of values is none, the simplest data structure is the **set**. The next simplest data structure is the **array**, a naturally occurring one in the sense that the **words** (or **bytes**) in any **random access memory** are stored sequentially, making the array operation of *indexing* to copy or update an individual word or byte trivially easy. The most complex data structure would be the complete **graph**, one in which every *node* (data value) is connected to every other. Most data structures can be implemented in more than one way. A particular representation of the data structure in **memory** is called a *storage structure*. Data structures such as one-, two-, and three-dimensional arrays (or higher) whose data values are contiguous are called *geometric* or *Cartesian* data structures in honor of René Descartes, the inventor of **analytic geometry**. Data structures such as **linked lists** and graphs whose nodes are connected by **pointers** and whose size can grow or shrink during processing are called *dynamic data structures*. The *persistent data structure*, invented by Robert Tarjan, Daniel Sleator, and James Driscoll in 1980, is one that preserves its old versions; that is, previous versions may be queried in addition to the latest one. In addition to **set**, **array**, and **graph**, other articles that define data structures are **database**, **deque**, **file**,

linked list, **pushdown stack**, **queue**, **record**, **string**, and **tree**. Of these, the **database** and the **record** (and hence the **file**, a sequence of records) may contain data of different **data type** and hence are *inhomogeneous*; all the rest are *homogeneous*. Implementation of an efficient **algorithm** is often possible only through use of a particular data structure, the point of view in Niklaus Wirth's classic 1975 textbook *Algorithms + Data Structures = Programs* and, of course, in the **Knuth textbooks**. Consistent with the definition of the lead sentence, each article that describes a data structure states the relationship among its data values and one or more operations that are typically performed on it.

Additional reading: *Encyclopedia of Computer Science*, pp. 512.

Data transmission   DEFINITIONS

Data transmission is a broad term that encompasses any means of sending information or **data** from one point to another. Conventional surface or air mail qualifies, as would **radio, telegraph, telephone**, and **television**, but the term more usually implies electronic transmission of data stored in the **memory** of one computer to the memory of another computer or device at a distant location. The best examples are **file** transfers over a **network**, particularly the **Internet**, and **facsimile transmission (fax)** either telephonically or over the Internet.

Data type   PROGRAMMING

In a **high-level language**, the entities that comprise a **data structure** have a specific *data type*. The simplest data type could be called a **bit** but is usually called *Boolean* because the only values that it can assume, 1 and 0, correspond to *true* and *false* in **Boolean algebra**. The next simplest could be called *byte*, but is usually called *character*, because a **byte** is just the right size to hold an **ASCII** or **EBCDIC** character (although two bytes are needed to hold a **Unicode** character). The data type **string**, then, is a one-dimensional **array** of **characters**. The most commonly used data types are *integer* and *real*. The set of integers usable on a **digital computer** is necessarily (because of **memory** limitations) a finite subset of the infinite **set** of integers of mathematics; each **programming language** supports some maximum and minimum integer that can (easily) be used. The real number axis in mathematics is continuous (between any two real numbers there are an infinite number of others) and is infinite in extent. In computing, this axis is simulated through use of **floating-point arithmetic**. Floating-point numbers form a discrete sequence of finite extent. Other data types supported in some, but not all, programming languages are *double-precision integer*, *double-precision real*, *complex*, and **record** (called *struct* in **C++**).

Additional reading: *Encyclopedia of Computer Science*, pp. 512–513.

Data visualization   COMPUTER GRAPHICS

Data visualization transforms numerical or symbolic **data** into geometric computer-generated images, often with liberal use of "false color" to enhance relationships that are otherwise difficult to discern. Data visualization in a general sense antedated widespread use of computers in that scientists and engineers have always sought to make elaborate charts and graphs that best illustrate possibly subtle relationships among seemingly disparate data. Perhaps the best-known example is the French engineer Charles Joseph Minard's famous depiction of the losses incurred by Napoleon's army during its retreat from Russia in 1812, a chart contained in Edward Tufte's classic 1983 book *The Visual Display of Quantitative Information*. Visualization tools allow scientists to observe and interpret the results of their computations in ways far superior to straightforward two-dimensional graphs or numeric tabulation. Generated images can be displayed statically or as animated sequences or movies shown in real time on a **personal computer** or **workstation**. Data visualization is closely related to the still-newer field of **data mining**. They differ in that visualization depicts relationships known to exist but that are not readily apparent when data is graphed conventionally, whereas data mining uses a blend of algorithmic and **heuristic** methods to uncover regularities and relationships not envisioned in advance, even by those who gathered the data being processed.

Additional reading: *Encyclopedia of Computer Science*, pp. 1550–1553.

Data warehouse

INFORMATION PROCESSING

The term *data warehouse* was coined in 1985 by Bill Inmon. In his book *Data Architecture* he defines the term as "a subject-oriented, integrated, time-variant, and nonvolatile collection of data in support of the management decision making process." By "subject-oriented" he meant information about a particular subject rather than just a record of a company's ongoing transactions. *Integrated data* is **data** gathered into the warehouse from a variety of sources and merged into a coherent whole. "Coherence" does not imply a common structure, however; warehouse data may be stored in a variety of formats and is much less structured than that used in a **relational database** or **management information system (MIS)**. This makes **software** used to query a data warehouse harder to write, but the advantage is that no one needs to decide in advance exactly what kinds of data will be warehoused. In Inmon's vision, data in the warehouse would be stable; additional data is constantly added, but none is ever removed. This has proved impractical, however; as warehouses approached terabyte size (trillions of **bytes**), it became necessary to limit the number of historical data generations stored. After a third year of use, perhaps, data for the oldest month is discarded and the newest month added.

Additional reading: Inmon, *Data Architecture*.

Database   DATA STRUCTURES

A *database* is a self-describing collection of interrelated **files** so arranged as to facilitate **information retrieval** in response to *queries*. When the term *database* is qualified by an adjective that implies the structure of the collection, the augmented terms meet the definition of **data structure**. Examples are *hierarchical database*, which is based on the **tree**, the *network database* in which informational nodes form a **graph**, and the **relational database**, which is based on two-dimensional **arrays** called *tables*. The concept of a database arose from the work of Charles Bachman, a **software** designer at **General Electric (GE)** and later **Honeywell**, who created a system called the *Integrated Data Store*. His vision was articulated in a 1973 paper called "The Programmer as Navigator" in which he described database technology as being a tool for "navigating" through stored **records** in search of the answer to a query.

Additional reading: *Computing Encyclopedia*, vol. 1, pp. 184–185.

Database Management System (DBMS)
INFORMATION PROCESSING

When a **database** is supported by a query language such as **SQL**, used extensively throughout a company, and maintained by a *database administrator*, it becomes a *Database Management System (DBMS)*, a particular form of **Management Information System (MIS)**. Following early experiments with the three types of database, current practice heavily favors the **relational database**. Significant **software** products based on that model and some sort of query language similar to SQL have been *dBase* and its successors, written by Wayne Ratliff of Ashton-Tate in 1979; **IBM**'s *R-Base* of 1982 under project leader Frank King; and *Oracle*, written by Ed Oates, Bruce Scott, and Robert Miner of Larry Ellison's **Oracle Corporation** and also released in 1982. Versions of Oracle are available for both **mainframes** and **personal computers**, but because of its much lower price and user friendliness, **Microsoft** *Access* is now the usual database software of choice for the home and small office.

Additional reading: *Computing Encyclopedia*, vol. 1, p. 185.

Dataflow machine
COMPUTER TYPES

A *dataflow machine* is a **parallel computer** that uses a **computer architecture** that reverses the operation of a conventional **von Neumann machine**. Instead of using a moving **pointer** called a *program counter* through a list of **instructions** that access and possibly modify static data values, the multiple processing nodes (*actors*) of a dataflow machine remain fixed,

and it is the **data** that moves (flows) to them for processing. Dataflow **algorithms** can be represented as directed **graphs** in which the arcs are data paths and the nodes are operations to be performed on the data tokens arriving on the incoming arcs. Multiple paths through such a graph represent those parts of a computation that can be executed in parallel. The names within the nodes of the graph indicate the operation to be performed. The act of performing an operation is called *firing* the node and results in the consumption of the input tokens and production of output tokens.

One of the first formal methods using dataflow was PERT / CPM (Project Evaluation and Review Technique / Critical Path Method) developed in the 1950s for project planning and control. Dataflow is also used by the simulation language **GPSS**. The **Burroughs Corporation** 1977 *Data Driven Machine 1* (DDM1) executed dataflow programs using a **tree** structure for organizing the atomic units and a switch at each node of the tree to distribute its output. A dataflow architecture first proposed by Jack Dennis and David Misunas in 1975 was implemented in the **Texas Instruments (TI)** *Distributed Data Processor* (DDP) of 1979. Each operation unit in the DDP had an **arithmetic-logic unit (ALU)** and a memory for instruction cells. Its nodes are connected by a **shift register** interconnection network. Both the operation units and instruction cells are addressable. The primitive operations correspond to the operations used in an **intermediate language** for compiling **Fortran** for TI's *Advanced Scientific Computer* (ASC). The LAU (Language à Assignation Unique) 32-processor system at CERT-ONERA in Toulouse, France, also implemented basic dataflow in 1979. Later work included the Manchester University "tagged-token" dataflow architecture of 1981. *Tagged-token* denotes a design that lets multiple **iterations** of a **loop** execute in parallel, labeling the data with tags to distinguish the values produced by different iterations. Tagged-token systems were also built at MIT in 1986 by David Culler and Arvind.

Additional reading: *Encyclopedia of Computer Science*, pp. 520–523.

DBMS. *See* **Database Management System (DBMS)**.

DEC. *See* **Digital Equipment Corporation (DEC)**.

DEC PDP series   COMPUTERS

The first of the *DEC PDP* (Programmed Data Processor) *series*, the 1960 PDP-1 designed by Ben Gurley, sold for $120,000. No PDP-2 was marketed, and successors PDP-3, -4, and -5 were small machines, but by 1962, the fifth year of **Digital Equipment Corporation (DEC)** operations, their cumulative success brought DEC to the level of $6.5 million in sales and net profits of $807,000. The PDP-6 of 1964, designed by Gordon Bell, was the company's first **mainframe**. It had a 36-bit **word**, a megabyte of main **memory**, and the first commercial **operating system** capable of **time sharing**. Two of the best known of the PDP series are the PDP-8, first sold in 1965 for $18,000, and the PDP-11 family, whose first system, the PDP-11/20, sold in 1970 for $11,000. The PDP-8 was the first highly successful **minicomputer**; over 50,000 were sold. The first version of **Unix** was written for the PDP-7. After it was rewritten in the language **C** in 1973, it became widely used on larger PDP-11s for teaching and research. PDP-11s were also used in several **multiprocessing** systems of the 1970s.

   Additional reading: *Encyclopedia of Computer Science*, p. 573.

DEC VAX series   COMPUTERS

The *DEC VAX series* of 1978 through 2000, the successor to the **DEC PDP series**, was the mainstay of most university computing centers for almost two decades. The VAX ran either the DEC VMS (Virtual Memory System) **operating system** or the Berkeley Software Distribution (BSD) version of **Unix**. The VAX used a 32-bit **computer architecture**. Its 16 general-purpose **registers** included a program counter, **pushdown stack** pointer, frame pointer, and argument pointer (*see* **Pointer**). **Instructions** varied in length from one **byte** to over 50. **Operation codes (opcodes)** were one or two bytes long. The number of **operands** varied from zero to six. Each operand was spec-

ified using a general operand specifier that allowed one of 13 addressing modes, including true postindexing. Several **data types** were supported, including 8-, 16-, and 32-bit integers; single, double, and quadruple precision **floating-point arithmetic**; decimal **string**, numeric string, **character** string, and 1- to 32-bit fields; and **queues**. The VAX was a classic **complex instruction set computer (CISC)**. It included a compatibility mode for **emulation** of PDP-11 user code. The original VAX architecture included 244 instructions when it was announced in 1977. Four instructions for manipulating queues in a multiprocessor system were added in 1978 and retrofitted to all VAX-11/780s in the field. In 1980, 56 new instructions were added to support a new extended range double precision data type and a quadruple precision data type. These new instructions were implemented as microcode options on the VAX-11/780 and VAX-11/750 (*see* **Microprogramming**). **Software** emulation was provided to achieve compatibility with systems that did not include the option. The *MicroVAX* was introduced in 1984 to allow single-chip **VLSI** implementation. In 1989, 63 new instructions were added to the VAX architecture to support register–based integrated vector processing. Although the VAX began as a **minicomputer**, the VAX series eventually included a wide range of architecturally compatible processors whose performance spanned a range from that typical of **workstations** through minis to **mainframes**.

   Additional reading: *Encyclopedia of Computer Science*, pp. 573–574.

Declaration   DEFINITIONS

A *declaration* is a **statement** used in **assembly language** or **high-level language** to specify the **data type** of a **constant** or **variable** and to allocate **memory** space for it or to provide other information to the **assembler** or **compiler** being used. A declaration may also be used to define a **macro instruction**. In an assembly language, a declaration is sometimes called a *pseudo-instruction* since its format mimics that of an executable instruction but uses a symbolic command that does not correspond to a command in the computer's **instruction set**. For example, the pseudo-instruction *CONST DEC* −*375* has the same format as a **hardware** instruction, but in this case the *DEC* just indicates that the programmer wants the (perhaps binary) equivalent of decimal −375 placed at a location called *CONST*. Since the **bit** pattern of −375 would not likely be a sensible **instruction**, this pseudo-instruction should not be placed relative to other instructions in such a way that the flow of control could ever pass through it.

Dell INDUSTRY

The *Dell* Corporation was founded in Austin, Texas, in 1984 by Michael Dell while still a college student. His vision was to assemble and sell **personal computers (PCs)** directly to end users via mail or phone order. The **Internet** has opened a third avenue and now accounts for $50 million per day in orders for "Wintel machines" (**platforms** consisting of **Microsoft Windows** and "**Intel** inside"). Dell offers a complete line of **servers**, desktops, notebook and **laptop computers**, and their accessories. The company holds almost a quarter of the U.S. market and 13% of the world market for PCs, first in both categories.

Additional reading: *Computing Encyclopedia*, vol. 1, pp. 190, 192.

Deque DATA STRUCTURES

A *deque* (pronounced "deck" and coined by Earl Schweppe in the 1960s) is a double-ended **queue**, a linear **data structure** to which items may be added or removed only at the *top* or the *bottom*. Thus a deque may be used as either a queue or a **pushdown stack** or, with more versatility and in keeping with its pronunciation, to simulate a deck of playing cards in which dealing from the bottom of the deck, legal or not, is easily done.

Additional reading: Reilly and Federighi, *Pascalgorithms*, pp. 442–446.

Desktop publishing SOFTWARE

Desktop publishing (DTP) is the creation and eventual printing of high-quality documents using **software** that supports the design of complex pages and allows their images to be displayed and edited. DTP arose during the late 1980s concurrent with the development of **per-**sonal computers (PCs) and **workstations** with a **Graphical User Interface (GUI)**, **laser printers**, and interactive software for document design and editing. So equipped, home and office users without formal training in document design can produce colorful well-formatted documents without the need to contract them to an outside publisher. In 1980 Les Earnest, who had earlier written the first **spelling checker**, founded IMAGEN to market the first DTP program geared to laser printer output. The first widely used DTP program was *PageMaker*, developed in 1984 by Paul Brainerd of the Aldus corporation for use on the **Apple Macintosh** and later ported to **Intel 80×86** systems. Some of the other popular DTP systems are *QuarkXPress*, *CorelDraw*, and *Microsoft Publisher*. Beyond what the usual **word processor** can do, these software packages support a myriad of text formatting facilities including word hyphenation, line justification, letter kerning, avoiding "widows" and "orphans" (single lines of a paragraph at the beginning or end of a page), fine control of line and **character** spacing, and typefonts in sizes ranging from 4-point to over 100-point at intervals of half a point or less. Simple text editing facilities for inserting and deleting text are provided, together with some form of cut-and-paste for moving blocks of text from one position in the document to another. A search-and-replace facility and a spelling checker are common. Text may be typed in directly or may be imported from a wide variety of word processors, **spreadsheets, database** programs, and other software. Desktop publishing image creation relies heavily on the use of Adobe's *PostScript* **Page Description Language (PDL)**.

Additional reading: *Encyclopedia of Computer Science*, pp. 532–535.

Difference calculation ALGORITHMS

One fertile source of new (or rediscovered) **algorithms** is *transformation of domain*, mapping a problem that is difficult to solve in one domain to another in which its solution is easier, followed by reverse transformation of the result. Examples that are the subjects of separate articles are **Horner's rule** and **logarithm**, both of which reduce or eliminate multiplication, and the **Fast Fourier Transform (FFT)**. Mathematical examples that are not discussed in this book are the *Laplace transform* (and others) that reduce solution of differential equations to a **table lookup** process and the *Schwartz-Christoffel transformation*, an ingenious method of solving certain problems in complex variables. A very historic example of transformation of domain is the basis for the Babbage and Scheutz **difference engines**. Calculation of a sequence of values of a polynomial such as $ax^3 + bx^2 + cx + d$ would seem to require several multiplications for each value of x of interest, but for regularly spaced values of x, the nth difference (as well as the nth derivative) of an nth degree polynomial is constant. This enables extension of a short table of values by addition only. For example, the triangular numbers 1 3 6 10 15 . . . , which correspond to the polynomial $(x^2 + x)/2$ for successive integer values of x, can be arranged with their first and second differences as

$$
\begin{array}{ccc}
1 & & \\
& 2 & \\
3 & & 1 \\
& 3 & \\
6 & & 1 \\
& 4 & \\
10 & & 1 \\
& 5 & \\
15 & &
\end{array}
$$

From this tableau, we can see that column one can be extended with just two additions: Add the first missing second difference, 1, to the first missing first difference, 6, and then add that 6 to the last item in column one to obtain 21, the next number in the sequence of triangular numbers. Instead of doing this ourselves, imagine, as Babbage did, that a special-purpose difference engine could be built to do so, one that could handle sixth-order differences to a precision of 20 digits. Calculation of tables through the method of differences is more generally applicable than might be first apparent because most **functions** of interest have **Taylor series** expansions which, though infinite series, can be truncated to polynomials of finite degree by omitting negligible high-order terms. Alter-

natively, an nth degree polynomial can be curve-fitted to selected values over a range of interest and then used as the basis for a *difference calculation*. The German mathematician Karl Weierstrass proved the fundamental theorem in this regard in 1885, showing that any function that varies smoothly (that is, continuously) over an interval can be approximated as closely as desired by a polynomial. (Modern practice is to compute the elementary functions of **subroutine** libraries through approximations that are ratios of polynomials.) Calculation by differences was known at least as early as 1624 when Henry Briggs made his table of **logarithms** and was discussed in Newton's *Principia* of 1687. Johann Müller described a rudimentary difference engine in 1786, long before Babbage, and the method of differences was used in the 1790s by Gaspard de Prony of France, who oversaw the labor of almost a hundred (human) **computers** to produce the 18-volume *cadastral tables* of logarithmic and trigonometric functions.

Additional reading: Lindgren, *Glory and Failure*, pp. 70–75.

Difference Engine   HISTORICAL DEVICES
A *difference engine* is a special-purpose **digital computer** that performs **difference calculations** in accord with the **algorithm** of the prior article. The first known attempt to construct a difference engine was that of Johann Müller in 1786. But as a proper noun, *Difference Engine* refers to the most famous attempts to mechanize this process, those made by Charles Babbage in the 1820s. Frustrated by the errors made by (human) **computers** preparing tables for him and urged on by Dionysius Lardner, a prolific writer on science and technology, Babbage produced a small demonstration model of a difference engine by mid-1822. This model did not survive, but it is reputed to have used six-digit **precision** and to have tabulated the polynomial $x^2 + x + 41$ (whose values for integer values of x from 1 to 39 happen to be **prime numbers**). Babbage then designed a much more extensive machine, intended to provide six orders of differences, each of 18 digits. But work on this *Difference Engine No. 1* ceased in 1833, at least partially because of

Babbage's deteriorating relations with his chief engineer, Joseph Clement. In 1847 Babbage produced a design for a *Difference Engine No. 2*, whose simpler design embodied ideas from his **Analytical Engine** of the mid-1830s. This machine was never completed by Babbage, but in 1991 the Science Museum in London completed a working Difference Engine No. 2 based on his drawings (*see* reference).

Inspired by Babbage's work, the father and son team of Georg and Edvard Scheutz of Sweden built a difference engine in the 1850s, but it never worked reliably, nor did a later nineteenth-century design by Martin Wiberg, also in Sweden. In the mid-1870s, American engineer George Barnard Grant built what was essentially an electrically powered difference engine the size of a grand piano and exhibited it at the Centennial Exposition in Philadelphia in 1876 (*see* **Grant calculators**). In the 1930s, Leslie Comrie of the British Nautical Almanac Office adopted **Burroughs Corporation** and **National Cash Register (NCR)** accounting machines and used them as difference engines. In 1933, French numerical analysis Louis Couffignal wrote a summary of calculating machines in which he showed complete familiarity with Babbage's work. He, too, aspired to build a programmable digital computer, but never received the financial support needed to carry his project any further than preliminary design.

Additional reading: Swade, *The Difference Engine*.

Differential Analyzer

HISTORICAL DEVICES

The *Differential Analyzer* was an **analog computer** described in a 1931 article by Vannevar Bush in the *Journal of the Franklin Institute*. The machine had been constructed by Bush and Harold Locke Hazen at MIT for the solution of ordinary differential equations. The Differential Analyzer was based on the use of mechanical integrators that could be interconnected in any desired manner. The integrator consisted of a variable-speed gear driving a small knife-edged wheel resting on a rotating horizontal disk. The use of mechanical integrators for solving differential equations had been suggested by William Thomson (Lord Kelvin—*see* **Kelvin tide**

Figure 13. A portion of Babbage's Difference Engine No. 1. (Bettmann/Corbis)

machine), and various special-purpose integrating devices were constructed at various times. Bush's Differential Analyzer was, however, the first device of sufficiently general application to warrant production in quantity. In 1945, Bush and Samuel H. Caldwell described an electrical, but still analog, differential analyzer. The first electronic differential analyzer was the **MADDIDA** (Magnetic Drum Digital Differential Analyzer) developed in 1949 at the Northrop Aircraft Corporation. Because of the hegemony of the general-purpose digital computer, no differential analyzers of any kind are now being built. Bush's more enduring legacy is his concept of a **memex**.

Additional reading: *Encyclopedia of Computer Science*, pp. 537–539.

Digital camera

HARDWARE

A *digital camera* captures images on what is essentially a solid-state retina instead of film. The artificial retina uses either a **charge-coupled device (CCD)** or CMOS (complementary metal oxide **semiconductor**) **arrays**. Images may be stored on removable and reusable **flash memory** cards or may be transferred to computer **memory** over a cable attached to a USB (Universal Serial Bus) port. Once uploaded, images may be edited with **software** tailored for **digital photography** and sent to friends and relatives via **electronic mail**. The digital camera dates to the mid-1970s when Kodak invented several solid-state image sensors that converted light to digital pictures for professional and consumer use. In 1986, Kodak scientists invented the first **megapixel** sensor, capable of recording 1.4 million **pixels** that could produce a 5" × 7" print of photographic quality. The first digital cameras designed for use with a **personal computer (PC)** were the **Apple** QuickTake 100 camera of 17 February 1994, the Kodak DC40 camera of 28 March 1995, the Casio QV-11 with a **liquid crystal display (LCD)** of late 1995, and the **Sony** Cyber-Shot Digital Still Camera of 1996. Current digital cameras use from two to five megapixels and are sold at a cost proportional to those resolutions.

Additional reading: *Computing Encyclopedia*, vol. 1, pp. 197–198.

Digital computer   COMPUTER TYPES

A *digital computer*, a term coined by George Stibitz, is a programmable **computer** whose logical operation deals with the discrete digits of a **positional number system**. The physical construction of a digital computer is irrelevant; most are now electronic, early ones were mechanical or electromechanical (*see* **Bell Labs relay computers**), and there are also **optical computers** and **DNA computers** that are digital. One of the three most significant milestones in this book was the invention of the **stored program** electronic digital computer, but there is contention as to which of several computers was the first to demonstrate its principles. *See* the Classification of Articles at the end of the book for a listing of the many articles on individual digital computers.

Additional reading: *Encyclopedia of Computer Science*, pp. 539–545.

Digital Equipment Corporation (DEC)   INDUSTRY

The *Digital Equipment Corporation (DEC)* was founded by Kenneth Olsen and Harlan Anderson in 1957. DEC began as a supplier of electronic circuit boards. By 1960, DEC was manufacturing the **DEC PDP series** of **digital computers**. By its thirtieth anniversary in 1987, DEC had a market value of $24 billion, tenth among all U.S. companies, thirty-eighth in the list of Fortune 500 companies, and second only to **IBM** among computer companies. Up through this time, the most successful DEC products were its DEC PDP-11 and the **DEC VAX series**, but the company had failed to introduce a viable **personal computer**. In January 1989, DEC announced its first **Reduced Instruction Set Computer (RISC)** based on a MIPS Computer Systems **microprocessor**. These systems ran Ultrix, DEC's version of **Unix**, and marked DEC's entry into the open systems market. These DECsystems and DECstations were used as **workstations** and **servers** in education and industry in the first half of the 1990s. But by the late 1990s, intense competition and DEC's failure to expand into personal computers proved fatal. In June 1998 the company was sold to the **Compaq** corporation for $9.6 billion, less than half of its 1987 value. The historic Digital Equipment Corporation continued, but as a division of its new parent company, one that was itself acquired (by **Hewlett-Packard**) in 2002.

Additional reading: Rifkin and Harrar, *The Ultimate Entrepreneur*.

Digital photography   APPLICATIONS

Digital photography is to **personal computers (PCs)** what **image processing** is to **mainframe** computers; that is, it brings to home and office users the capability to modify, store, and transmit over the **Internet** photographs of interest to them and their correspondents. The beginnings of digital photography are correlated with the perfection of the **optical scanner** in the 1970s. Even before most PCs had their own scanner, the amateur photographer was and still is able to ask photo developers such as Kodak to scan their prints and deliver them on **floppy disks** or post them on an accessible **World Wide Web** site. The development of reasonably priced **digital cameras** in 1995 provided a still more direct way to produce digital images that can be uploaded to a PC and processed by any one of a number of **software** products. These programs allow such operations as modifying contrast, color hue, and intensity, sharpening an image, and changing its size and orientation. One of the first popular packages that allowed these operations was *Adobe Photoshop*. A very versatile image editor called *IrfanView* written by Irfan Skiljan of Bosnia is available as **free software** over the Internet. There are now inkjet printers that use six colors (instead of the usual four) and may be used to make 19" × 13" prints of high-resolution images on special photo paper. The resulting prints are indistinguishable from those made by a professional photo finisher.

Additional reading: *Computing Encyclopedia*, vol. 1, pp. 201–202.

Digital Signal Processor (DSP)

COMPUTERS

A *Digital Signal Processor (DSP)* is a special-purpose **hybrid computer** that is optimized to convert incoming analog signals to digital form and process them in real time. John C. Murtha and James A. Ross, Jr., of Westinghouse obtained a patent for a programmable DSP on 21 May 1974. A DSP is about ten times faster than a general-purpose **microprocessor** made with the same technology. DSPs are now essential to the operation of any **real-time system** for which noise suppression is vitally important. The principal **algorithms** implemented by a DSP are the **Fast Fourier Transform (FFT)** and the **Kalman filter**. DSPs are embedded in every automobile, **cellular telephone**, **digital camera**, and **personal computer** sound card (*see* **Embedded system**). Digital signal processing can restore vintage music recordings to their original clarity, erase static from long-distance phone lines, and enable satellites to resolve terrestrial objects as small as a baseball (*see* **Global Positioning System [GPS]**). In automobiles, DSPs create digital audio "surround sound" and monitor active suspension systems that adjust automatically to road conditions. In cell phones, DSPs squeeze more information into limited **bandwidth** and scramble signals to thwart eavesdropping. The leading producers of DSPs are **Motorola** and **Texas Instruments (TI)**, whose latest models are based on **micro-**processors that use **Very Long Instruction Word (VLIW)** architecture.

Additional reading: *Computing Encyclopedia*, vol. 1, pp. 202–203.

Digital Subscriber Line (DSL)

COMMUNICATIONS

A *Digital Subscriber Line (DSL)* is used as "last-mile technology" to connect home users to the **Internet** over existing copper wires but using digital rather than analog data transmission. Although the technology was developed in the early 1990s, DSL service has only recently become widely available. A DSL provides Internet access at speeds about half that of a **fiber optics** cable **television** connection, but DSL service is available only to those who live within a limited distance to the nearest **telephone** switching center. The monthly cost of the two high-speed connections is about the same, and neither interferes with normal voice or other signals carried by the same medium.

Additional reading: *Computing Encyclopedia*, vol. 1, pp. 203–204.

Digital television

COMMUNICATIONS

Digital television means *all-digital* television, in contrast to conventional **television** in which analog signals are transmitted and converted to digital images for display at the receiver. All-digital television was made possible by a breakthrough made in the United States by Korean immigrant Woo Paik of the General Instrument Company in 1990. But much had to happen before prospects for digital television were propitious. The aspect ratio (ratio of width to height) originally used for television screens was developed by the French engineer W. K. L. Dickson in 1889 while working in Thomas Edison's laboratory in the United States. Dickson was experimenting with a motion-picture camera called a *Kinescope*, and he made his film one inch wide with frames three-fourths inches high, and this 4:3 aspect ratio became the standard for the film and motion-picture industry. In 1941, when the NTSC (National Television Standards Committee) proposed standards for TV broadcasting, they adopted the same ratio as the film industry. But the most common theater screen aspect ratio is 16:9, or 1.778, which is closer to the *Golden Ratio* (1.618) that is allegedly most pleasing to the human eye. U.S. High-Definition Television (HDTV) uses the same aspect ratio. Besides a wider screen, an HDTV picture has greater resolution and is thus sharper than conventional television.

On a standard television screen, the electron beam has about 256 levels of intensity for each of three colored phosphors—red, green, and blue (RGB)—giving each **pixel** a spectral range of about 16.8 million colors. The old NTSC format uses rectangular pixels that are slightly taller than they are wide. The HDTV format is composed of smaller but square pixels, just like some computer **monitors**. In the area of a single standard NTSC TV pixel, HDTV will pack 4.5 pixels, and the greater the density, the better the picture. Older NTSC television sets use a display of 720 pixels wide by 525 pixels high, lower than computer monitors whose resolution is often 800×600 or better. HDTV can have a resolution of up to $1,920 \times 1,080$ (2,073,600 pixels), six times greater than the older resolution.

HDTV is not necessarily digital; Japanese HDTV is broadcast as an analog signal. But **data compression** can pack more digital than analog information into the same **bandwidth**. Video on digital TV is compressed using a scheme called MPEG-2, which provides a bit-reduction of about 55:1. The method is "lossy"; that is, some information is discarded, but the eye doesn't notice. The human ear, however, is much more sensitive to subtle changes in sound. HDTV sound uses the Dolby Digital AC-3 audio encoding system, the same system used in most movie theaters, DVDs, and many home theater systems since the early 1990s. It can include up to 5.1 channels of sound—three in front (left, center, and right), two in back (left and right), and a subwoofer bass (the one-tenth of a channel) for a sound one can feel as well as hear. Digital sound is close to compact disc (CD) quality, including frequencies lower and higher than most people can hear.

Additional reading: Fisher and Fisher, *TUBE: The Invention of Television*, pp. 340–351.

Digitizing tablet   I/O DEVICES AND MEDIA

A *digitizing tablet* is a **computer graphics** input device that translates movement detected on its surface into numeric coordinates used to form the lines and vertices of a stored **graph**. Points of the tablet are mapped to fixed points on a **monitor**, in contrast to **mouse** or **trackball** input, which is relative to the current position of a cursor. The *puck* used to trace outlines on a digitizing tablet looks something like a mouse but has a transparent center containing crosshairs that enable selection of a particular point on the tablet by clicking a button on the puck. Digitizing tablets are used by surveyors, by artists for freehand drawing, and by engineers using **CAD / CAM** systems. The earliest digitizing tablets were Tom Dimond's *Stylator* of 1957 and the 1964 *Graphicon* ("RAND tablet") of Malvin Davis and Thomas Ellis of the **RAND Corporation**, both used for research in handwritten character recognition.

Additional reading: *Computing Encyclopedia*, vol. 1, p. 208.

Directory   OPERATING SYSTEMS

A *directory* (catalog) is a list of the **files** stored on a given volume of **auxiliary storage**, most often a **hard disk**, by a particular **operating system**. Directories are hierarchical; that is, they have the **data structure** of a **tree**. Nodes are usually called *folders*, subdirectories that may contain either file names (leaves) or the names of other folders, and so on, to any depth. In keeping with tree nomenclature, the highest-level folder in a **Unix** directory is called the *root*. The concept and need for a hierarchical file directory is traceable to Hal Draper's 1961 short story "Ms. Fnd in a Lbry" in *Fantasy and Science Fiction* in which he wrote:

These were the innocent days before the problem [file location] became acute. Later, Index runs were collected in Files, and Files in Catalogs—so that, for example, $C^3F^5I^4$ meant that you wanted an Index to Indexes to Indexes to Indexes, which was to be found in a certain File of Files of Files of Files, which in turn was contained in a Catalog of Catalogs of Catalogs. Of course actual superscripts were much greater; the structure grew exponentially.

The first hierarchical directories were implemented in the 1970s on systems such as Unix,

CP / M, MS-DOS, and the operating system for the **Apple II**. **Microsoft Windows** includes *Windows Explorer* (not to be confused with the *Microsoft Explorer* Internet **browser**), which, upon invocation, displays portions of the directory of any of several auxiliary storage devices and makes it easy to navigate through its tree to find and activate a particular file, no matter how deeply nested.

Additional reading: *Encyclopedia of Computer Science*, pp. 583–585.

Discrete mathematics   MATHEMATICS

Early in their studies, students of **computer science** must pass a course in *discrete mathematics*, the mathematics of individual countable objects, and they usually find this endeavor more difficult than the study of **calculus**, which deals with continuous quantities. Discrete mathematics is the foundation for digital computation in the same way that calculus is the foundation for analog computation. The differential equations of calculus, which are rather naturally dealt with by **analog computers**, must be discretized to form approximating *difference equations* before they can be solved on a **digital computer** (*see* **Numerical analysis**). Discrete mathematics, much of which dates back to antiquity, is essential to the **analysis of**

algorithms. The topics of greatest utility in such analysis are the mathematics of *probability* traceable to Blaise Pascal and Pierre Fermat in 1654; proof by *induction*, first thoroughly enunciated by Pascal; *combinatorics*, the enumeration of combinations and permutations of discrete objects studied extensively by Leibniz in 1666; and the *binomial theorem*, which was known to James Gregory in 1670 before it was later stated and proved by Newton. The binomial theorem concerns the algebraic expansion of $(x + y)^n$ for any value of n; for example, $(x + y)^2 = x^2 + 2xy + y^2$. The coefficients of such expansions, 1 2 1 in the specific example given, can be arranged in what is known as *Pascal's triangle* even though the triangle was known to the Chinese scholar Jia Xian of the eleventh century, to Yang Hui in 1261, and to Peter Apian in 1527, all well before the birth of Pascal. The first seven rows (for $n = 0$ through $n = 6$) are:

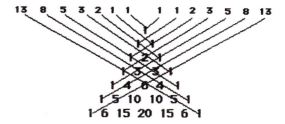

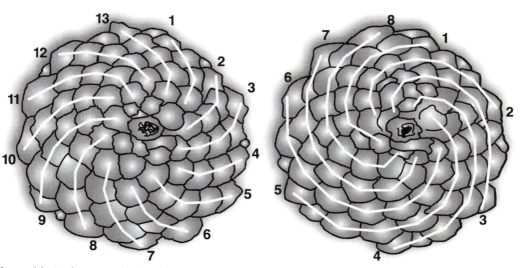

Figure 14. A pine cone that has eight spirals running clockwise and 13 running counterclockwise. Different pine cones would have differing numbers of spirals, but because of the way a cone grows, the spiral counts in the two directions will always be consecutive Fibonacci numbers.

The triangle, the production of which is frequently assigned as homework for programming students, has several noteworthy properties. Each row sums to 2^n. The *triangular numbers* 1 3 6 10 15 . . . can be seen along either of two diagonals. And the *Fibonacci sequence* 1 1 2 3 5 8 13 . . . (in which each integer starting with 2 is the sum of the prior two integers) is obtained by adding the numbers along the skewed diagonals identified by the straight lines. Finally, the limit of the ratio of the nth Fibonacci number to the prior one (which has reached $^{13}/_8 = 1.625$ as of the seven lines of the triangle shown), approaches $(1 + \sqrt{5})/2 = 1.618033989 . . .$, the *Golden ratio* that occurs in many natural objects (*see* Figure 14).

Additional reading: *Encyclopedia of Computer Science*, pp. 587–593.

Figure 15. The successive rotations needed to bring a randomly oriented Rubik cube back to its canonical position (all faces of uniform color) can be deduced from group theory, a topic in discrete mathematics. (From the collection of the author)

Diskette. *See* **Floppy disk**.

Distributed system

INFORMATION PROCESSING

A *distributed system* is a computing system whose **data** and possibly some of its **programs** are stored at multiple sites (nodes) of a **network**. Nearly all large **software** systems are necessarily distributed. The largest and best-known distributed system is the **Internet**, but the first large-scale distributed systems were the **SAGE** and **SABRE** projects of the 1950s. The most common distributed systems are networks that implement **client-server computing**. In a distributed environment, no single computer is responsible for controlling the system as a whole—distributed systems are federations of autonomous agents. Distributed systems rely on a common set of basic **algorithms** and **protocols** that are used to solve problems such as the detection of termination of a distributed computation, election of a leader for a group of nodes, synchronization of redundant computations performed for the sake of fault tolerance (*see* **Fault-tolerant computer**), coordination of **database** transactions, and mutually exclusive access to shared resources. More specialized algorithmic problem domains include distributed **simulation**, **electronic commerce**, digital libraries, distributed multimedia, and collaborative **groupware**.

Additional reading: *Encyclopedia of Computer Science*, pp. 595–602.

DNA computer   COMPUTER TYPES

DNA computing is the use of biological operations on DNA molecules to solve certain classic computational problems efficiently. One can think of a *DNA computer* as a massively **parallel computer** where each DNA strand serves as a separate processor. A DNA strand is a sequence (polymer) of four types of nucleotide distinguished by the bases they contain, bases denoted A, C, G, T (adenine, cytosine, guanine, and thymine). When each strand contains fewer than 10,000 nucleotides, 10^{18} strands of DNA can be dissolved in a liter of water. DNA processors are extremely slow, requiring several hours to complete the simplest operations of **Boolean algebra**. Nonetheless, a parallel computer that can perform 10^{18} such operations every several hours is a significant computational resource.

DNA computing began in 1994 when Leonard Adleman showed how DNA can be used to solve the *directed Hamiltonian path* (DHP) graph problem (*see* **Graph theory**). A Hamiltonian path is one that visits all nodes in a **graph** exactly once. Testing whether a particular graph has a Hamiltonian path is an **NP-complete problem**. In 1995, Richard Lipton showed that SAT (satisfiability for Boolean **expressions** in conjunctive normal form) can be done by the DNA method in time linear in the size of the formula. The number of strands needed is 2^n, where n is the number of variables in the formula. Adleman did such a computation for seven variables shortly thereafter and extended this to 20 variables in 2002. In 1996, a research group led by Daniel Boneh discovered how to simulate a nondeterministic **Turing machine** using DNA. Boneh's group also showed how to decipher messages enciphered by the **Data Encryption Standard (DES)**. The massive parallelism of a DNA computer is used to test all possible keys until the correct one is found. In 1998, Erik Winfree showed how complicated DNA patterns can be used to simulate a **cellular automaton**.

Additional reading: Siegfried, *The Bit and the Pendulum*, pp. 95–113.

DRAM
MEMORY

Development of the *DRAM* (Dynamic Random Access Memory), or more completely *MOSFET DRAM* where *MOSFET* is an acronym for *Metal Oxide Semiconductor Field Effect Transistor*, is a double milestone. The first DRAM using one **transistor** per **bit** (and hence four per four bits) was invented by **IBM** Fellow Robert H. Dennard in 1966 and patented in 1968. In 1969, William Regitz of **Honeywell** canvassed U.S. **semiconductor** companies looking for someone to share in the development of a 256-bit DRAM based on a 3-transistor per 4-bit cell that he and his co-workers had invented. **Intel** showed interest, and Regitz worked with that company's Joel Karp and Ted Hoff, later coinventor of the first **microprocessor**, the **Intel 4004**, to achieve a circuit-packing density four times as great as

had been anticipated. After Intel engineer John Reed worked to increase the yield of good chips produced during fabrication, Intel was able to produce a 1,024-bit (1K) DRAM for $10.24 per chip, a penny a bit, and announced the product in October 1970. The author of the reference calls the breakthrough "a turning point in the history of the computer industry" because this second and arguably greater milestone marked the beginning of the end of **magnetic core** memory for **digital computers** of all levels. Transistors had been used for **logic circuits** for over a decade before the DRAM, the first semiconductor-based **solid state memory**, was ready for widespread use.

Additional reading: *Encyclopedia of Computer Science*, pp. 1132–1135.

DSL. *See* **Digital Subscriber Line (DSL)**.

DTSS. *See* **Dartmouth Time-Sharing System (DTSS)**.

DVD-ROM. *See* **Optical storage**.

Dynamo
LANGUAGES

Dynamo, the first and still the dominant continuous-event **simulation** language, was developed in 1959 by Phyllis Fox and Alex Pugh of MIT. Dynamo is an outgrowth of Richard Bennett's SIMPLE (Simulation of Industrial Management Problems with Lots of Equations). Dynamo runs on **IBM PC**–compatible **personal computers** under either **MS-DOS** or **Microsoft Windows**. The language provides an equation-based development environment for system dynamics models and can be used to simulate a general-purpose **analog computer**. In 1970 Jay Forrester ran a Dynamo program called the "World Dynamics Model" to help the mayor of Boston, John Collins, manage his large city.

Additional reading: Bergin and Gibson, *History of Programming Languages*, pp. 389–391.

E

EBCDIC   CODING THEORY
EBCDIC (Extended Binary-Coded Decimal Interchange Code) was an 8-bit **character** code
developed by **IBM** in 1964 for use with the
IBM 360 Series and later IBM and IBM-
compatible **mainframe** computers. Although
this 8-bit code allowed representation of four
times as many characters as did the 6-bit BCD
code used on the **IBM 700 Series**, EBCDIC
proved awkward to use. Its uppercase and lowercase alphabets were not assigned consecutive
binary values, and its collating sequence (the
relative values of digits versus letters, for example) did not correspond to what seemed natural to most users. For these reasons, EBCDIC
has now been superseded by **ASCII**, even on
IBM computers, a reversal of what was once
the common practice that other vendors followed the lead of IBM.

    Additional reading: Petzold, *CODE*, pp. 295–
297.

E-book. *See* **Electronic book (e-book)**.

Eccles-Jordan trigger circuit. *See*
Flip-flop.

**Eckert-Mauchly Computer
Corporation (EMCC)**   INDUSTRY
In 1946, J. Presper Eckert and John W.
Mauchly, fresh from their success with **EN-
IAC**, formed the Electronic Control Company

and renamed it the *Eckert-Mauchly Computer
Corporation (EMCC)* a year later. EMCC was
the first company founded for the express purpose of making **digital computers** and no other
product. Under the EMCC name, the **BINAC**
was built, and then after purchase by **Remington Rand** in 1950, EMCC became the Univac
Division of that company and produced the
UNIVAC.

    Additional reading: Ceruzzi, *A History of Mod-
ern Computing*, pp. 13–16.

E-commerce. *See* **Electronic commerce
(e-commerce)**.

Editor. *See* **Text editor**.

EDSAC   COMPUTERS
The *EDSAC* (Electronic Delay Storage Automatic Calculator) was, despite its name, a **digital computer** built in England during the late
1940s at the University (of Cambridge) Mathematical Laboratory (UML). The project leader
was Sir Maurice Wilkes, who designed the
computer in conjunction with William Ren-
wick. Others who worked on the project were
Stanley Gill, David Wheeler, Douglas Hartree,
Ben Noble, Don Willis, Eric Mutch, Phil Farmer, Roy Piggott, Sid Barton, Gordon Stevens,
and John Bennett. The EDSAC is one of four
computers that might validly be called the first
stored program computer. Ron Fisher pub-

lished the first scientific paper containing results from a digital computer, a genetics paper based on an EDSAC program. EDSAC was a serial binary computer with an **ultrasonic memory**. The mercury tanks were configured as two "batteries" of 16 tanks. A battery could store 256 numbers of 34 **bits** plus a sign bit. Addressing was done in terms of 17-bit half-words, and a half-word could hold either a short integer or any one of 17 one-address instructions. The **instruction set** included multiplication but not division. Five-channel **punched paper tape** was used for **input / output (I/O)**. The I/O **instructions** called for the transfer of five bits from the tape to the **memory**, or vice versa. Operation of the machine began when a bootstrap loader known as the **initial orders** written by Wheeler was read into main memory from a **read-only memory (ROM)** consisting of a set of rotary **telephone** switches. Programming for the machine was documented in the now-classic 1951 text *The Preparation of Programs for an Electronic Digital Computer* by Wilkes, Wheeler, and Gill. The EDSAC did its first calculation on 6 May 1949. Its design, only slightly modified, became the basis for **LEO**, the first commercial computer used for extensive **data processing**. EDSAC 2, which began operation early in 1958, the same year that the original EDSAC ceased operation, was a **vacuum tube** computer designed by the same team that had built EDSAC 1. It was the first computer to have a microprogrammed control unit (*see* **Microprogramming**).

Additional reading: Martin Campbell-Kelly, "Past into Present: The EDSAC Simulator," in Rojas and Hashagen, *The First Computers: History and Architectures*, pp. 397–416.

EDVAC
COMPUTERS

The *EDVAC* (Electronic Discrete Variable Automatic Computer), one of four candidates for having been the first **stored program computer**, was a direct outgrowth of work on the **ENIAC**. During the design and construction of the ENIAC in 1944 and 1945, the need for more storage than its 20 10-digit numbers was acutely felt. Experience with **ultrasonic memory** led to the concept of recirculating storage of digital information. The group at the Moore School of Electrical Engineering at the University of Pennsylvania started development work on mercury delay lines for such storage and initiated the design of the EDVAC. The principal scientists responsible for the EDVAC were J. Presper Eckert, John W. Mauchly, Herman Goldstine, Arthur Burks, John von Neumann, and T. Kite Sharpless. The EDVAC had about 4,000 tubes and 10,000 crystal diodes. It used a 1,024-word **ultrasonic memory**, consisting of 23 lines, each 384 microseconds long. The words were 44 **bits** long, and instructions consisted of a 4-bit **operation code (opcode)** and four 10-bit addresses. The arithmetic unit did both fixed and floating-point operations at a clock frequency of 1 MHz. Input and output were via **punched paper tape** and **punched cards**. Although the conceptual design of the EDVAC was complete in 1946 and was delivered to the Ballistic Research Laboratories at Aberdeen, Maryland, by 1950 the entire computer had not yet worked as a unit and was still undergoing extensive tests. The delay was primarily due to the exodus of computer people from the Moore School in 1946. Eckert and Mauchly resigned to build the **UNIVAC**. Herman Goldstine and Arthur Burks went to Princeton to work with von Neumann, and Harry Huskey left to work with Turing in England. The EDVAC finally became operational as a unit in 1951.

Additional reading: *Encyclopedia of Computer Science*, pp. 626–628.

EFT. *See* **Electronic Funds Transfer (EFT)**.

Electronic book (e-book)
GENERAL

At its simplest, an *electronic book (e-book)*, a term first used by the American expert in **computer graphics** Andries van Dam, is just a stored **file** that contains the complete text of a book that was initially published, or could have been published, in the usual printed form. The origin of e-books can be traced to a chance happening in 1971 when Michael Hart was given a computer account with $100 million of computer time by the operators of the Xerox Sigma V **mainframe** at the Materials Research Lab at the University of Illinois. Hart decided that the best thing to do with it was to encourage the typing, and later scanning, of books and manuscripts to be preserved as computer files. Thus began what is now called Project Gutenberg. The Project continues to this day, with e-books stored at many sites, the principal one being at the University of North Carolina. With the same intention, the University of Virginia now maintains an **Internet**-accessible archive of over 2,000 titles of classic books that are out of copyright. Over 4 million copies of these e-books have been downloaded since inception of the service in 2000. Among the most popular accessions are Lewis Carroll's *Alice in Wonderland* (with or without illustrations), Jules Verne's *Around the World in Eighty Days*, and Arthur Conan Doyle's *The Hound of the Baskervilles*. Some publishers now post e-books that never had print editions and make them accessible for a fee. **Microsoft** and other **software** firms now offer **programs** for **personal computers** and for **handheld computers** such as the *Palm Pilot* and *Pocket PC* that make it easier to navigate through e-book files, and there are special-purpose book-size devices specifically designed to facilitate reading e-book material.

Additional reading: *Computing Encyclopedia*, vol. 2, pp. 12–13.

Electronic calculator
CALCULATORS

An *electronic calculator* is a calculator that uses **transistor** logic. While there might have been **vacuum tube** calculators, in practice there were not; the expense and heat generation of vacuum tube technology led to large devices that qualified as **computers** rather than mere **calculators**. The earliest electronic calculators to attain widespread use were the **HP-35** of **Hewlett-Packard** and the **TI SR-50** of **Texas Instruments**. These devices cost a few hundred dollars when introduced, but small handheld electronic calculators of equivalent or greater functionality are now available from companies such as Casio that sell for as little as $10 or $15.

Electronic commerce (e-commerce)
APPLICATIONS

Electronic commerce (e-commerce) enables the execution of financial and commercial transactions between two or more parties over a **network**. Historically, such networks comprise one or more of POTS (plain old **telephone** service), cable TV, leased lines, and wireless, and now, most significantly, the **Internet**. The earliest form of e-commerce, dating to the 1960s, was **electronic funds transfer (EFT)** between banks over secure private networks. During the late 1970s, e-commerce became widespread on private **intranets** in the form of electronic messaging technologies such as electronic data interchange (EDI) and **electronic mail (e-mail)**. In the 1990s, the advent of the **World Wide Web** on the Internet represented a turning point in e-commerce by providing an easy-to-use technological solution to the problem of information publishing and dissemination over a public network. Significant milestones were the establishment of Amazon.com by Jeff Bezos in July 1995 to sell books (and later CDs, toys, and gifts), and the founding of the popular eBay auction Website by Pierre Omidyar in October 1995. Amazon did not exist prior to the Web, but older companies such as Barnes and Noble quickly set up competitive "click and order" sites to augment existing chains of traditional brick and mortar stores. Somewhat recursively, **Dell** uses **computers** to sell computers. There are now very few businesses of any size who are *not* willing to accept orders over the Internet.

Additional reading: *Encyclopedia of Computer Science*, pp. 628–634.

Electronic Data Systems (EDS)
INDUSTRY

Electronic Data Systems (EDS) was founded in 1962 by former **IBM** employee and later U.S. presidential candidate H. Ross Perot. The story as usually told is that he did so with a $1,000 loan from his wife, but his alleged need for capital does not jibe with his reputation as having been the most successful salesman in IBM history. The business plan for EDS was based on Perot's perception that there would be many companies who did not want to establish **data processing** departments of their own but would be willing to pay handsomely for long-term contracts to a service bureau. EDS's first major customer was Frito-Lay of Plano, Texas, where Perot later built his impressive world headquarters. EDS became extraordinarily successful, but Perot chose to sell the company to General Motors (GM) in 1984 for $2.5 billion and join its board of directors. But Perot was not successful in telling GM what to do, and the arrangement was short-lived. Perot went on to dabble in politics and eventually founded a company similar to EDS but named for himself. In the meantime, EDS continued to prosper and now has 120,000 employees worldwide and annual revenues of about $20 billion.

Additional reading: *Computing Encyclopedia*, vol. 1, p. 16.

Electronic Funds Transfer (EFT)

APPLICATIONS

Electronic Funds Transfer (EFT) is the movement of funds between financial institutions over an electronic **network**. There are two major worldwide EFT networks: the Clearinghouse Interbank Payments System (CHIPS) and FedWire, the oldest EFT system in the United States. In 1997, these networks moved over $2 billion each banking day. A third major network, the Society for Worldwide Interbank Financial Telecommunications (SWIFT), is capable of handling nearly 2 million messages per day. The original expansion of EFT was stimulated by the standardization of **magnetic ink character recognition (MICR)** technology in the mid-1950s. EFT has progressed through four various forms: **automatic teller machines (ATMs)**, automated clearing houses (ACH), electronic funds transfer point-of-sale systems (EFTPOS) and debit cards, and electronic funds transfer electronic data interchange systems (EFT-EDI). A successful EFT system relies on **fault-tolerant computers**. The U.S. **Data Encryption Standard (DES)** has traditionally been the encryption **algorithm** of choice for secure EFTPOS systems. **Smart cards** provide an added element of security.

Additional reading: *Encyclopedia of Computer Science*, pp. 635–637.

Electronic mail (e-mail)   NETWORKS

Electronic mail (e-mail) is the transfer of a message from one **digital computer** user to another over a **network**. The message travels through a series of computer systems until it reaches its final destination, where it is stored for retrieval at the leisure of the intended recipient. E-mail is now the world's most common form of communication, supplanting traditional mail ("snail mail"), **facsimile transmission (fax)**, and to the consternation of advocates of personal interaction, **telephone** conversation. Electronic mail began as a simple messaging scheme in the first **time sharing** systems, **CTSS** at MIT and the **Dartmouth Time-Sharing System (DTSS)** of the 1960s. The sender would merely place a message in a **file** accessible by another user. It was not until elements of the **ARPAnet** were created at **Bolt, Beranek, and Newman (BBN)** in 1969 that e-mail messages could be sent between different computers. The milestone event was a successful 1971 experiment at BBN by Ray Tomlinson, who gained both fame and notoriety as the inventor of the now standard e-mail address format based on the @ sign. The spread of e-mail from this point on was rapid. Other early networks such as UUCPnet and Bitnet quickly adopted e-mail systems. Proprietary mail systems were developed mainly to work within a **local area network (LAN)** for a university or company. Systems such as Lotus ccMail, Microsoft Exchange, and QuickMail all were designed to allow transfer of mail within **intranets**. In 1988, Jarkko Oikarinen of Oulu University in Finland invented Internet Relay Chat (IRC), a form of rapid-exchange e-mail that lets young people gather in virtual "chat rooms." The 1990s saw origination of public online services such as CompuServe, Prodigy, and **America Online (AOL)**, all of which supported e-mail transfer. AOL also devised a rapid exchange of messages between parties known to be simultaneously online called *Instant Messenger*, and other vendors have followed suit. Spam, the electronic equivalent of junk mail, has grown in both volume and notoriety as use of e-mail has become ubiquitous. Technological and legislative responses to unwanted e-mail began to appear in the late 1990s, but the problem continues to worsen. But e-mail is here to stay, and its development is a most significant milestone.

Additional reading: *Encyclopedia of Computer Science*, pp. 637–642.

Electrostatic memory — MEMORY

Electrostatic memory consists of tiny **bits** of charge stored as a two-dimensional **array** on the inside face of a **cathode ray tube (CRT)**. The first such device, the *Williams tube* of 1947, was named after its English inventor, Frederic C. Williams of Manchester University. The Williams tube was a bit-serial device. To write information on an initially uniformly charged tube, the electron beam is deflected along a horizontal line, and at each point where the beam is turned off, a residual positive charge remains. To read the information, an electrode is placed on the outside of the face of the CRT. As the beam again sweeps over a line, the change of potential on the inside face is detected by the electrode. Bits of charge below a certain threshold are read as 0s, and those above it as 1s. After sweeping a line, the whole array must be systematically regenerated. Typically, regeneration took place during odd-word times and information access during even-word times. By changing the vertical or word deflection, several different numbers can be stored on one CRT; thirty-two 32-bit words was typical. Among the computers that used electrostatic storage were the **SWAC**, the **IAS computer**, and the IBM 701 (*see* **IBM 700 series**). Electrostatic memory and ultrasonic delay lines were the principal forms of **memory** used with first-generation **vacuum tube** computers, but these technologies were soon replaced by use of **magnetic core** in second-generation machines and **solid state memory** in third-generation machines of the present day.

Additional reading: *Encyclopedia of Computer Science*, pp. 1851–1853.

Eliza — ARTIFICIAL INTELLIGENCE

Eliza is a simple but ingenious artificially intelligent **program** written by Joseph Weizenbaum of MIT in 1966. The program, essentially a **Turing test** administered in a very restrictive domain, emulates a psychiatrist counseling a patient. "Patients" type their "problems" at a computer terminal, and Eliza responds principally be reflecting what they say back to them in modified form. For example, if the patient enters, "I am sad," Eliza might respond, "Is it because you are sad that you came to see me?" Since, unlike an **expert system**, Eliza has no stored *knowledge base*, it tends to answer questions with questions and to change the subject if semantically confused. For example, if the patient has mentioned a father, mother, sister, or brother, Eliza might stall by saying, "Tell me more about your family." But the technique is amazingly effective—for a limited session, at least—and the program continues to fascinate into its fourth decade. Eliza-like programs are available on the Web; *see* http://i5.nyu.edu/~mm64/x52.9265/january1966.html.

Additional reading: Weizenbaum, *Computer Power and Human Reason*.

Elliott 803 — COMPUTERS

During the 1960s, the principal **digital computer** used at many university and college computing centers in Great Britain was an *Elliott 803*, built by Elliott Brothers Limited of London and introduced in 1962. The machine was compact, requiring only about 400 square feet of floor space, had the undemanding power requirement of only 3.5 kilowatts and 10 kilowatts of air conditioning, and supported hardware **floating-point arithmetic** as an option, so the Elliott could be used as a low-cost scientific machine. Its **computer architecture** was similar to that of the Ferranti Pegasus, a 1956 **vacuum tube** machine. At least 250 Elliott 803s were delivered, and one has been restored to use by members of the Computer Conservation Society working under the direction of John Sinclair. The most surprising feature of the machine is its **magnetic tape** subsystem, which comprises modified film handlers of the kind used in movie studios instead of the more familiar half-inch tape decks. The principles are the same: Oxide-coated celluloid tape passes near a multichannel read/write head, but the "tape" is conventional 35mm film stock coated with oxide, and the mechanics include sprockets and spring-loaded tension arms. The films were specially made by Kodak and formatted at the factory, although service engineers could effectively reformat them by using a special film copy **program** to replicate a known good tape. The 1,000-foot films could hold over 7 million characters (*see* Figure 16). The machine is built from germanium **transistors**, and a large number of ferrite core logic elements are used, not as **memory** but as logic gates. The 803 is a bit-serial machine based on a single-address, single accumulator **instruction set** and a 39-bit **word length**. An auxiliary **register** was used for multiply and divide instructions. Its 19-bit **instructions** are packed two to a word, the 39th bit being a so-called B-bit. Within an instruction, three bits specified the group, three bits the instruction type, and the remaining 13 bits were used for addressing or specifying an **input / output (I/0)** function. Hence the direct addressing capability was 8,192 words. When the B-bit of a word was set, the contents of the location addressed by the first instruction in the word were added to the second instruction before execution, allowing indirect and other more esoteric forms of addressing and instruction modification at some cost in comprehensibility. Simple arithmetic instructions all executed in 576 microseconds. Jump instructions required 288 microseconds and floating-point division up to 9.792 milliseconds. Elliott supplied a very compact 8,192-word one-pass **compiler** for **Algol 60** written by Anthony Hoare, a classics graduate hired as a programmer in August 1960. Although a typical half-page Algol program took a half hour to compile and execute, the Elliott 503—the successor to the 803 despite its lower model number—was 60 times faster. His first task at Elliott had been to implement the Shell sort for the new Elliott 803 (*see* **Sorting**), but instead he invented a much better **algorithm** that he called **Quicksort**. Until the completion of his Algol compiler allowed succinct expression of his algorithm using **recursion**, he had not been able to convince others of its superiority; now it is acknowledged as a classic. Hoare went on to dominate many aspects of **computer science**, inventing, among other things, the *case statement* (*see* **Control structure**) and a theoretical basis for parallel computing. Hoare recounted some of his experiences at Elliott in his 1980 ACM Turing Award lecture.

Additional reading: Anthony Hoare, "The Emperor's Old Clothes," *Communications of the ACM*,

24, 2 (February 1981), online at http://lambda.cs.yale.edu/cs422/doc/hoare.pdf

Figure 16. A detail of the magnetic tape unit of the Elliott 803 of 1963, an early transistorized digital computer. The tape units were unique in that they used 35mm film stock coated with magnetic material. (National Museum of Photography, Film & Television/Science & Society Picture Library)

E-mail. *See* **Electronic mail (e-mail)**.

Embedded system HARDWARE

An *embedded system* is a device that contains a nonaccessible **microcomputer** chip to control its function. Typical embedded systems include traffic lights, vending machines, automobiles, home and office appliances, **cellular telephones**, toys, medical equipment, and unmanned space vehicles. The first embedded system in the automotive industry was the Volkswagen 1600 of 1968, which used a **microprocessor** in its fuel injection system. An embedded system implies use of a **program** stored in nonvolatile **read-only memory**

(ROM), a program that must be thoroughly de-bugged because it can never thereafter be mod-ified. As the end of the twentieth century approached, part of the **Y2K problem** was con-cern that embedded systems around the world might suddenly fail because of reliance on faulty date calculation **algorithms**, but the fear proved unfounded.

Additional reading: *Encyclopedia of Computer Science*, pp. 646–647.

EMCC. *See* **Eckert-Mauchly Computer Corporation**.

Emulation COMPUTER ARCHITECTURE

Emulation, a term coined by Larry Moss and Stuart Tucker of **IBM** in 1964, is the ability of one **digital computer** to interpret and execute the **instruction set** of another computer of dif-ferent architecture. The control unit of a digital computer can consist of either *hard-wired logic* (special-purpose logic circuitry for each **oper-ation code**) or microprogrammed control (*see* **Microprogramming**). On a machine with hard-wired logic, emulation in the sense of this article cannot be done. On a machine having a *writable control store*, its **instruction set** con-sists of sequences of **microinstructions** stored in **read-only memory (ROM)**. Emulation is then implemented through use of an alternative ROM with the microinstruction sequences needed to mimic the instruction set of the target machine. The technique was used by Moss and Tucker in the **IBM 360 series** to facilitate con-version of **programs** written for the **IBM 1400** and **IBM 700 series**. Emulation was common until widespread use of **high-level languages** with efficient **compilers**. **Software** can now be moved to new systems by recompilation, so emulation is now less important as a means of running old programs on new **hardware**.

Additional reading: *Encyclopedia of Computer Science*, pp. 647–648.

Encapsulation DEFINITIONS

Encapsulation is that property of an **object-oriented programming (OOP)** system whereby a **procedure**, called a *method*, is com-bined with the **data structure** upon which it operates to form a *module* in such a way that details of how that structure is implemented is kept as a *secret of the module*. Encapsulation is thus the epitome of **information hiding**, the principle that users are given access to only what is needed to solve their problems, leaving maintenance and possible upgrade of methods to systems programmers.

Engineering Research Associates (ERA) INDUSTRY

The historic firm *Engineering Research Asso-ciates (ERA)* was founded in 1946 in St. Paul, Minnesota, by Howard Engstrom, Ralph Meader, and William Norris. Seymour Cray be-came an early and most valuable employee. Its first major product was the **ERA 1101** of 1951. A year later, ERA was purchased by **Reming-ton Rand**, which merged with Sperry Gyro-scope, but the ERA group remained sufficiently intact to produce the *Bogart* in 1954. Bogart, of which the first was shipped to the National Security Agency in Washington, D.C., was the first computer built with solid state diodes, and it also used **magnetic core** memory. It was not named for the actor but rather for John B. Bo-gart, city editor of the New York *Sun*. Reming-ton Rand merged with Sperry Gyroscope to form **Sperry Rand** on 30 June 1955. Ninety days later, on 1 October, ERA became part of the Univac Division of Sperry Rand and under that banner built the **Univac 1103**. In 1957, William Norris left Sperry Rand to found the **Control Data Corporation (CDC)**.

Additional reading: Erwin Tomash, "The Start of an ERA," in Metropolis, Howlett, and Rota, *The His-tory of Computing in the Twentieth Century*, pp. 485–495.

English Electric KDF9 COMPUTERS

The *English Electric KDF9*, first delivered in 1963, and the **Burroughs B5000** of late 1962 were the first two **digital computers** that used zero-address (stack-based) **computer architec-ture**. The principal designers of the KDF9 were A. C. D. Haley and R. H. Allmark. Like the B5000 and the **Philco Transac S-2000**, the KDF9's 48-**bit word** provided good **floating-point** numerical **precision**. The machine had up to 32KB (kilobytes) of **magnetic core** memory and used **magnetic tape** as **auxiliary storage**.

Its hardware **pushdown stack**, known as a "nesting store," had a maximum size of 16 words. Arithmetic operations were carried out on the top few values in the stack (nest), replacing the **operands** by the results. A second nesting store was capable of holding up to 16 return addresses for **subroutines**. The operating system, called the "Director," was one of the first to implement fully preemptive process switching. Control returned to the Director whenever an attempt was made to use a busy device, to overflow either nesting store, or to execute an illegal **instruction. Interrupts** were also generated from the console and the internal clock. Two **compilers** for **Algol 60** were supported. One, the *Kidsgrove* compiler, was a highly optimized version intended for production use. The second, the *Whetstone* compiler, was intended for fast compilation and debugging. About 29 KDF9 computers were sold between 1963 and 1969.

Additional reading: Lavington, *Early British Computers*, pp. 76–77.

ENIAC   COMPUTERS

The *ENIAC* (Electronic Numerical Integrator and Computer) was designed and built at the Moore School of the University of Pennsylvania by J. Presper Eckert and John W. Mauchly over the period 1943 to 1946. John Grist Brainerd served as administrative supervisor, and others involved in the project were Arthur Burks, Joseph Chedaker, Chuan Chu, Richard Clippinger, Herman Goldstine, T. Kite Sharpless, and Robert Shaw. ENIAC is often described as the "world's first computer," but a more accurate claim would be "world's first *automatic, general-purpose, electronic, decimal, digital* computer." Omitting any of the italicized adjectives grants priority to some other computer. ENIAC was not a **stored program computer** and not easily programmable in the modern sense, which may or may not be the reason why so many sources claim that the "C" of ENIAC stands for **Calculator**. But ENIAC was *automatic* in that once a computation was started, it ran to completion without manual intervention, and it was *electronic* in its use of **vacuum tubes**. Each of ENIAC's 20 **registers** (*accumulators*) could add or subtract signed 10-

digit numbers in 200 microseconds. Multiplication involved 6 accumulators and took 2.6 milliseconds. The ENIAC programming manual was written by Herman Goldstine's wife Adele. Stanley Frankel and Nicholas Metropolis of the Los Alamos Scientific Laboratory ran the first **program** on the ENIAC in late 1945. ENIAC was dedicated at the Moore School in February 1946 and then moved to the U.S. Army's Aberdeen Proving Grounds in Maryland, where it remained in operation from 1947 until it was ceremoniously turned off on 2 October 1955. Parts of the machine are on display at the Smithsonian Institution in Washington, D.C. Eckert and Mauchly had filed for a U.S. patent on ENIAC in 1947 and obtained it in 1964, but it was invalidated on 19 October 1973 in the famous case **Honeywell** v. **Sperry Rand**. The judge's controversial ruling declared that the ENIAC principles were based on earlier work of John V. Atanasoff even though his **ABC** was a special-purpose device.

Additional reading: McCartney, *ENIAC: The Triumphs and Tragedies of the World's First Computer.*

Figure 17. Eckert and Mauchly's ENIAC of 1945 is, perhaps, the most historic early electronic digital computer, but not a **stored program computer**. See the article of that name for the contenders for the first of that kind. (Bettmann/Corbis)

Enigma. *See* Bombe.

ERA. *See* Engineering Research Associates (ERA).

ERA 1101   COMPUTERS

The *ERA 1101*, designed by Arnold Cohen and built by **Engineering Research Associates**

(ERA) in 1950, was the first commercially produced **stored program computer**. A one-megabit **magnetic drum** was used as its main **memory**. The first of an eventual ten ERA 1101s that were produced was delivered to the U.S. Navy in fulfillment of the specifications of its Task 13. Forgetting or foregoing the origin of 1101, the binary representation of 13, successive models were numbered 1102, 1103, and 1104, although these did not form an upward compatible family. The most noteworthy of these models became known as the **Univac 1103**, also designed by Cohen, after ERA was purchased by **Remington Rand** in 1952.

Additional reading: Ceruzzi, *A History of Modern Computing*, p. 37.

Error-correcting code   CODING THEORY

Error-correcting codes that introduce redundant **bits** into a data stream so that transmission errors can be detected and sometimes corrected are essential parts of most forms of digital communication and storage. Proper operation of the **Internet, modems**, and compact discs (CDs) would be impossible without them. The simplest error-detecting code is the simple **parity** check that uses a single redundant bit and can detect but not correct an odd number of errors in a group of bits. In interactive communication, it is usually sufficient to detect errors and request retransmission. In other situations, such as data stored on a CD or DVD, rereading erroneous data just results in the same error, so the errors must be corrected before use even though the disc itself cannot be (unless it is rewritable).

An error-correcting code developed by Richard Hamming in 1950 uses multiple parity checks in a very clever way to simplify error correction. A parity check is assigned to those positions in the **code** that have a 1 in the rightmost position of their binary representation, a second parity check for those positions that have a 1 in their second-to-right position, and so on. Thus, when a single error does occur, exactly those parity checks will fail for which the binary expansion of the position of the error has 1s. Moreover, the failures collectively point directly to the position of the error. It is then easy to reverse that bit and correct the error.

Reed-Solomon codes based on abstract algebra are the most powerful error-correcting codes known. They were invented in 1959 by Irving S. Reed and Gustave Solomon, then staff members at MIT's Lincoln Laboratory, and described in a famous five-page paper that appeared the next year in the *Journal of the Society for Industrial and Applied Mathematics*. This paper marks the beginning of **coding theory** as a significant mathematical endeavor. Actually, Reed and Solomon showed only how to encode in accord with their **algorithm**; not until 1970 did Elwyn Berlekamp, a professor of electrical engineering at the University of California at Berkeley, invent an efficient algorithm for *de*coding the Reed-Solomon code and thus make it practical for use in digital transmission. Other important error-correcting codes are the (Marcel) Golay code and the BCH code, named for its inventors Raj Chandra Bose and Dijen(dra) Kumar Ray-Chaudhuri of India and Alexis Hochquenghem of France.

In applications such as CDs where burst errors (long sequences of errant bits) are especially common, **interleaving** is used during recording to disperse the bits from one code word over a long portion of the data stream. When the data stream is deinterleaved, the burst of errors will appear as a large number of code words each with exactly one correctable error. The combination of interleaving and use of Reed–Solomon codes can correct bursts of over 4,000 errors, which is why accidentally scratching a CD will not necessarily distort its music.

Additional reading: *Encyclopedia of Computer Science*, pp. 674–677.

Ethernet   NETWORKS

The *Ethernet* data transmission **protocol** and its implementation in **hardware** was invented in 1973 by Robert Metcalfe, David Boggs, and Charles Thacker at **Xerox PARC**. Ethernet was designed to interconnect a number of desktop **personal computers** and **laser printers** at a data transmission rate of 2.94 Mbps (million **bits** per second). In 1980, Xerox, **Digital Equipment Corporation**, and **Intel** published specifications for a 10 Mbps Ethernet, and these were used as the basis for the IEEE 802.3 standard released in 1983 and subsequent variations. Ethernet has now evolved to include speeds of 10, 100, and 1,000 Mbps, as well as a variety of transmission media including thin coaxial cable, twisted copper pairs, and optical fibers. The IEEE classifies Ethernet using a **code** of the form `<speed> <baseband or broadband> <physical medium>`. The first part indicates the speed in Mbps, while the second part indicates whether baseband or broadband transmission is used. In baseband transmission, only one device can transmit at a time, and the entire **bandwidth** of the medium is available for transmission. In broadband transmission, the bandwidth of the medium is divided into two or more **channels** of differing frequency to support multiple simultaneous transmissions over a single cable. Cable **modems** use broadband transmission. The third part, physical medium, is either a number, in which case it refers to the longest allowable Ethernet segment length in hundreds of meters, or a letter used to denote a particular transmission medium. For example, 10Base5 refers to 10 Mbps baseband transmission using a thick coaxial cable whose maximum segment length is 500 meters, 100BaseT refers to 100 Mbps baseband transmission using unshielded twisted pairs, while 1000BaseSX refers to 1000 Mbps baseband transmission (Gigabit Ethernet) using a short wavelength **laser** over two multinode fibers. A typical personal computer configuration now includes an Ethernet card and a cable modem to provide fast **Internet** connection. Metcalfe founded the **3Com** corporation in 1979. *See also* **Metcalfe's law**.

Additional reading: *Computing Encyclopedia*, vol. 1, pp. 38–39.

Euclidean algorithm   ALGORITHMS

The author of the **Knuth textbooks** calls the *Euclidean algorithm* for finding the greatest common divisor (gcd) of two positive integers "the oldest nontrivial algorithm which has survived to the present day." Knuth devotes 46 pages to the subject, one of the longest sections in volume 2. Euclid described the **algorithm** in his *Elements* of 300 B.C., although it may date back to Eudoxus of 375 B.C. The integers 63 and 99 are both divisible by 3, but their gcd is 9 because 9 is the *greatest* of the divisors com-mon to both. One version of the Euclidean algorithm uses **modular arithmetic**, but another is based on repetitive subtraction: Successively replace the larger of the two initial numbers with their difference until the resulting numbers are the same. Thus gcd(63,99) ⇒ gcd(63,36) ⇒ gcd(27,36) ⇒ gcd(27,9) ⇒ gcd(18,9) ⇒ gcd(9,9) = 9. But the chain stemming from gcd(7,12) leads to gcd(1,1), indicating that 7 and 12 have no common divisor; they are said to be *relatively prime*. The Euclidean algorithm can also be stated (and programmed) using **recursion**:

gcd(a,b):
If a = b then return(1) else if a > b then return(gcd(a − b,b)) else return(gcd(a,b − a).

The Euclidean algorithm is indispensable to the analysis of algorithms and to implementation of many cryptographic algorithms (*see* **Data encryption**; **RSA algorithm**). A version of the Euclidean algorithm that is particularly well suited for use with the **binary number system** was discovered but not published by Roland Silver and John Terzian in 1962 and rediscovered and published by Josef Stein in 1967.

Additional reading: Knuth, *The Art of Computer Programming*, vol. 2: *Seminumerical Algorithms*, pp. 293–338.

Event-driven programming

PROGRAMMING METHODOLOGY

Event-driven programming is what must be done to create a **program** that is continually ready to respond to *events*—**interrupts** from an external source or internal *traps* that signal a request for service or invocation of an error-handling routine. In the era of **batch processing**, a program, once initiated, either ran to completion or halted prematurely because of some error condition. Even in the era of **operating systems** such as **MS-DOS**, there was little that a **personal computer** user could do but to either watch a program run or interrupt it with a **keyboard** control key. But in the era of highly **interactive computers** inaugurated by the **Graphical User Interface (GUI)**, a running program is subject to many different kinds of interruption or redirection, events that vary depending on where the monitor's **cursor** is positioned when a **mouse** or **trackball** is clicked. Thus the birth of event-driven programming as a milestone is coincident with the invention of the GUI, more particularly with the GUI developed for the **Apple Macintosh**. The GUI itself, a **browser** surfing the **Internet**, a **spreadsheet**, and a full-featured **word processor** are all examples of interactive programs that must be written in event-driven form. Such programs need a program structure that allows them to respond to a multitude of possible inputs, any of which may arrive at unpredictable times and in an arbitrary sequence. Under the event-driven programming model, the program structure is divided into two groups, *events* and *services*. An event represents the occurrence of something interesting. A service is what is done in response to the event. The event-driven program executes by constantly checking for possible events and, when one is detected, executing the associated service. In order for this approach to work, the events must be checked continuously and often, and services must execute quickly, so that the program can resume readiness to cope with still more events. Event-driven programs can be written in any **programming language** but some, such as **C++** and **Java**, were designed to facilitate their creation through natural integration with their support for **object-oriented programming (OOP)**.

Additional reading: *Encyclopedia of Computer Science*, p. 1856.

Evolutionary programming. *See* Artificial Life (AL); Genetic algorithm.

Expert system   ARTIFICIAL INTELLIGENCE

An *expert system* (ES) is a computer **program** that uses **heuristic** reasoning in conjunction with a *knowledge base* to solve complex problems with a proficiency close to that of a human expert in the domain of interest. Examples are programs that diagnose illnesses based on symptoms, and programs that analyze loan applications to assess the credit worthiness of applicants. In 1965, Edward Feigenbaum, Joshua Lederberg, and Bruce Buchanan of Stanford University created a program called DEN-

DRAL to interpret the mass spectrum of an organic molecule in terms of a hypothesis as to its structure. In doing so, they found that the problem-solving methods of **artificial intelligence (AI)** were useful but not sufficient; knowledge of chemistry and mass spectroscopy was more important. That knowledge was added to DENDRAL in collaboration with scientists at the Stanford Mass Spectrometry Laboratory. The effort to codify the knowledge of specialists for use in expert systems came to be called *knowledge engineering*, and the first association of the term *expert* with a computer program was in a publication describing DENDRAL experiments. In 1968, Saul Amarel wrote a seminal paper called "On Representations of Problems of Reasoning About Actions" that stressed the importance of *knowledge representation* in artificial intelligence, and his ideas were incorporated into an early expert system for medical diagnosis. Extensive research into knowledge engineering was subsequently conducted by Ryszard Michalski of George Mason University in Virginia. The two best-known industrial expert systems of the 1970s were XCON from **Digital Equipment Corporation (DEC)** and the Dipmeter Adviser from Schlumberger Ltd. For DEC, XCON optimized the configuration of **minicomputers** with respect to their planned workload about 300 times faster than human engineers. Schlumberger's Dipmeter Adviser interprets the data taken from instruments lowered into bore holes during the search for oil and gas. It offers hypotheses about the tilt, or "dip," of the rock layers far beneath the earth's surface. Knowing the dip of each of the hundreds of rock layers in a bore hole is valuable in oil exploration.

Additional reading: *Encyclopedia of Computer Science*, pp. 684–689.

Expression   DEFINITIONS

In both mathematics and in **high-level language** syntax, an *expression* is a symbolic representation of a rule for calculating a value. To the extent that the concept of an expression is a milestone, its invention was necessarily coincident with that of the concept of a **variable** by François Viète in 1591; until that date, expressions were limited to combinations of numbers. In mathematics, variables are traditionally represented by a single character, so multiplication can be implied through juxtaposition. For example, the volume enclosed between concentric spheres of radius r_1 and r_2 $(r_2 > r_1)$ is described by the expression

$$4\pi(r_2^3 - r_1^3)/3.$$

But there are several problems in trying to model this mathematical expression with an expression in a high-level computer language. Greek letters are not usually available; subscripts and superscripts must be linearized in some way; and since variables may have two or more characters, explicit multiplication **operators** must be used. Thus the expression in the example might be written in **Fortran** as

$$4*\text{pi}*(r(2)**3 - r(1)**3)/3$$

(which assumes that the radii are stored in a one-dimensional **array** of two elements). An expression may appear to the right of the replacement symbol (usually = or := or ←) in **statement**-oriented languages such as **Pascal** or Fortran or may stand alone and be evaluated immediately to yield a particular value in expression-oriented **functional languages**. A *statement-oriented language* is one in which sentencelike **statements** calculate and save intermediate values but (except for specific **input / output [I/O]** statements) do not print them. In Pascal, for example, the statement

$$p := a*b+c$$

is composed of an **identifier** (variable name) p, a *replacement symbol*, $:=$, and the *expression* a*b+c. Such an expression makes sense to the Pascal **compiler** if all of its identifiers have been previously declared as to **data type** (real, integer, etc.) and if it is *well formed* according to the grammatical rules of the language. An *expression-oriented language* is one in which expressions may stand alone such that, when encountered during program flow, their values are calculated and printed immediately (*see* **Functional language**). Thus, if the expression

$3+4$ is presented to **APL** at an interactive terminal session, APL will respond immediately by outputting 7. Since complex expressions may be built up recursively from simpler ones, the rules for recognizing a well-formed (syntactically valid) expression in any given language may be stated quite rigorously in a notation such as **Backus-Naur Form (BNF)**.

Extensible language   LANGUAGE TYPES

Any **high-level language** is extensible through use of **subroutines** or **macro instructions**, but the term at issue here has a more specialized meaning. An *extensible language* is one that allows a programmer to enrich or modify its syntax so smoothly that someone who does not know the base language and examines an extended program cannot tell customized constructs from those originally supported. One way of providing extensibility is to translate **source code** with a *syntax-directed **compiler***, one that parses **statements** in accord with a grammatical description such as, for example, **Backus-Naur Form**. That description would encompass the language's original grammar plus the additional grammatical definitions supplied by the programmer, who must be careful not to introduce ambiguities. **Lisp** and **Algol 68** are extensible to some degree. Extensible languages that were the subject of research in the early 1970s include Jan Garwick's GPL (General Purpose Language) of 1969, Tom Standish's PPL (Polymorphic Programming Language) of 1969, an extension of **APL**, and Ben Wegbreit's EL1 (Extensible Language 1) of 1974. The most widely used extensible language still in use is **Forth**. Interest in the concept has waned since even the mainstream language **C++** includes the ability to define combinations of new **data types** and **data structures** called **classes** and use them in conjunction with *overloaded operators*, **operators** whose meaning depends on context. The original C++ has already been greatly extended through provision and documentation of a large number of useful classes in a package called the *Standard Template Library* (STL).

   Additional reading: *Encyclopedia of Computer Science*, pp. 691–692.

F

Facsimile transmission (fax)

COMMUNICATIONS

The development of *facsimile transmission (fax)*, certainly a communications milestone, became a milestone in computer applications as soon as the combination of an **optical scanner** and a **laser printer** permitted a **personal computer** to double as a free-standing fax machine. In 1840, the Scot Alexander Bain proposed a fax machine using synchronized pendulums to scan an image at the transmitting end and send electrical impulses to a matching pendulum at the receiving end to reconstruct the image. In 1863, Italian physicist Giovanni Caselli received a U.S. patent for a similar fax machine called the "pantelegraph" that he had invented in 1855. Service between Paris and Lyons, France, operated between 1865 and 1870, but the Franco-Prussian War ended the experiment. In 1902, German scientist Arthur Korn developed the first practical fax machine using selenium cells to deconstruct pictures into components, then reconstructing them at the receiving end. In 1906, Max Dieckmann used an early **cathode ray tube (CRT)** to build a fax machine, and in 1925 he and his assistant Rudolph Hell built a better one. The fax machine as we now know it is essentially the device invented by Edouard Belin in 1913. It is not surprising that facsimile transmission antedates **television** because it is certainly more difficult to transmit moving images rather than digitized pictures and documents, but the fax machine, and the ability of a PC to mimic one, is a central feature of any modern office.

Additional reading: *Computing Encyclopedia*, vol. 2, pp. 52–53.

Fairchild Semiconductor

INDUSTRY

Fairchild Semiconductor was founded in Mountain View, California, on 1 October 1957 by eight scientists who left Shockley Semiconductor—Gordon E. Moore, C. Sheldon Roberts, Eugene Kleiner, Robert N. Noyce, Victor H. Grinich, Julius Blank, Jean A. Hoerni, and Jay T. Last. The company is named for the founder of the parent company, *Fairchild Camera and Instrument, Inc*. The founders used $3,500 of their own money to develop a method of mass-producing silicon **transistors** that previously had to be manufactured one at a time. The Fairchild Camera and Instrument Corporation—founded in 1925 by Sherman Fairchild, son of George Winthrop Fairchild, who had been the first chairman of the board of **IBM**—then invested $1.5 million in return for an option to buy the company within eight years. In 1958, Robert Noyce developed the monolithic **integrated circuit (IC)** at about the same time that Jack St. Clair Kilby invented a similar one at **Texas Instruments (TI)**. Noyce's colleague Jean Hoerni took the idea a step further and put a collector, base, and emitter all on one plane, creating the planar transistor and a new industry along with it. Decades later, the planar process is still the primary method for producing transistors. In 1968 Noyce and Gordon Moore left Fairchild to found **Intel**, but Fairchild continued to develop innovative **semiconductor** products. In 1979 Fairchild Camera and Instrument became a subsidiary of Schlumberger Limited, a global petroleum services and electronics company founded by the brothers Conrad and Marcel Schlumberger in 1920. Schlumberger sold its subsidiary Fairchild Semiconductor to **National Semiconductor (NS)** in 1987, but National kept control for only a decade, selling it in 1997 for $550 million to investors who now operate Fairchild Semiconductor as an independent company. In 1999 Fairchild acquired Samsung's Power Device Division and became a publicly traded company. In 2000 Fairchild acquired QT Optoelectronics, the world's largest company engaged in that emerging technology and one that Fairchild expects to become a major focus of its activity.

Additional reading: http://www.fairchildsemi.com/company/history_1950.html

Fast Fourier Transform (FFT)

ALGORITHMS

The *Fourier Transform*, due to the French mathematician Jean-Baptiste-Joseph Fourier in 1807, is widely used in physics and engineering to convert a function of time or space into a function of frequency. This is an integral transform, but since signal **data** is normally discrete, the transform of most practical interest is the *Discrete Fourier Transform* (DFT) that involves summation rather than integration. The straightforward application of the **algorithm** has running time that is proportional to the square of the problem size (number of signal values to be transformed), but in 1965, John Tukey and James Cooley published an equivalent algorithm called the *Fast Fourier Transform (FFT)* whose running time is O(n log n). (*See* **Analysis of algorithms** for a discussion of "Big O" notation.) In the reference, Cooley describes why the discovery was actually a rediscovery; the origin of the FFT can be traced back to Carl Friedrich Gauss in 1866. Cooley credits the first FFT program that actually achieved O(n log n) running time to Philip

Rudnick of the Scripps Institution of Oceanography in San Diego, California, who said that he wrote it (presumably in the 1950s) based on a 1942 publication of G. C. Danielson and Cornelius Lanczos in the *Journal of the Franklin Institute*. Lanczos worked in general relativity and corresponded extensively with Albert Einstein. Somewhat interesting is that one of the key steps in computing the FFT is the need to reverse a binary number **bit** by bit. Doing so by **table lookup** is discussed in the article of that name.

Additional reading: James W. Cooley, "How the FFT Gained Acceptance," in Nash, *A History of Scientific Computing*, pp. 133–139.

Fault-tolerant computer

COMPUTER TYPES

A *fault-tolerant computer* is one that continues to operate satisfactorily even in the presence of most **hardware** failures. The first fault-tolerant computer was the SAPO relay computer of 1954, built in Prague, Czechoslovakia, by a team led by Antonin Svoboda. The processor used *triple modular redundancy* (TMR), circuitry that is triplicated and voted. Operation continues so long as two of three circuits agree with each intermediate result. SAPO used a **magnetic drum** with automatic reread when an error was detected but could not be corrected through use of **error-correcting codes**. The U.S. National Aeronautics and Space Administration (NASA) was an early sponsor of fault-tolerant computing. In the 1960s, its first fault-tolerant machine was the onboard computer for the Orbiting Astronomical Observatory (OAO). The Jet Propulsion Laboratory STAR (Self-Testing-and-Repair) computer was developed for NASA in the late 1960s for a ten-year mission to the outer planets. The STAR, designed by Algirdas Avižienis, was the first computer to employ dynamic recovery throughout its design. Ultradependable **real-time systems** must be designed for applications such as aircraft control, mass transportation systems, and nuclear power plants where an error or even a serious delay can be catastrophic. The applications justify massive investments in redundant hardware, **software**, and testing. An example is the Saturn V guidance computer, developed in the 1960s, which contained a TMR processor and duplicated memories. More recently, fault-tolerant computing has become the dominant mode in transaction processing for banks, airline reservations, and so on. Fault-tolerant **servers** are in great demand to support the needs of the **Internet** and **local area networks (LANS)** for uninterrupted service. Server manufacturers offer systems that contain redundant processors, disks, and power supplies and automatically switch to backups if a failure is detected. Examples are the **Sun Microsystems** ft-Sparc and the **Hewlett-Packard** Stratus Continuum 400. For use as scientific **supercomputers**, emphasis is increasing on fault-tolerant **clusters** in which redundant nodes can assume the work of a failed machine.

Additional reading: *Encyclopedia of Computer Science*, pp. 698–702.

Fax. *See* Facsimile transmission (fax).

Ferranti, Ltd.   INDUSTRY

Ferranti, Ltd. of Great Britain was founded in 1882 as an electrical products company by the 18-year-old Sebastian Ziani de Ferranti. For over a century, the company's fortunes alternately prospered and waned under the leadership of the founder, his son Sir Vincent de Ferranti, and his sons Basil and the second Sebastian. For the most part, the firm enjoyed high growth and engineering success, pioneering significant breakthroughs in many fields from power station practice through to **digital computers**, microelectronics, and avionics. Dietrich Prinz joined Ferranti's instrument department in about 1947 and became involved in the firm's work in computers. Prinz wrote one of the earliest **computer chess** programs and became an expert in linear programming (*see* **Simplex method**). In 1948 he visited the United States to assess computer developments on behalf of Ferranti. The *Ferranti Mark I*, a production version of the **Manchester Mark I**, was the first commercially available digital computer. Manchester's involvement with Ferranti continued through the 1950s with the design of a Mark II computer whose 1957 production version was known as the *Ferranti Mercury*, and with the **Atlas** of 1962. In the

tradition of names like Mercury and Atlas, other Ferranti computers were named Orion, Perseus, and Sirius. After selling a total of 97 large computers, Ferranti withdrew from the **mainframe** computer business in 1963 but continued to make electronic components. Severe financial difficulties plagued Ferranti in the early 1990s, and the historic company closed its doors in 1994 after 111 years of business.

Additional reading: Wilson, *Ferranti: A History*.

Fiber optics   COMMUNICATIONS

Fiber optics technology uses *optical fibers*, thin flexible strands of clear glass or plastic, that transmit light instead of electrical signals. Optical fibers are the backbone of cable **television** networks that can be used to support high-speed **Internet** access and more broadly to support an *integrated services digital network* (ISDN) in which voice, video, and **data** can be simultaneously transmitted over other public and private networks. This fiber optics medium is capable of carrying up to several billion **bits** (gigabits) per second over long distances. An optical fiber transmitter uses a **light emitting diode (LED)** or **laser** to convert electrical signals to light signals, and the receiver uses a photodiode to effectuate the reverse conversion. An optical fiber consists of two concentric layers, an inner core and an outer cladding that has a refractive index smaller than that of the core. Light propagation through the fiber then depends on its diameter, its construction, its refractive index profile, and the nature of the light source.

Spun glass fibers were made as early as 1713 by René de Reaumur, and in the 1790s Claude Chappé invented an "optical **telegraph**" in France. In 1854 John Tyndall demonstrated to the Royal Society that light could be conducted through a curved stream of water, proving that a light signal could be bent. In 1880 Alexander Graham Bell invented a "Photophone" that transmitted a voice signal on a beam of light. On 15 October 1926 John Logie Baird applied for a British patent on an array of parallel hollow tubes to carry images in a mechanical television system. On 30 December 1926 Clarence W. Hansell outlined the principles of a fiber optic imaging bundle in his notebook at the **Radio Corporation of America (RCA)** Rocky Point Laboratory on Long Island. In 1930, German medical student Heinrich Lamm was the first to assemble a bundle of optical fibers to carry an image. In July 1952 Harold Hopkins applied for and received a grant from the Royal Society in London to develop bundles of glass fibers for use as an endoscope at the Imperial College of Science and Technology and hired Narinder Kapany as a graduate assistant. In 1954 Hopkins and the Dutch scientist Abraham Van Heel wrote separate papers on imaging bundles. In the summer of 1955 Kapany completed a doctoral thesis on fiber optics under Hopkins that is now considered to be the defining document in the long history of fiber optics development. Production and installation of practical fiber optic systems had to await development of the **semiconductor** diode laser by Robert Hall of **General Electric (GE)** in 1962. In 1970, Corning Glass researchers Robert Maurer, Donald Keck, and Peter Schultz invented fiber optic wire that they called "Optical Waveguide Fibers" capable of carrying 65,000 times more information than copper wire. Twenty-five million kilometers of the cable they designed has been installed worldwide, and more than 85% of the world's long-distance traffic is now carried over optical fiber cables.

Additional reading: Hecht, *City of Light*.

File

DATA STRUCTURES

A *file* is a variable-length one-dimensional **array** of **records** in **auxiliary storage**. Before **digital computers** (and still, to some extent) "auxiliary storage" was a file cabinet containing files in the form of manila folders that contained paper records. Now, files are typically stored on **magnetic tape** or **hard** or **floppy disks**. An individual file is typically accessed by only a small number of persons at a single location, whereas a **database** consists of an integrated collection of multiple files accessed by many people, perhaps over the **Internet** or an **intranet**. The first common type of file organization was called BSAM (Basic Sequential Access Method) because to reach and retrieve a record from an inherently sequential magnetic tape the preceding records must be skipped

over. A variant was QSAM (Queued Sequential Access Method). Among the various access methods that have been developed for use with (relatively) random access storage devices are DAM (Direct Access Method), ISAM (Indexed Sequential Access Method), and VSAM (Virtual Storage Access Method).

Additional reading: *Encyclopedia of Computer Science*, pp. 5–8, 708–711.

File server. *See* **Server**.

File Transfer Protocol (FTP)

NETWORKS

The *File Transfer Protocol (FTP)* is a **protocol** used to upload or download data **files** to and from an **Internet** host. FTP was developed in 1972 by a team led by Abhay Bhushan at **Bolt, Beranek, and Newman (BBN)**. The protocol was first used on the **ARPAnet** 20 years before the appearance of the **World Wide Web**, whose built-in file transfer routine has now largely displaced the need for FTP, but it is still the method of choice for "publishing" (uploading) **HTML** or **XML** files to be used as Web pages.

Additional reading: Hafner and Lyon, *Where Wizards Stay Up Late*, pp. 174–175.

Finite element method

ALGORITHMS

The *finite element method* of **numerical analysis** is an **algorithm** for transforming partial differential equations into difference equations that are solved by piecewise linear approximation over a network of polygons, usually triangles. The method is reminiscent of the way the mathematicians of ancient Greece estimated the value of π by noting that the perimeter of a polygon inscribed in a circle approximates its circumference (*see* **Pi [π]**). The earliest work that presages modern finite element techniques was that of Karl Schellbach in 1851. In 1939, Gabriel Kron named his own variant of the Schellbach technique *diakoptics*, or "the method of tearing." In his view, large structures should be figuratively torn apart, solved in small pieces, and the partial results reassembled. In the late 1950s, Eugene Wachspress of the (then GE) Knolls Atomic Power Laboratory (KAPL) used a similar method to combine results from the analysis of two-dimensional (2D) planes of a nuclear reactor into an approximate 3D solution through what he called *flux synthesis*. In 1959, E. D. Reilly and Jerry Fletcher of the same laboratory were the first to solve reactor diffusion problems over a 2D mesh of uniformly spaced congruent equilateral triangles, and Bruce Kellogg, Richard Varga, and J. H. Cadwell of the (then Westinghouse) Bettis Atomic Power Laboratory in Pittsburg extended the method to a grid tiled with triangles of arbitrary shape and size. But the first complete elaboration of the finite element method as now applied was contained in a 1956 paper by Mark J. Turner, Ray W. Clough, Harold C. Martin, and L. J. Topp, but the term *finite element method* was not coined (by Clough) until 1960. Their papers are an important milestone because the finite element method has become the method of choice for the engineering analysis of large structures of irregular shape.

Additional reading: J. Tinsley Oden, "Historical Comments on Finite Elements," in Nash, *A History of Scientific Computing*, pp. 152–166.

Finite State Machine (FSM) THEORY

The *Finite State Machine (FSM)*, or *finite automaton*, is the lowest (least capable) of the four types of machine in the **Chomsky hierarchy**. An FSM merely accepts an input while in one state, advances to another state depending on the value of that input, and (possibly) emits an output value as a result. Perhaps the simplest example is a coin-operated vending machine. While in its *initial state*, the machine accepts a sequence of coins and for each one advances to another state until it reaches its *final state* (also called an *accepting state*), one that will dispense the product (and change) and move back to the initial state. A finite automaton is said to *accept* any sequence of symbols (coin values in the example given) that brings it to its final state. An FSM can recognize only *regular expressions*, those that conform to a very restricted set of rules. For every regular expression, there is a corresponding FSM that will accept the language symbolized by that expression.

Additional reading: Dewdney, *The Turing Omnibus*, pp. 8–13.

Firewall NETWORKS

A *firewall* is a system that prevents unauthorized access to a proprietary **network** or, inversely, to a **personal computer** or **workstation** connected to any kind of network. The first firewall was invented at Bell Labs in the mid-1980s. Firewalls are frequently used to prevent unauthorized **Internet** users from accessing private networks connected to the Internet, especially **intranets**. All messages entering or leaving the intranet pass through the firewall, which examines each message and blocks those that do not meet the specified security criteria. Firewalls can be implemented in **hardware** or **software** or a combination of both. The *packet filter* examines each data packet entering or leaving the network and accepts or rejects it based on user-defined rules. The *application gateway* applies security measures to specific applications, such as **file transfer protocol (FTP)** and **Telnet**, whereas the *circuit-level gateway* applies them when a **TCP / IP** connection of any kind is established. A *proxy server* intercepts all messages entering and leaving the network, effectively hiding the true network addresses. A firewall is merely the first line of defense in protecting private information; for greater security, **data encryption** is more reliable.

Additional reading: *Computing Encyclopedia*, vol. 2, pp. 72–73.

FireWire COMMUNICATIONS

FireWire, also known as i.LINK, is a high-speed serial **input / output (I/O)** technology for connecting digital devices such as **optical scanners**, printers, camcorders, and **digital cameras** to **personal computers (PCs)**. FireWire connections transfer data at up to 400 Mbps (million bits per second), more than 30 times the **bandwidth** of a Universal Serial Bus (USB—*see* **Bus**). FireWire was developed by **Apple** in the mid-1990s but became commercially available only at the turn of the century. FireWire is now the cross-**platform** industry standard IEEE 1394. The combination of low-cost, high-quality camcorders and built-in FireWire allows the creation of broadcast-quality video. FireWire has revolutionized video editing within the **television** and film industry. Because of its impact, Apple received a 2001 Primetime Emmy Engineering Award for the development of FireWire from the U.S. *Academy of Television Arts and Sciences.*

Additional reading: *Computing Encyclopedia*, vol. 2, p. 73.

Firmware
DEFINITIONS

Early in the history of digital computing, the useful distinction was made between **hardware**, the tangible components of a computing system, and **software**, the collection of **instructions** that directed what was to be computed. Software had to be recorded on some tangible medium, such as **IBM** cards (*see* **Punched card**) or **magnetic tape** (early) or, later, hard or **floppy disks**; the software itself was considered to be pure information and hence intangible. Supporting the "softness" of this interpretation was the fact that when recorded on a magnetic medium, software could be modified with ease. The question arose as to what to call **programs** recorded indelibly on a medium such as **read-only memory (ROM)** or embodied in hard-wired computer circuitry. The result, no longer "soft" enough to be modified, was still functionally "software," but had it become, though still executable, "hardware"? The term coined by Ascher Opler in 1967 to solve the dilemma was *firmware*. Given adequate **memory**, any general-purpose **digital computer** can execute programs written for any other through use of a *simulator*, a program that interprets each target machine instruction and executes whatever sequence of host machine instructions is needed to do the same thing, **bit** for bit. Interpretation is naturally slow, but **emulation**, implementation of key parts of the simulator as firmware, provides a significant increase in speed of execution.

Additional reading: *Encyclopedia of Computer Science*, pp. 713–714.

Flash memory
MEMORY

Flash memory is a form of nonvolatile **solid state memory**; that is, it retains its content even when its power is switched off. Because flash memory is writable only in blocks, not a **byte** at a time, it is not suitable for use as main **memory**, but it is ideal for use in **cellular tele-**phones, **digital cameras**, **MP3** players, and videogames. A version of flash memory based on the NAND function of **Boolean algebra** was invented in 1984 by Fujio Masuoka. EPROM tunnel oxide (ETOX) flash memory, invented by Stefan K. Lai of **Intel** in 1988, is now the industry standard. This variant uses the NOR function to emulate EPROM (Erasable Programmable **Read-Only Memory**).

Additional reading: http://www.howstuffworks. com/flash-memory1.htm

Flat-panel display. *See* Liquid Crystal Display (LCD).

Flip-flop
HARDWARE

A *flip-flop* is a bi-stable device whose input can switch (or "toggle") it from one state (representing "0") to the other (representing "1"). Thus a flip-flop is capable of storing one **bit** of information. Flip-flops can be, and have been, realized by optical, pneumatic, biochemical, hydraulic, mechanical, electromechanical, and electronic means—even by Tinkertoys®. The first electronic flip-flop was built in 1919 by W. H. Eccles and F. W. Jordan who used two interconnected triode **vacuum tubes**. One type of flip-flop, the *JK flip-flop*, has two inputs traditionally labeled *J* and *K* either arbitrarily or, according to some sources, because they are the initials of Jack (St. Clair) Kilby of **Texas Instruments**, a coinventor of the **integrated circuit (IC)**. Current flip-flops are made from **transistors**, but regardless of construction, a flip-flop is still sometimes referred to as an *Eccles-Jordan trigger circuit*. Their invention is a milestone because, once invented, multiple flip-flops could be used to implement the **logic circuits** needed to build electronic **digital computers**.

Additional reading: Petzold, *CODE*, pp. 155–179.

Floating-point arithmetic
GENERAL

Just as those who once used a **slide rule** to do multiplication and division had to keep track of the decimal point, early programmers had to keep track of the decimal or binary point when calling for a sequence of inherently integer arithmetic operations. This was done through a tedious and error-prone process known as *scaling*. *Floating-point arithmetic* automates this process through use of numbers that are essentially stored in scientific notation, a concept known to the Chinese of 100 B.C. Examples are 1.273×10^{-6} and -4.32×10^5 in decimal (radix 10) and 1.00101×2^{110} in binary (radix $2 = 10_2$). Floating-point numbers are an abstraction of the mathematical real number axis. In the memory of a binary computer, a floating-point number is stored in three parts: Its *magnitude* (size aside from sign and exponential scaling) is stored as a sequence of **bits** called the *mantissa*, the sign of the number is assigned one bit (0 for plus and 1 for minus), and the exponent of the radix plus a constant *bias* (to avoid negative exponents) is stored in a sequence of bits called the *characteristic* (or *biased exponent*). Since the radix never changes, it need not be stored explicitly. The Z3, a **Zuse computer** of the early 1940s, was the first to feature floating-point **hardware**. On early **microprocessors**, floating-point arithmetic was either simulated through **software** or emulated through use of a coprocessor (*see* **Emulation**; **Firmware**). Almost all current computers feature hardware floating-point **instructions** that conform to an ANSI/IEEE standard promulgated in 1985.

Additional reading: *Encyclopedia of Computer Science*, pp. 85–88.

Floppy disk   MEMORY

The *floppy disk*, or *diskette*, invented by David Noble of **IBM** in 1967, is the primary removable **auxiliary storage** medium for a **personal computer (PC)**. Diskettes were originally called "floppy disks" in distinction to the *hard disk* because the earliest such disks were intrinsically flexible and remained so when encased in their square cardboard protective envelopes. Diskettes are made by depositing a ferromagnetic metallic oxide material on a mylar substrate. The 8-inch disks used with early **CP/M** personal computers are now obsolete, as are the 5.25-inch diskettes used with the earliest **MS-DOS** and **Apple** systems. The most commonly used version is a 3.5-inch diskette encased in hard plastic (and hence no longer "floppy").

Diskettes are inexpensive and easy to mail and carry. Information is recorded on both sides of the diskette in *double-sided format*. Therefore, each floppy disk drive has two read/write heads. Information is organized in concentric *tracks* that are either prerecorded when the diskette is manufactured or magnetically encoded by a formatting program that comes with the **operating system**. Each track is subdivided into *sectors* and each sector into **bytes**. The 3.5-inch diskette has 80 tracks per side, 18 sectors per track, and 512 bytes per sector, yielding a capacity of 1.44MB (megabytes). "Thick" diskettes are now available that hold up to 250MB, but their associated disk drive is not compatible with those used with the small diskettes.

Additional reading: *Computing Encyclopedia*, vol. 2, pp. 80–81.

Flowchart   GENERAL

As used in computer science, a *flowchart* is an organized pictorial representation of the logic of a **program** for a **digital computer**. In that sense, the first flowchart may have been the elaboration of the formulas to be used for computing successive values to be assumed by **variables** in Ada Augusta Byron's 1842 "Diagram for the Computation by the Engine of the Numbers of Bernoulli." This meticulously detailed diagram is part of Ada's famous addendum to Luigi Menabrea's description of Charles Babbage's **Analytical Engine** (*see also* **Program**) and is included as a foldout in an appendix to Bowden's *Faster Than Thought*. The style incorporated in her diagram looks very much like the programs of 60 years later written in **Plankalkül** by Konrad Zuse for his **Zuse computers**. But with the advent of the **stored-program computer** in the mid-1940s, it became much easier to tell a computer to make decisions and alter its flow of control depending on the outcome. At this point, flowcharts assumed the form of a directed **graph** in which nodes had specialized shapes such as rectangles for most imperative actions, diamonds or hexagons for *predicates* (decisions having either a *true* or *false* outcome), and rhombi for **input/output (I/O)** operations. Once the move to **structured programming** gained momentum, however, the programs corresponding to such flowcharts began to be denigrated as "spaghetti code." In response, two new forms of flowchart were devised. In 1973, Isaac Nassi and Ben Shneiderman invented what they called *structured flowcharts* but which are now called *Nassi-Shneiderman diagrams*, and in 1974 J. D. Warnier proposed use of an alternative system later refined by Kenneth Orr and now called *Warnier-Orr diagrams*. Nassi-Shneiderman diagrams use variously shaped outlines that fit together like jigsaw pieces that "tile the plane" (leave no gaps), where the boundary of the plane in question is an enclosing rectangle. When implemented in a structured **high-level language** such as **Pascal** or **C++**, the code that corresponds to this kind of flowchart contains no distracting GOTO statements. This is also the case with the Warnier-Orr diagram, whose basis is a linear layout of successive actions with their alternative actions, if any, stacked underneath.

Additional reading: *Encyclopedia of Computer Science*, pp. 714–716, 1704–1706.

Flynn taxonomy

COMPUTER ARCHITECTURE

The *Flynn taxonomy*, devised by Stanford University **computer science** professor Michael J. Flynn in 1966, is a way of classifying **digital computers** into four categories with regard to the way they qualify, if at all, as being **parallel computers**. The earliest digital computers were *uniprocessors* incapable of any degree of parallelism other than, perhaps, the **multiplexing** of I/O devices and the *buffering* (simultaneous execution) of **input / output (I/O)** and computation. Flynn classifies the uniprocessor as being an SISD machine, which stands for *Single Instruction Single Data* stream computer. That means that as **data** is ingested, it is processed by the one and only available processor. A *Single Instruction Multiple Data* (SIMD) computer such as the **ILLIAC IV** used a two-dimensional **array** of identical processors, each of which simultaneously performed the same operation of a grid of data belonging to the same program; that is, it performed as a **cellular automaton**. A *Multiple Instruction Single*

Data (MISD) machine such as the **CDC 6600** had multiple functional units of variously different kinds that could be applied to different portions of the data stream emanating from the same program after ascertaining through **instruction lookahead** that it was safe to do so. Pipelined computers that operate like an assembly line are also MISD machines, as are **Very Long Instruction Word (VLIW)** computers. Finally, a *Multiple Instruction Multiple Data* (MIMD) machine uses multiple processors, each of which operates on a data stream emanating from different concurrently running programs. There are *shared memory* MIMD machines that contend for access to a single main **memory**, and there are *distributed memory* MIMD machines in which each processor has an independent memory of its own. A **dataflow machine** is a shared memory MIMD. Despite its simplicity, the Flynn taxonomy has nonetheless remained useful for almost four decades.

Additional reading: *Encyclopedia of Computer Science*, p. 1351.

Forth

LANGUAGES

Forth is a **high-level language** invented by Charles H. Moore in 1970 for **real-time systems** applications such as the control of **radio** telescopes and orbiting satellites. Early Forth implementations included an **interpreter**, a **compiler**, a disk **operating system** (DOS), a **multitasking** executive, a **text editor**, and an **assembler**, all resident in less than 16 kilobytes of **memory**. Members of the Forth Interest Group (FIG) formed in 1978 have developed a model implementation for common **microprocessors**. Forth is an **extensible language** that may be either interpreted (the usual practice) or compiled. It uses **Polish notation** and implicit parameter passing by way of a **pushdown stack**. A Forth development environment typically includes a large set of predefined commands called *words*. Any word may be executed from the **keyboard** or combined with others to define a new word. Associated with each word is an *action*, and the meaning of a sequence of words is the sequence of actions implied by their order.

Additional reading: *Encyclopedia of Computer Science*, pp. 721–723.

Fortran   LANGUAGES

Fortran (FORmula TRANslator) was the first widely used **high-level language** for scientific and engineering computing. John Backus of **IBM** proposed its development to his boss, Cuthbert Hurd, in 1953. The 1954 team that began design of the language included Sheldon Best, Harlan Herrick, Robert Nelson, Roy Nutt, Peter Sheridan, Richard Goldberg, Lois Haibt, David Sayre, and Irving Ziller. The team's principal design goal was that the Fortran **compiler** would generate very efficient **object code**. The goal was achieved so well that its level of performance in this regard was not exceeded until the development of optimizing compilers more than ten years later. On 20 September 1954, Harlan Herrick ran the first successful compilation and execution of a Fortran program. Fortran was distributed in April 1957 for use on the IBM 704. The first **bug** in its compiler reported by a customer was that of Herbert Bright of the Westinghouse Bettis Atomic Power Laboratory near Pittsburgh, Pennsylvania. The first Fortran textbook, called simply *Fortran*, was written by Daniel McCracken and published in 1961. Fortran II, distributed in 1958, featured improved diagnostics and a **subroutine** facility. Fortran III of the same year allowed the combination of high-level and **machine language** code, but this version was never distributed outside IBM. Fortran IV of 1963 added COMMON storage, double-precision and logical **data types**, relational **expressions**, and a DATA **statement** for initializing **variables**. Fortran IV was the basis for the first Fortran standard, ANSI Standard X3.9 in 1966 and the later ISO Standard 1539 in 1972. Fortran 77, a decade in development, became the official Fortran standard in April 1978. In a last-minute decision by the designers of Fortran 77, an if-then-else **control structure** had been added as a minimal concession to the growing number of advocates of **structured programming**, and the tepid reaction prompted a "five-year" plan to produce a truly structured successor. Five years stretched into 12, and the product was released as Fortran 90 (F90). Again, a five-year

schedule for revisions was adopted, and Fortran 95 was released on schedule. Fortran 2000 is now scheduled for 2003, the major goal being interoperability with **C** and **C++**. Separately from these efforts, a consortium with Charles Koelbel of Rice University as executive director developed a 1993 extension of Fortran 90 called HPF (High Performance Fortran) for use on **parallel computers**.

Additional reading: Lohr, *Go To: The Programmers Who Created the Software Revolution*, pp. 11–34.

Fractal   MATHEMATICS

Fractal is a term coined in 1975 by the Polish mathematician Benoit Mandelbrot to describe planar curves that are so highly irregular that they seem to have a dimensionality greater than one but less than two; that is, a "fractal" dimension that is not an integer. It is characteristic that such curves are "self-similar"; that is, no matter how much they are magnified, the section viewed continues to be topologically equivalent. For example, the classic von Koch "snowflake curve" depicted in the reference has a *self-similarity dimension* of 1.262. Fractals arise naturally in the study of **chaos theory**, but their attraction to computer scientists is that they make excellent models of the irregular terrain—a mountain or a plowed field, perhaps—that is needed in **computer graphics** images. It is also the case that when "false color" is used to depict the various differently shaped regions of space generated during fractal **iteration**, the created pattern is an attractive example of computer-generated art.

Additional reading: *Encyclopedia of Computer Science*, pp. 725–732.

Free software   SOFTWARE

Free software, in contrast to **freeware**, is **software** that is not necessarily free of nominal cost but is distributed subject to something called a General Public License (GPL) by the creator of the concept, Richard Stallman. In particular, the GPL requires that software that incorporates other GPL software must itself be subject to a GPL, which thus, by **recursion**, prohibits charging more than the cost of distribution for any modification to a GPL product. Stallman created the not-for-profit *Free Software Foundation* (FSF) in 1985 to promote the idea that **source code** to all **programs** should be freely available but distributed more restrictively than general **open source software** so that it would remain freely available indefinitely. Through employment of both paid programmers and volunteers, the FSF promoted the creation of GNU (Gnu's Not Unix) software that includes, among other programs, a GNU equivalent of **Unix** whose kernel became **Linux**, GNU **C++**, and the GNU Emacs **text editor** that Stallman wrote himself.

Additional reading: *Encyclopedia of Computer Science*, pp. 732–733.

Freeware
SOFTWARE

The term *freeware* was first used and trademarked by Andrew Fluegelman, the founding editor of *PC World* magazine. He said that the idea for freeware came to him one day in 1983 while driving across the Golden Gate Bridge in San Francisco. Fluegelman had developed a product called *PC-Talk* but didn't have the time or resources to market his product. Instead of the usual distribution methods, he decided to make the compiled (binary) version of his **program**, but not its **source code**, freely available to anyone who wanted it. Thus **open source software**—in which "open" *does* imply availability of source code—is freeware, but the converse is not true. (*See also* **Free software** and **Shareware**.) In the summer of 1985, Fluegelman had been missing for about a week when his car was found parked near the toll plaza on the Marin County side of the Golden Gate Bridge. His family held a memorial service in New York the following Sunday, but his body was never found.

Additional reading: *Computing Encyclopedia*, vol. 2, pp. 90–92.

Friden calculator
CALCULATORS

The Friden Calculating Machine Company was founded in the United States by Swedish engineer Carl Mauritz Fredrik Friden on 11 January 1934. The company's business plan was to market a high-quality calculator, an improved version of one that he had invented and sold to his employer while working at Marchant

Calculating Machine Company. The company's first product was the electrically operated *Friden Automatic Calculator* with automatic division. Its design incorporated the Friden roto-flow one-way drive principle, which allows the actuating mechanism to rotate continuously at all times, eliminating the interruptions to which reversible calculating mechanisms were subjected. Among the many other design principles that Friden devised were the return clear key; fully automatic, continuous nonstop division and multiplication; selective dial accumulator locks; **keyboards** with individual key locks; and selective automatic half-cent adjustments. By 1930, Friden held more calculating machine patents than any other living engineer. By the 1950s, four-function motor-driven mechanical calculators had become common, with Friden and Monroe being the market leaders. In 1952 Friden introduced the *Friden SRW*, an imposing calculator weighing 42 pounds. The carriage contained two **registers**. The upper register showed totals, subtotals, products, and so on, and the lower register showed the number of items added or subtracted, multipliers, and quotients. The **keyboard** contained well over 100 keys including both full and ten-key keyboards. But its major innovation was a square root key that invoked a square root routine that delivered its result in just a few seconds.

Additional reading: http://www.fridenites.com

FTP. *See* **File Transfer Protocol (FTP)**.

Fujitsu, Ltd.
INDUSTRY

The origin of *Fujitsu* dates back to 1876 when the German company Siemens (*see* **Siemens-Nixdorf**) set up an office in Tokyo as the sole supplier of electrical equipment to the Furukawa cartel. After World War I, Siemens and Furukawa established Fuji Electric as the sole importer of Siemens products. In 1935, Fuji Electric created a subsidiary that was first called Fuji Communications Corporation and then reorganized as *Fujitsu, Ltd.* after World War II. In April 1954, Fujitsu developed Japan's first commercial **digital computer**, the FACOM 100. In February 1960, the company began making **transistors** and in 1961 produced the transistorized FACOM 222. Several additional FACOM models followed. By 1965, Fujitsu had developed profitable product lines in **automatic control** devices and **robotics**. In the 1970s, Fujitsu became the leading Japanese computer manufacturer by virtue of arrangements with the **Amdahl Corporation** and introduction of its own M series of **IBM**-compatible **mainframes**. Fujitsu built Japan's first **supercomputer**, the VP-100/200, in 1982. By 1990, Fujitsu owned 43% of Amdahl and 80% of Britain's **International Computers Limited (ICL)**. Its initial entry into the **workstation** and **personal computer** market in 1981 had been unprofitable, but after investments in **Sun Microsystems** RISC chips and **Advanced Micro Devices** (AMD) 80×86 chips, Fujitsu made a successful reentry in the late 1990s and is now a leader in application of those technologies. The company has now moved heavily into **handheld computers** and very light **laptop computers** called *notebooks*. Fujitsu employs about 187,000 people and has annual sales of about $40 billion.

Additional reading: Chandler, *Inventing the Electronic Century*, pp. 196–198.

Function PROGRAMMING

A *function* as used in a **high-level language** is comparable to a single-valued function in mathematics. Syntactically, a function is a form of **subroutine** that returns a single value that may be used in an **expression** (formula). In most languages, an expression like $5.7*sin(x) - cos(y - 2.1)$ is valid provided x and y have reasonable values at the time the *library functions* for the sine and cosine are invoked. Similarly, the expression $3.7*zilch(8)$ would be valid provided the programmer has defined the homemade function *zilch* so as to make it applicable to a range of numbers that includes 8.

Functional language LANGUAGE TYPES

A *functional language*, sometimes called an *applicative language* or *expression-oriented language*, is a **high-level language** in which the only **control structure** is the application of a **function** to its arguments. Functional languages, as do **spreadsheets**, emphasize *what* is to be computed rather than *how*, as *imperative*

languages like **Pascal**, **Ada**, and **C** do. In a functional language, functions are invoked by **expressions**, with the final value of the expression or the value of a stand-alone function being printed, displayed, or *bound* (associated with) a name supplied as a **parameter**, the equivalent of the more usual *assignment statement*. Instead of conditional statements such as `if-then-else`, *conditional expressions* are used. For example, the statement if `(a>b)` then `write 10 else write 4` can be rendered as `4+6*(a>b)`. This works because `(a>b)` produces 1 (true) or 0 (false). The best-known early functional languages were **Lisp** and **APL**, which has an assignment statement but is otherwise a functional language, as is its successor *J*. Even **C++**, otherwise an imperative language, has one functional construct; the statement discussed above can be written `cout << (a>b ? 10 : 4)`. One of the earliest advocates of *functional programming* was John Backus, the very person who led the group that produced the world's most widely used imperative language, **Fortran**. He advanced the advantages of the concept though his language FP, which he developed and described in 1978. Subsequent functional languages that have gained favor are David Turner's *Miranda* of 1985 and **Haskell** of 1987, which is based on the **Lambda calculus**.

Additional reading: *Encyclopedia of Computer Science*, pp. 736–739.

Functional programming. See Functional language.

Fuzzy logic MATHEMATICS

In **Boolean algebra**, **variables** may have only one of two values, 1 (true) or 0 (false), and in mathematics, an element from the domain of a **set** is either *in* or *not in* any given set of such elements. But *fuzzy logic* and *fuzzy set*, terms introduced by Lofti Zadeh in 1965, deal with variables with values in the interval [0, 1]. In a fuzzy set, transition between membership and nonmembership is gradual rather than abrupt; each element is given a degree of membership between 0 (nonmembership) and 1 (full membership), such as 0.18 or 0.73. Fuzzy sets naturally arise when a finite set of terms has to be mapped to a linguistic scale. For instance, if a temperature scale is chosen to be the real interval [0, 100] in degrees Celsius, to which a set of terms {cold, cool, warm, hot} corresponds, then it is difficult to find exact thresholds a, b, and c such that, for instance,

cold = [0, a), cool = [a, b),
warm = [b, c) and hot = [c, 100]

because the predicates *cold*, *cool*, *warm*, and *hot* are imprecise. Fuzzy arithmetic is well suited to engineering design. Poorly specified **parameters** can be represented by fuzzy numbers and adjusted as necessary until **simulation** indicates acceptable system performance, at which point their bounds are incorporated into final specifications.

Additional reading: Sangalli, *The Importance of Being Fuzzy.*

G

Game of Life GENERAL

The *Game of Life* is a **cellular automaton** invented by John Horton Conway that captured the public's imagination when described in Martin Gardner's *Mathematical Recreations* column in the October 1970 and successive issues of *Scientific American*. This one-person game is so simple that it can be played with pencil and quadrule paper or with counters on a chessboard or GO board, but it has maximum effect when programmed such that its evolving results are displayed on the **monitor** of a **digital computer**. Consider a uniform two-dimensional **array** (grid) in which some small number of cells are filled in as a starting position. This population of "organisms" is then followed from generation to generation until it either dies, becomes a stable "still life," oscillates, collectively drifts out of range, or grows without limit. The genetic laws that determine the configuration of a new generation based on the preceding one are as follows: (1) Each counter with either two or three neighbors (of the eight it might have had) survives into the next generation. (2) Each counter having four or more neighbors dies from overcrowding, and each counter with one or no neighbor dies from isolation. (3) Each empty cell having exactly three adjacent neighbors spawns a new organism born at that site as part of the next generation. These rules are applied simultaneously to all cells of one generation to produce the next.

The effects produced by such simple rules are amazing; they provide the illusion of "creatures" that move, change shape, and evolve. Some initial configurations die out, some evolve into oscillating patterns, and some expand indefinitely. The first pattern that does the latter—one that emits a "glider" that moves off to infinity every 30 iterations—was discovered by William Gosper in 1970. Thus the Game of Life is a milestone example not only of a cellular automaton but also of **artificial life (AL)**.

Additional reading: Federighi and Reilly, *Weighting for Baudot and other problems for you and your computer*, pp. 175–176.

Game theory THEORY

Game theory is a branch of mathematics applicable to highly complex interactive systems. Although more usually applied to economics and strategic military planning, game theory has recently proved invaluable in the theoretical analysis of the behavior of large **distributed systems**. In November 2001, the Center for Discrete Mathematics and Theoretical Computer Science (DIMACS) held a workshop entitled *Computational Issues in Game Theory and Mechanism Design* at Rutgers University in New Jersey. Although there were antecedents, the effective origin of game theory was the 1944 Princeton University Press publication *Theory of Games and Economic Behavior* by John von Neumann and Oskar Morgenstern. As used therein, a "game" is a metaphor for a wide range of human interactions in which the outcomes depend on the respective strategies of two or more persons, countries, or organizations that have opposing motives. In 1950, while addressing an audience of psychologists at Stanford University as a visiting professor, Albert W. Tucker created the *Prisoners' Dilemma* to illustrate the difficulty of analyzing certain kinds of games. The situation he used as an example concerned the separate interrogations of two prisoners believed to have been involved in the same burglary. Either would probably be treated more leniently if he were to confess, but the pair would be better off if neither did so. What should they do? Tucker's simple explanation has since given rise to a vast body of literature in subjects as diverse as philosophy, biology, sociology, ethics, political science, economics, and, of course, **computer science**. After almost 50 years out of sight and mind of even most scientists, interest was reinvigorated by the award of the 1994 Nobel Prize in Economics jointly to John Forbes Nash and John Harsanyi of the United States and Reinhard Selten of Germany. The award to Nash in particular caught the public's imagination seven years later with the appearance of the Academy Award–winning motion picture *A Beautiful Mind*, based on Sylvia Nasar's book of the same name.

Additional reading: Holland, *Emergence: From Chaos to Order*, pp. 28–80.

Gate. *See* **Logic circuit**.

Gateway   NETWORKS

Gateway is usually considered to be a synonym for a **hardware** device, the **router**, but some—including the influential Tim Berners-Lee, inventor of the **World Wide Web**—consider that a gateway is the **software** that allows a **digital computer** to act as a **server** to **Internet** clients (*see also* **Apache**). The distinction is minor because servers need routers to operate.

Additional reading: Berners-Lee and Fischetti, *Weaving the Web*, pp. 49–50.

Gateway, Inc.   INDUSTRY

Gateway, once *Gateway Country* and *Gateway 2000*, was founded in 1985 by Ted Waitt at his family's farm in Iowa. Like **Dell**, Gateway makes and sells **IBM**-compatible **personal computers (PCs)** by phone, mail order, and **Internet**. But unlike Dell, Gateway also supports a national network of stores that display merchandise, take orders that are filled by mail, and repair machines as necessary. Since 1993 when the company went public, Gateway headquarters have been in San Diego, California. In 2002, Gateway held about 6% of market share for IBM-compatible PCs, tied for fourth place with **IBM**. Dell was the leader at 14%, with **Compaq** and **Hewlett-Packard** *(hp)* (now merged) at 10% and 8%, respectively.

Additional reading: *Computing Encyclopedia*, vol. 2, p. 105.

GE. *See* **General Electric (GE)**.

GE 200 series   COMPUTERS

General Electric (GE) entered the **mainframe** computer business in June 1959 with delivery of the first member of its *GE 200 series*, the GE 210, designed by Arnold Spielberg (Steven's father) and Charles Prosper. The series ultimately consisted of three additional models, the 225 of March 1961, an intermediate model 215 in June of 1963, and the top of the line, the 235 in June 1964. The computers were limited by an unusually short **word length**, 20 **bits**, the equivalent of only six decimal digits, so its "double" **precision** arithmetic was less precise than the single-precision arithmetic of contemporary computers such as the **CDC 1604** and **Philco Transac S-2000**. The combination of a GE 235 processor and of a Datanet 30 communication processor was called the *GE 265* and used as the basis for the **Dartmouth Time Sharing System (DTSS)**. The DTSS software was designed in 1963–1964 at Dartmouth College, principally to support instruction in **Basic**, but **Algol 60** and **Fortran** were also supported. Most of the DTSS **operating system** was resident in the Datanet front-end processor; only user **programs** and their environment were executed in the GE 235.

Additional reading: http://perso.club-internet.fr/febcm/english/ge200.htm

GE 600 series

COMPUTERS

The first model of the *GE 600 series*, the 635, was delivered to MIT in November 1964. Like the IBM 7094 (*see* **IBM 700 series**), the GE 635 had a 36-bit **word**, 18-bit addresses, eight **index registers**, and programmable data **channels** with direct **memory** access. The main innovation of the GE 600 series was its use of a *Symmetric Multiprocessing Platform* (SMP), the first ever built. The origin of the 600 series was the M236 computer developed by **General Electric (GE)** for the U.S. Air Force Cape Canaveral Missile Range and installed at Eleuthera in the Bahamas. The M236 was a 36-bit computer designed to meet the computational requirements of radar tracking and the need to exchange data with an IBM 7094 located at the Range. The chief architect of the M236 was John Couleur, who became the technical leader of subsequent large GE systems. The MIT 635 was used to support a simulated GE 645 until that successor was delivered in 1968. The operational speed of the 645 was about half a MIPS (million instructions per second), somewhat disappointing compared to competitive machines of that era, but it is historic in that it was the first to support **Multics**, an **operating system** capable of **time sharing** more than a hundred terminals. GE went on to deliver a higher-performance 655 in March 1971 and brought out less costly and lower-performance models of the series, the 625 in April 1965 and the 615 of June 1969.

Additional reading: http://perso.club-internet.fr/febcm/english/gecos_to_gcos8_part_1.htm

General Electric (GE)

INDUSTRY

The Edison Electric Light Company, the earliest predecessor of the *General Electric Company (GE)*, was founded by Thomas Edison on 15 October 1878. In 1886, Edison established the Edison Machine Works in Schenectady, New York. On 15 April 1892, General Electric was incorporated in Schenectady through merger of Edison's two companies and the Thomson-Houston Company. The GE Research Laboratory, the first industrial laboratory in the U.S. dedicated to research, was founded in 1900. The 12-year period (1958–1970) in which GE attempted to be a major manufacturer of **digital computers** is only 10% of its long history, but the company's scientists made many contributions to computer technology before and after that time. GE engineer Ernst Alexanderson invented several devices that facilitated the development of **television**. In 1891, GE's Charles Proteus Steinmetz discovered the law of hysteresis, which forms the basis for the ferrite cores used in **magnetic core** memory. Steinmetz had also showed how to apply complex arithmetic to the analysis of alternating current electrical systems, the basis for many applications of the **Bell Labs relay computers**.

In 1962, Robert Hall of the GE Research Laboratory built the first **semiconductor** injection **laser**. Charles Bachman's Integrated Data Store of 1964 is regarded as the first **Database Management System (DBMS)**. In 1977, a very successful full-body **computerized tomography (CT)** scanner was developed by J. M. Houston and W. Rey Whetten of the same Laboratory. Before GE began to market the **GE 200 series** of computers in 1959, the company had built the special-purpose scientific computer *OMIBAC* and, under technical director Jerre Noe, the business computer *ERMA* for Bank of America. In 1964, GE acquired a two-thirds interest in **Machines Bull** of France and full acquisition of the computer division of **Olivetti** of Italy. The **GE 600 series** of the mid-1960s was reasonably successful, but the company sold its computer division to **Honeywell** in 1970. For a time in the 1970s, the GE Corporate R & D Center in Niskayuna, New York, was home to a group of internationally renowned theoretical computer scientists including Richard Stearns and Juris Hartmanis, Turing Award winners of 1993, and Richard Lewis and Daniel Rosenkrantz, who joined Stearns in developing the definitive theory of how a **compiler** might best analyze ("parse") programmed **expressions** so as to generate good **object code** in the least time. The **Radio Corporation of America (RCA)**, which had its own flirtation with computer manufacture (*see* **RCA Spectra series**) began as a GE subsidiary in 1919, became independent in 1930, and was reabsorbed by GE in 1987. As recently as 2002, GE announced that it was making a major ex-

pansion of its Corporate R & D Center, renamed the GE Global Research Center, as a
prelude to intensive research in the emerging
field of **nanoelectronics**. GE is the only company in the Dow Industrial Index that was in
the first Dow Index of 1886.

Additional reading: Chandler, *Inventing the Electronic Century*, pp. 15–27, 101–103.

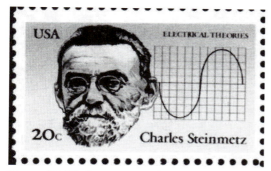

Figure 18. "Steinmetz was the most valuable piece
of apparatus General Electric had—until he wore out
and died" (John Dos Passos, *The 42nd Parallel*).
(From the collection of the author)

Genetic algorithm ALGORITHMS

A *genetic algorithm* is an **algorithm** that embodies a computerized version of evolution to
create new inventions. The creation of **programs** that use genetic algorithms, known as
genetic programming, was invented in 1987 by
John R. Koza of Stanford University. The process, outlined in **artificial life (AL)**, starts with
a population of thousands of randomly generated test objects and then selects the most
highly adapted species of "individual" after mu-
tation and recombination introduce additional
variations into thousands of successive generations. Here, "most highly adapted" means the
closest approximation to achievement of a
mathematically stated criterion. Naturally occurring operations that follow a genetic algorithm are the likely basis for the evolution and
gradual optimization of the genetic code that is
essential to life on earth. Genetic programming
is beginning to compete favorably with human
inventors in the field of engineering design, especially the design of specialized electronic circuits. As a result of the work of Koza and his
team using a **Beowulf** cluster of their own construction (*see* reference), genetic algorithms

have reproduced the design of 15 **logic circuits**
patented between 1917 and 2001, and have created nine other circuits, some of which may
well be patentable (if, that is, a patent can be
granted to a **digital computer**). Among these
are algorithms for **quantum computing**, the
NAND circuit of **Boolean algebra**, and digital-
to-analog (D to A) and analog-to-digital (A to
D) converters (*see* **Analog computer**). Koza's
cluster can follow the evolution of a population
of 100,000 individuals for 100 generations in
about seven days, and the fastest current **supercomputers** could reduce that time to about
an hour. Of the three principal attempts to
achieve **artificial intelligence (AI)**, genetic
programming holds the greatest promise for
eventual development of a computer that can
pass the **Turing test**. And this would be the
very way that Turing himself envisioned that it
might happen in his famous paper of over 50
years ago (in, appropriately, the journal *Mind—
see* **Listing of References of Interest** for the
complete citation).

Additional reading: John R. Koza, Martin A.
Keane, and Matthew J. Streeter, "Evolving Inventions," *Scientific American*, **288**, 2 (February 2003),
pp. 52–59.

Geographic Information System
(GIS) INFORMATION PROCESSING

A *Geographic Information System (GIS)* is a
specialized **database management system
(DBMS)** that analyzes and displays geographic
data. Geographic information describes the locations, characteristics, and boundaries of both
physical and human-made features on the surface of the Earth. Traditionally, such information has been produced and disseminated in the
form of paper maps and atlases. Increasingly,
however, use of instruments capable of sensing
the Earth's surface from space have enabled
the capture of relevant information in precise
digital form. GIS requires specialized **input /
output (I/O)** devices, including map-sized **digitizing tablets**, high-resolution **optical scanners**, and pen plotters. The development of
such **peripherals** in the 1960s was a major im-
petus, but significant use of the GIS concept
began only in the early 1980s, following the
development of **relational database** manage-

ment systems and **workstations** equipped with high-resolution display **monitors**. GIS data is organized in *layers*, each of which defines a representation of some specific class of objects over the geographic area of the database. The ability to combine and display information from different layers is an important characteristic of a GIS. For example, a city planner may wish to direct a GIS to display a multicolored map of a given neighborhood that shows the locations of underground utilities derived from one layer with land parcel zoning information derived from another. The first GIS was the *Canada Geographic Information System* of the mid-1960s. The current leading producers of GIS software are MapInfo of Troy, New York, and ESRI of Redlands, California, whose major products are called MapInfo and ArcInfo, respectively.

Additional reading: *Encyclopedia of Computer Science*, pp. 748–750.

Global Positioning System (GPS)

APPLICATIONS

The *Global Positioning System (GPS)* is a satellite-based digital information system that enables the precise location of objects on the surface of the Earth, within a few yards for civil use, within a few centimeters for military use. GPS is funded by and controlled by the U.S. Department of Defense (DoD). GPS provides specially coded satellite signals that can be processed by a ground receiver that contains a **microprocessor** and, because incoming data must be corrected for relativistic effects (*see* reference), a very high precision atomic clock. Four GPS satellite signals are used to compute positions in three dimensions and a time offset with respect to the receiver clock. The spatial segment of the system consists of GPS satellites whose *transponders* emit **radio** signals. The GPS *Operational Constellation* consists of 24 satellites that orbit the Earth at an altitude of 10,900 nautical miles every 12 hours. As the Earth turns underneath, the satellites repeat the same track and configuration over any point every 24 hours, but four minutes earlier each day. There are six orbital planes with four satellites in each, equally spaced at 60 degrees apart and inclined at about 55 degrees with re-

spect to the equatorial plane. This constellation provides the user with between five and eight satellites visible from any point on the Earth. GPS systems are now available as optional equipment in some luxury automobiles that use them to gather the information needed for display of a map of current position and heading on a dashboard screen.

Additional reading: Neil Ashby, "Relativity and the Global Positioning System," *Physics Today*, **55**, 5 (May 2002), pp. 41–47.

Gödel incompleteness theorem

MATHEMATICS

In 1931, the Austrian logician Kurt Gödel astounded the mathematical community by proving what is now known as the *Gödel incompleteness theorem*. Contrary to what had been believed for centuries, the theorem states that in any "sufficiently rich axiomatic system" such as the *Peano postulates* (axioms) used to prove theorems in **number theory** or the Russell-Whitehead axioms of symbolic logic, there are theorems that are true but which cannot be proved to be true from the axioms. The implication became significant to **computer science** when Alan Turing proved an equivalent corollary, namely, that the **Halting problem** is undecidable.

Additional reading: Casti, *The Cambridge Quintet*.

GOLEM computers. *See* WEIZAC.

GPSS

LANGUAGES

GPSS (General Purpose Simulation System) is a discrete-event **simulation** language designed and implemented in 1961 by Geoffrey Gordon of **IBM** with programming assistance from Robert Barbieri and Richard Edron. The initial version of GPSS was written for the **IBM 700 series** in **assembly language**, as were all subsequent versions for IBM computers. GPSS is not a **block-structured language** with regard to syntax but rather deals with blocks as **data structures**, each block describing what happens at that stage of the chain of blocks that define the discrete system being modeled. Only one continuous-event language was ever widely used, **Dynamo**, but many discrete-event languages have been defined, GPSS, **Simscript**, and **Simula 67** being the usual languages of choice in this area.

Additional reading: Wexelblat, *History of Programming Languages*, pp. 403–437.

Grant calculators   CALCULATORS

After exhibition of his electrically powered piano-sized **difference engine** in 1876, American engineer George Barnard Grant turned his attention to building smaller **calculators**. He produced two models, the *Centennial* and the *Rack and Pinion* of which 125 were sold. The latter was simpler to set up and operate than earlier calculators and did much to build public confidence in the reliability of machine calculation.

Additional reading: Evans, *The Making of the Micro*, p. 49.

Graph   DATA STRUCTURES

A *graph*, or more precisely a *connected graph*, is a **data structure** that consists of a collection of data *nodes* and **pointers** that connect each node to one or more other nodes of the same graph. A **tree** is a restricted graph that has no circuits. Graphs are useful in modeling **networks**, in which nodes have arbitrary degrees of connectivity. In a directed graph, a distinction is made between a node's *in-degree*, the number of other nodes that point to it, and its *out-degree*, the number of other nodes that it points to. These terms correspond exactly to what electrical engineers call the *fan-in* and *fan-out* of a **logic circuit**. In 1847 Russian physicist Gustav Kirchhoff, a student of Gauss, may have been the first to apply graphs to scientific or engineering problems when he devised his **algorithm** for solving for currents and voltages in closed electrical circuits by applying the laws of energy conservation at nodes. Some early **Database Management Systems** were based on graphical models, but these have largely been supplanted by **relational database** systems. Although graph **algorithms** have been of interest to mathematicians for centuries (*see* **Graph theory**), very few computer languages have been devised that facilitate graph navigation. The earliest may have been that of Douglas Ross of MIT in 1961, who called a graph a *plex* (from com*plex*) and went on to use the same name to describe a whole philosophy of science. Most recently, *Xpath*, a language for addressing parts of an **XML** document, was created in 1999 by James Clark and Steven DeRose of Brown University.

Additional reading: Knuth, *The Art of Computer Programming*, vol. 1: *Fundamental Algorithms*, pp. 362–372, 377–378.

Graph theory   THEORY

Graph theory is a branch of mathematics that has proved to be a fertile source of problems in **computational complexity** that require an extensive **analysis of algorithms**. The origin of graph theory is considered to be the work of Leonard Euler in solving the problem called the *Seven Bridges of Königsberg*. Königsberg, once a city in Eastern Prussia, is now the Russian city of Kaliningrad. The river Pregel (now Pregolya) flows through the city, and it contains two islands connected to themselves and the opposite river banks by a network of seven bridges (*see* Figure 19). Citizens of the city speculated for years as to whether one could start at one point (a *node* of an equivalent **graph**) and follow a path that crosses each bridge (*edge* of the graph) exactly once and return to the starting point. They could not find one, and in 1736 it was finally proved by Euler that none was possible. More generally, he discovered the rule for determining whether there existed such an *Eulerian circuit* in a given graph of arbitrary complexity. When a closed path can be found that visits each node (rather than edge) only once, the path is called a *Hamiltonian circuit* in honor of the great Irish mathematician William Rowan Hamilton, although he was not the first to study such a problem. In 1994, Leonard Adleman showed how to use a **DNA computer** to find such circuits. Some graph problems are quite tractable. For example, in 1956 the Dutch computer scientist Edsger Dijkstra published an **algorithm** for finding the shortest path from a given vertex of a connected graph to any other vertex in the graph, and in 1971, John Hopcroft and Robert Tarjan published an algorithm for determining whether two planar graphs were isomorphic (topologically equivalent). **Java** *applets* can be downloaded from the **World Wide Web** that let one watch the running of Dijkstra's algorithm on a **monitor**. Another graph problem of great practical interest is the *Traveling Salesman Problem*, trying to find the circuit through a given list of cities that visits each city only once and minimizes the total distance traveled. Unfortunately, solving the problem for *n* cities with assurance that one has indeed found the optimal solution appears to require time that is an exponential function of *n*, and the problem has been proved to be **NP-complete**.

Additional reading: Biggs, Lloyd, and Wilson, *Graph Theory*.

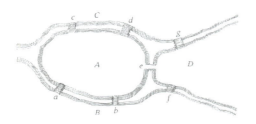

Figure 19. The famous "Seven Bridges of Königsberg" (Prussia), over which the good bergers could find no Eulerian circuit. (From Euler's 1736 paper, reproduced in Biggs, Lloyd, and Wilson, *Graph Theory*.)

Graphical User Interface (GUI)

OPERATING SYSTEMS

A *Graphical User Interface (GUI)* is a **monitor** screen presentation that consists of small *icons*, *toolbars*, and possibly multiple *windows* in which text or graphical images are displayed. The functionality of a GUI depends on use of a pointing device such as a **mouse** or **trackball** that is moved by hand to position a **cursor** at a chosen point on the screen. Clicking a mouse button at that location will activate the program associated with an icon on the "desktop" (central portion of the screen) or, more particularly, select a toolbar option associated with the currently running **program**. On **personal computers (PCs)** and **workstations**, but not on **mainframes**, GUI-based **operating systems** have almost completely displaced command-line systems such as **MS-DOS**. Even **Unix**, the quintessential command-line operating system, can now be used with an optional GUI, and doing so is now popular with the **Linux** version of Unix. Use of a GUI is an essential part of modern **word processing**, **spreadsheet**, **database**, and **desktop publishing** systems. The GUI is the outgrowth of 1960s work in **computer graphics** by Ivan Sutherland and David Evans. The first GUI to attract the attention of computer professionals was used with the **Alto** at **Xerox PARC** in 1973, and the first to capture the imagination of the public at large was that of the **Apple Macintosh** of 1984. Despite a late start **Microsoft** has been able to capture and hold a dominant share of the GUI market with successive versions of **Microsoft Windows**.

Additional reading: *Computing Encyclopedia*, vol. 2, pp. 118–119.

Graphics. *See* **Computer graphics**.

Grosch's law
GENERAL

In 1953, H. R. J. ("Herb") Grosch of **IBM** noted that spending twice as much for a new computer as an older version of the same model would generally yield four times the processing speed; that is, processing speed rises as the square of the cost. The observation became known as *Grosch's law* and proved accurate for at least three decades. Either in fulfillment of an inherent "law" or because IBM marketers chose a conforming price structure, Grosch's law modeled the performance of individual members of the **IBM 360 series** quite well. Over the last two decades, however, **multiprocessing** and **multitasking** have made it very hard to measure intrinsic processing speed, and the law is much less quoted than heretofore. See also **Moore's law**.

Additional reading: *Encyclopedia of Computer Science*, pp. 960–963.

Groupware
APPLICATIONS

Groupware is **software** technology designed to facilitate the work of two or more people. Groupware, whose origin goes back to the 1970s, is used to communicate, cooperate, coordinate, solve problems, compete, or negotiate. The term is ordinarily used to refer to a specific class of technologies relying on computer **net-**

works, such as **electronic mail**, newsgroups, **bulletin board systems**, instant messaging, and chat rooms. The goal of groupware use is to increase the gains that can arise from group work while limiting potential losses. Sources of gain include the fact that a group has greater knowledge than any one individual, stimulation from exchanges among participants, and the opportunity for a participant to learn from the behavior of others. Potential losses include limited speaking time among participants, reluctance to share ideas, excessive socializing, and the tendency of some members not to carry a fair share of the work. Groupware technology is typically categorized in accord with whether participating users are working at the same time (*synchronous groupware*) or different times (*asynchronous groupware*) and whether they are working at the same place (*colocated groupware*) or in different places (*remote* or *distant groupware*). Groupware vendors now offer software that blurs the distinction between the corporate **intranet** and the **Internet** by offering products that function equally well in either environment. Several of the best-known groupware packages that do so are *Lotus Notes*, *Microsoft Exchange*, *Netscape SuiteSpot*, and *Novell GroupWise*.

Additional reading: *Computing Encyclopedia*, vol. 2, pp. 123–125.

H

Hagelin machine HISTORICAL DEVICES

The *Hagelin machine* is a cryptographic device invented in 1925 by the Russian-born Boris Hagelin, an employee of a Swedish company started by Arvid Damm. Hagelin's machine was not the first multiple-rotor cryptographic machine, but it was the first commercially successful one. It typified the rotor-based machines that were so widely used by both sides during World War II that they were one of the two driving forces behind that era's rapid development of **digital computers**, the other being the need for high-speed calculation of equations related to nuclear fission. Additionally, all cryptographic machines were special-purpose digital computers in their own right, because **data encryption** is a quintessentially digital process. A rotor-based crypto machine uses several rotors (wheels) with scrambled letters around their circumference that are used to encipher messages according to a polyalphabetic encryption **algorithm**. Then U.S. Secretary of State Thomas Jefferson invented one with 25 rotors in 1792 but most unfortunately was talked into using a much less secure system. The twentieth-century machines use only three to six rotors, but the way they are ratcheted one with respect to another in accord with day-to-day lug settings created an enormous number of key combinations that made the devices extremely secure.

Rotor machines were devised by four different inventors from four different countries within a short time of one another. The first was that of American Edward Hebern in 1918, but it did not prove reliable. In 1919, Hugo Koch of the Netherlands patented a rotor machine three days before Arvid Damm of Sweden. In 1923, a machine called the *Enigma* was patented by Arthur Scherbius of Germany, who also acquired the rights to the Koch machine. Hagelin's contribution, while he was still working for Damm, was to use retractable pins near the rims of his rotors, and their variable in-or-out settings provided an effective key length of 100 million characters, enough to encipher at least 25 copies of *War and Peace* before cyclic repetition. The first version of his machine, the B-211, now produced by his own company, was about the size of an attaché case and weighed 37 pounds. He then designed a six-rotor pocket-sized version, the C-36, and sold 5,000 to the French government in 1935. In 1940, the U.S. government adopted the full-sized version as its M-209, and Hagelin licensed Smith-Corona to produce 400 per day. Over 140,000 were produced, and historian David Kahn calls it "the most ingenious mechanism in all cryptography." An M-209 in good working order is a prized collector's item.

Germany's use of the rotor-based Enigma machine and Japan's use of a rotor machine for its Purple code during World War II prompted an intensive Allied effort to break the intercepted messages enciphered in those systems. The story of the British effort to crack the Enigma is told in the article **Bombe**, a project that employed the talented Alan Turing. Since the early 1940s, the first version of every new U.S. **supercomputer** is delivered to the National Security Agency (NSA) at Fort Meade, Maryland, and subsequent versions to the U.S. national laboratories dedicated to work in nuclear weapons or reactors. Without question, development of the Hagelin machine and others like it constitutes a most important milestone in the history of digital computer development.

Additional reading: Kahn, *The Codebreakers*, pp. 410–434.

HAL 9000 COMPUTERS

HAL (Heuristic ALgorithmic computer), is, of course, the fictional **digital computer** in Stanley Kubrick and Arthur C. Clarke's 1968 movie (and later Clarke novel) *2001: A Space Odyssey*. Clarke claims that it is only a coincidence that when each letter of HAL is replaced by the next higher one in sequence, the result is **IBM**. The novel, if not the computer, is historic because it tried to assess the state of **artificial intelligence (AI)** 33 years beyond the year in which it was written. HAL appeared to be able to think and almost certainly would have been able to pass the **Turing Test**. HAL was completely aware of its surroundings; it could differentiate and recognize several different people both visually and through their voices, it could read their lips, and it could speak quite distinctly (or rather the unseen actor Douglas Rain could). But the reality is that **computer vision** is still not well developed, nor is **speech recognition**, because of the great difficulties of **pattern recognition**. The one anthropomorphic attribute of HAL that can now be done reasonably well is **speech synthesis**—we can make a computer speak clearly. Telling it what to say in order to let it uphold its end of a conversation is quite another matter, of course.

Additional reading: Clarke, *2001: A Space Odyssey*.

Halting problem THEORY

The *halting problem* was not really a "problem" but rather speculation as to whether a **digital** **computer** could inspect a **program** and its **data** and, without running the program, decide whether it would ever halt. Interest in the question intensified in 1931 after proof of the **Gödel incompleteness theorem** showed that there must be true theorems that cannot be proved from the usual axioms of mathematics. The British mathematician and computer scientist Alan Turing, suspecting a relation between Gödel's theorem and the halting problem, proved in 1936 that no **algorithm** exists that a **Turing machine** (and hence any conceivable **computer**) can use to decide whether a program of *arbitrary* complexity will halt or, alternatively, **loop** indefinitely. (Trivially, a program could be written to decide that another program consisting of a single HALT **instruction** will indeed halt.) Turing's result is considered the most profound in theoretical **computer science** because, first, it showed that given that one particular problem is undecidable, others may be too. By now, many others are known. Second, Turing's result has deep mathematical significance because it answered the *Entscheidungsproblem*, the second of a famous set of 23 problems posed on 8 August 1900 by David Hilbert: There can be no algorithm capable of deciding whether each and every mathematical conjecture it is asked to examine is true or not. In fact, if Turing's result had been obtained first, it could have been used to prove Gödel's incompleteness theorem.

Additional reading: Davis, *The Universal Computer: The Road from Leibniz to Turing*.

Hamming code. *See* **Error-correcting code**.

Handheld computer COMPUTER TYPES

Literally, a *handheld computer* is a (**digital**) **computer** than can be held in the hand. The earliest form of handheld computer was, despite its terminology, the **programmable calculator**. When **Apple** introduced the *Newton* in 1990, it was called a **personal digital assistant (PDA)**. That term is still sometimes used to refer to more recent handheld computers that are also called *pen computers* because data is entered into them with a stylus, something like writing on an Etch A Sketch®. Using a handheld computer equipped with a wireless connection to the **Internet** is called **mobile computing**. The first handheld computer was a pen computer, the GRiDPAD, designed in 1989 by Jeff Hawkins. The device measured 9 × 12 × 1.4 inches and ran on a 10 MHz **Intel 80×86** processor using **MS-DOS**. The GRiDPAD had a CGA (Color Graphics Array) 640 × 400 **pixel** display **monitor** and sold for $2,370 without **software**. Data storage was via 256 or 512KB (kilobyte) battery-backed **random access memory (RAM)** cards.

Additional reading: http://www.digital-doc.com/why.html

Hard disk   MEMORY

A *hard disk* is a high-capacity, high-speed rotational **auxiliary storage** device. The hard disk was invented in the mid-1950s by an **IBM** team led by Reynold Johnson, who had also invented the **test scoring machine**. Johnson's inspiration was a conceptual proposal of Jacob Rabinow. In January 1955, the first hard disk, the IBM 350, was delivered as part of the IBM 305 **RAMAC**. Hard disks were once called *Winchester drives*, a name derived from a 1969 IBM hard drive that stored 30MB (megabytes) of information on each of two spindles, a 30–30 arrangement that reminded its inventors of the Winchester 30–30 rifle. A hard disk consists of a rigid aluminum alloy disk that is coated with a magnetic oxide material, much like a diskette (*see* **Floppy disk**). Because the disks are rigid, they can endure much greater rotational speed than a floppy, up to 10,000 rpm or more. Their read/write heads float on a cushion of air very close to the disk surface, the gap being about 1/100,000 of an inch. Large drives contain multiple platters mounted on a rotating spindle. Each platter surface has its own read-write head. **Mainframe** hard disks can hold up to a terabyte—a trillion **bytes**. Hard disks used with **personal computers (PCs)** typically range in capacity from 20 to 100GB (gigabytes—billions of bytes). Hard disks store data in magnetically encoded *tracks*. There may be over 10,000 tracks on a high-capacity hard disk. The collection of tracks with the same number arranged vertically on all platters is called a *cylinder*. Tracks are divided into *sectors*. When a

file is written to a hard disk, it is not written to consecutively numbered sectors. Sectors are organized as **linked lists** in order to minimize the rotational delay (*latency*) involved in retrieving their information. A **directory** records the size of the file and when it was created.

Additional reading: Brian Hayes, "Terabyte Territory," *American Scientist*, **90**, *3* (May–June 2002), pp. 212–216.

Hardware   DEFINITIONS

Long before **computers** and their components, *hardware* was used as slang to denote firearms and then later more broadly for any military equipment, in contrast to the soft bodies of cowboys and soldiers. In computing, *hardware* is used to denote the tangible devices that comprise a physical computer, devices such as its **integrated circuits**, **memory**, **auxiliary storage**, and **input / output (I/O)** devices. The term is not really a milestone because its coinage cannot be traced to a particular person; it is defined here only for the sake of contrast with **software** and **firmware**.

Additional reading: *Encyclopedia of Computer Science*, pp. 768–780.

Hardware Description
Language (HDL)   LANGUAGE TYPES

A *hardware description language (HDL)* is a **high-level language** that facilitates the design, **simulation**, and documentation of a **digital computer**. A digital computer can be described at various levels from the broad system level through the **computer architecture** and **instruction set** level down to the **logic circuit** (gate) level as a network of elements of varying degrees of complexity. Languages that may be used include the general-purpose languages **Ada**, **APL**, and **C++** and those created specifically as HDLs such as ISP, Esterel, Hardware C, Verilog, and VHDL. The first HDL was ISP, invented by C. Gordon Bell and Allen Newell at Carnegie Mellon University and described in their book *Computer Structures* in 1972. ISP was also the first to use the term *register transfer level*, which derives from the use of ISP in describing the behavior of the PDP-8 (*see* **DEC PDP series**) as a set of **registers** and logical functions describing the transfer of data from source register to destination register. Standard or predefined networks of gates and **flip-flops** are often used as building blocks at the register transfer level. Typical components are registers, multiplexers, and **arithmetic-logic units**. At the gate level, the elements of design are primitive blocks—the basic logic gates. Three industry-supported efforts of the mid-1980s sought to standardize the format of HDLs, leading to the creation of the *Very high-speed integrated circuits Hardware Description Language* (VHDL), Verilog, and the Electronic Design Interchange Format (EDIF). In VHDL, the hardware entities are modeled as abstractions of the real hardware, called *design entities*. Verilog functions are similar to VHDL at the lower levels of abstraction, but VHDL's modeling capability also extends to higher levels. Because of strong support from the U.S. Department of Defense (DoD), VHDL and Verilog are now the most widely used HDLs.

Additional reading: *Encyclopedia of Computer Science*, pp. 768–773.

Harvard Mark I (ASCC) COMPUTERS

In 1936, Howard Hathaway Aiken was a graduate student in applied physics at Harvard who perceived the need for an electromechanical device to solve some equations relating to his research in the design of **vacuum tubes**. After failing to receive support from the Monroe Calculating Machine Company, he asked James Bryce of **IBM** to intercede with IBM president Thomas Watson. Watson provided Aiken with an initial $15,000 and much more as design progressed. The result was an enormous machine, 51 feet wide but only 2 feet deep. Its capacity was only 72 numbers, and it could perform only three additions or subtractions per second, with multiplication taking as long as six seconds. Its IBM name was ASCC (*Automatic Sequence Controlled Calculator*), but it was more commonly known as the Mark I, or, because "Mark I" was a popular designation for other computers, the *Harvard Mark I*. The Mark I was readied for public dedication on 7 August 1944, seven years after Aiken's first contact with IBM. But in preparation for the day, Watson had commissioned a noted industrial designer to envelop the machine in a sleek, futuristic outer skin of stainless steel and glass, ignoring the fact that Aiken preferred the accessibility of its electrical circuits. In protest, Aiken ignored IBM officials at the ceremony and gave every impression that he was its sole creator. Watson, fuming, was not pleased.

Additional reading: Cohen, *Howard Aiken: Portrait of a Computer Pioneer*.

Figure 20. The Harvard Mark I, aka the IBM ASCC, of 1944. Note the sleek encasement that super-salesman Tom Watson liked but engineer Howard Aiken did not, a difference of opinion characteristic of their lukewarm relationship. (Courtesy of IBM)

Haskell LANGUAGES

Haskell is a **functional language** based on the **lambda calculus** and named for the logician Haskell Brooks Curry. Haskell, an extension of an earlier language called *Miranda*, was designed by a committee from the functional programming community in April 1990. It features static **data types**, a **class** system, user-defined algebraic data types, primitive operations for **pattern matching**, and separate compilation of **subroutines**.

Additional reading: *Computing Encyclopedia*, vol. 2, p. 134.

Hayes modem. *See* Modem.

Heuristic DEFINITIONS

Heuristic, used as either a noun or an adjective, means "of or pertaining to a method that is likely but not certain to achieve success." In simpler terms, a heuristic is a rule of thumb, an educated guess. Heuristics are widely used in

artificial intelligence (AI), an example being the built-in tendency of a **computer chess** program, other factors being equal, to make moves that keep control of the center of the board.

Hewlett-Packard (*hp*)

Hewlett-Packard (hp) was founded in 1938 by William Hewlett and David Packard to make electronic analytical instruments. By starting in Hewlett's garage—now a California National Historic Site featured in *hp*'s TV ads—the two engineers started a trend that was followed by several later computer entrepreneurs, including Steve Jobs and Steve Wozniak of **Apple**. After selling **electronic calculators** such as the **HP-35** in the 1960s and the HP-65 **programmable calculator** a few years later, *hp* developed a successful **minicomputer**, the HP 3000, in 1972. In the 1980s, *hp* began to sell its still highly profitable line of *LaserJet* printers. In 1989, *hp* acquired Apollo, which had once been the leading manufacturer of **workstations** but had lost most of its market share to **Sun Microsystems**. In the mid-1990s through the present, *hp* has offered a successful line of **IBM**-compatible **personal computers**. On 19 March 2002, *hp* stockholders approved, by a razor-thin margin, a proposal that the company purchase **Compaq**. Three of William Hewlett's children, led by son Walter B. Hewlett, had mounted a $100 million effort to defeat the merger, and its passage was seen as a major victory for *hp* CEO Carleton S. Fiorina, who strongly advocated approval. The new company is the largest ever to be led by a female CEO. Michael Capellas, former CEO of Compaq, was appointed president on 7 May 2002.

Additional reading: Anders, *Perfect Enough: Carly Fiorina and the Reinvention of Hewlett-Packard*.

Hexadecimal number system

The base 16, or *hexadecimal number system*, is just one of an infinite number of possible **positional number systems**. Just as the decimal number system needs ten different symbols for its constituent digits, the hexadecimal system needs 16. Any 16 could be used, but the usual choice is to borrow the representations of the first ten decimal digits and the first six capital letters: 0, 1, 2, 3, 4, 5, 6, 7, 8, 9, A, B, C, D, E, and F. Thus 2AC5 is a perfectly good hexadecimal number whose decimal value can be computed as $2 \times 16^3 + A (=10) \times 16^2 + C (=12) \times 16^1 + 5 \times 16^0 = 10,949$. Since, like the base 8 of the **octal number system**, 16 is a power of 2, hexadecimal numbers make a very useful shorthand notation for expressing binary values in computers whose **word length** is a multiple of four. For example, the four **bytes** (8 nibbles) that comprise the tediously written 32-bit binary integer 1101 0110 1001 1010 1111 0001 1011 0100 can be expressed much more succinctly in hexadecimal as D69AF1B4.

Hexadecimal notation was used at least as far back as the **IAS**-class computers **ILLIAC** and **ORDVAC** of the early 1950s. (With those computers, the peculiarities of the modified **punched paper tape** telegraph code forced use of the letters K, S, N, J, F, and L for the values 10 to 15 instead of A through F. Programmers used the catchphrase "King Sized Numbers Just For Laughs" to remember them.) But hexadecimal notation was not widely used until the advent of the **IBM 360 series** in 1964. In that series, hexadecimal was used not only as shorthand for binary, but, somewhat unfortunately, 16 was used as the implied base of the exponent (mantissa) used in its **floating-point arithmetic** representation. This resulted in a loss of almost a full equivalent decimal digit of **precision** as compared to the more conventional representation used in competitive systems such as the **DEC VAX series**. The defect was corrected in the IBM 370, 390, and later successors to the 360 through support for the ANSI/IEEE floating-point standard of 1985.

Additional reading: Petzold, *CODE*, pp. 180–189.

High-level language

High-level language is a relative term. Early in the history of **programming languages, assembly language** was considered high level, and certainly it is of higher level than **machine language**. By now, however, a high-level language is generally understood to be either a **procedure-oriented language** or a **problem-**

oriented language (which share the acronym POL). Machine language is called a low-level language, but it is the lowest-level language only on **digital computers** that implement their **instruction set** in *hard-wired logic*. On computers that do so through underlying **microinstructions**, the syntax of those instructions constitutes a still lower-level language, although it may or may not be made accessible to the programmer. Assembly languages, then, could be called *intermediate-level languages*, but that term is not often used, perhaps because **intermediate language** has a quite different connotation.

Hitachi INDUSTRY

Namihei Odaira founded *Hitachi* in 1910, naming the company for his hometown in Japan. Initially, Hitachi made locomotives, transformers, elevators, pumps, and other electrical and mechanical goods. In 1956, Hitachi began to market **radio** and **television** sets and other consumer products such as washing machines, electric fans, and air-conditioning units. In the 1960s, Hitachi marketed nuclear reactors and equipment for experimentation in nuclear fusion. Hitachi introduced its first line of **mainframe** computers in 1974, and its first **supercomputer** in 1982. By the late 1980s, Hitachi had secured licenses to make **semiconductor** products and soon did so on a par with **Toshiba** and the **Nippon Electric Corporation (NEC)** and had also licensed **client-server computing** technology from **Hewlett-Packard**. Hitachi is a leader in **fiber optics**, **CAD / CAM** systems, and applications of **fuzzy logic**. Hitachi, Ltd. now has 1,069 subsidiaries, including 335 overseas corporations, with annual sales that are approaching $70 billion.

Additional reading: Chandler, *Inventing the Electronic Century*, pp. 205–208.

Hollerith machine HISTORICAL DEVICES

The first practical electric **punched card** tabulating system was developed and patented by Herman Hollerith in the late 1880s and used to process the U.S. Census of 1890. Hollerith, a young Census Bureau employee, reported to John Shaw Billings, who guided and supported his work. Fifty *Hollerith machines* were used to tally and cross-tabulate population, dwelling counts, and other characteristics for nearly 63 million persons. Until some of the workload was assumed by a **UNIVAC**, the Census Bureau continued to use later versions of the Hollerith tabulating machine for the next 60 years. Hollerith also developed a number of auxiliary *unit record* punched card machines to support his tabulator, including a keypunch, a reproducer, a collator, and a sorter. To capitalize on his success, Hollerith formed the Tabulating Machine Company in 1896, a company that merged with two others in 1911, the Computing Scale Company of America and the International Time Recording Company, to form the Computing-Tabulating-Recording Company (C-T-R). In 1924, the name of C-T-R was changed to International Business Machines (**IBM**). The Hollerith machines that IBM perfected evolved first into the **Card Programmed Calculator (CPC)**, then into the **stored program computers** that the company sold so successfully over many years. Thus Hollerith's achievement of 1890 is one of the more important milestones in all of computing.

Additional reading: Austrian, *Herman Hollerith: Forgotten Giant of Information Processing*.

Holography COMPUTER GRAPHICS

Among other applications, *holography* is an optical method of generating three-dimensional (3D) **computer graphics** images of a quality suitable for use in the implementation of **virtual reality**. Hungarian scientist Dennis Gabor developed the theory of holography in 1947 and received the Nobel Prize in Physics for his conceptual work in 1971. Gabor coined the term *hologram* from the Greek words *holos*, meaning "whole," and *gramma*, meaning "message" in 1947, but actual construction of a device for making holograms had to await the 1960 invention of the **laser**, whose pure, intense monochromatic light was ideal for making holograms. In 1962 Emmett Leith and Juris Upatnieks of the University of Michigan were the first to realize and demonstrate that holography could be used as a 3D visual medium. In 1972 American physicist Lloyd Cross developed multiplex holograms by combining white-light transmission holography with conventional cinematography to produce 3D motion pictures.

Holograms can also be made with sound waves and electromagnetic waves other than laser light. Holograms made with X-rays or ultraviolet light can record images of particles smaller than visible light, such as atoms or molecules. Microwave holography detects images deep in space by sensing their **radio** wavelength emissions. Acoustical holography uses sound waves to "see through" solid objects. **Optical scanners** read bar codes (*see* **Universal Product Code [UPC]**) by using a holographic lens system to direct laser light onto product labels. Holograms can be used to make an **associative memory** for use in **information retrieval**, a memory from which desired information can be retrieved through knowledge of its content rather than its address. Three-dimensional holographic **television** may be available within the next ten years, and **optical computers** will be able to process terabits (trillions of **bits**) of information faster than any current **supercomputer**.

Additional reading: *Computing Encyclopedia*, vol. 2, p. 144.

Homebrew Computer Club   GENERAL

The *Homebrew Computer Club* met for the first time in 1975 in the garage of Gordon French in Menlo Park, California. The name of the club reflected the interest of its members in assembling, or *homebrewing*, a **personal computer (PC)** from **Intel 8080** kits or parts acquired individually. Members liked to meet to discuss progress in building or upgrading their computers or to show off videogames or other interesting programs that they had written. They considered themselves "hackers" in the good sense; most of their programs were written in **machine language** in attempts to coax the last ounce of performance from their machines. The Club is significant in that so many of its members went on to found computer companies. Steve Wozniak and Steve Jobs were members who founded **Apple**. Adam Osborne founded a company to market the **Osborne portable** computer, having hired fellow member Lee Felsenstein to design it. Felsenstein and Homebrew member Bob Marsh later designed the **SOL**

computer. Harry Garland and Roger Melen founded *Cromemco*, an early producer of **S-100 bus** PCs, and George Morrow founded Microstuf (later, Thinker Toys). Although the Club is no longer active, Gordon French was still listed as its secretary as of 2001.

Additional reading: Freiberger and Swaine, *Fire in the Valley*, pp. 118–124.

Honeywell Corporation   INDUSTRY

Honeywell, once called *Minneapolis Honeywell*, traces its roots back to 1885 when American inventor Albert M. Butz patented a furnace regulator and alarm. He formed the Butz Thermo-Electric Regulator Company of Minneapolis, Minnesota, on 23 April 1886. The Consolidated Temperature Controlling Company, Inc. acquired Butz's patents and business and by 1893 had been renamed the Electric Heat Regulator Company (EHR). In 1912, the name of the company changed again to Minneapolis Heat Regulator Company (MHR). In 1906, engineer Mark Honeywell formed the Honeywell Heating Specialty Co., Inc. (HHS), specializing in hot water heat generators. In 1927, MHR and HHS merged to form the Minneapolis-Honeywell Regulator Co. and became the largest producer of high-quality jeweled clocks. Honeywell entered the **digital computer** business in 1955 through a joint venture with Raytheon, the Datamatic Corporation. The company's first computer, the Datamatic 1000, weighed 25 tons, occupied 6,000 square feet, and cost $1.5 million. In 1960, Honeywell bought Raytheon's interest and called its new subsidiary Electronic Data Processing (EDP). In 1963, Honeywell capitalized on the success of the **IBM 1400 series** by making the compatible but faster H-200 series. Existing 1400 software was ported to the H-200 series by a program called the "Liberator," a name that Honeywell found amusing but, not surprisingly, **IBM** did not. In 1970, Honeywell merged its computer business with that of **General Electric (GE)** to form Honeywell Information Systems, which performed well in **mainframe** markets when Honeywell assumed production of the former **GE 600 series**. In 1986, the company formed Honeywell Bull, a joint venture with Compagnie des **Machines Bull** of France and the **Nippon Electric Corporation (NEC)** of Japan to produce **personal computers (PCs)**. But Honeywell's level of support gradually decreased until, in 1991, Honeywell left the computer business in order to concentrate on its traditional field of **automatic control**.

Additional reading: *Computing Encyclopedia*, vol. 2, pp. 147–148.

Horner's rule   ALGORITHMS

By *Horner's rule*, a polynomial such as $a_0 + a_1x + a_2x^2 + a_3x^3$ can be rewritten as $a_0 + x(a_1 + x(a_2 + xa_3))$, that is, when generalized to an nth order polynomial, evaluation of the parenthesized version for a particular value of x reduces what would have required $2n - 1$ multiplications (and n additions) to just n multiplications (and n additions). The shorter algorithm is also less subject to roundoff error. Horner's rule is named for William George Horner, who described the algorithm in 1819, but it can be found in a 1711 publication of Isaac Newton. The rule represents one of the earliest efforts in the **analysis of algorithms**, although a **difference calculation** would eliminate time-consuming multiplication altogether. More significant to **numerical analysis** is *Horner's method* for extracting the roots of algebraic equations, but that, too, has proved to be a rediscovery, as it was known to the fifteenth-century Persian mathematician Jamshid al-Kashi.

Additional reading: Knuth, *The Art of Computer Programming*, vol. 2: *Seminumerical Algorithms*, pp. 422–423.

HP-35   CALCULATORS

In January 1972, **Hewlett-Packard** introduced the 9-ounce $395 *HP-35*, the first handheld **electronic calculator**. Designed by David Cochran and Barney Oliver, it was given the designation "35" because it had 35 keys arranged in a 7×5 **array**. The machine proved very popular among scientists and engineers despite their having to learn reverse **Polish notation** (RPN) to use it. Its principal competitor, the **TI SR-50** from **Texas Instruments**, used normal infix *algebraic notation*. The two calculators and their ever-cheaper successors spelled the end of the **slide rule**.

Additional reading: *Computing Encyclopedia*, vol. 2, pp. 137–138.

HTML   LANGUAGES

HTML (HyperText Markup Language) is a **Page Description Language (PDL)** devised by Tim Berners-Lee of England in 1990 as a medium for the composition and later display of "pages" on the **World Wide Web**. HTML is a subset of an older PDL called SGML (Standard Generalized Markup Language). HTML **files** are expressed in standard **ASCII** character format, so they can be created and edited with an ordinary **text editor** or **word processing** program. An HTML document may contain several different kinds of *tag*, information delimited by $<$ (for "begin") and $/>$ (for "end") characters. Tags are used to specify the insertion of images or music segments; to select typefont styles, sizes, and colors; and to define links to other remote or local Web pages. An encoded document is interpreted and executed by a **browser** that converts its information to a properly formatted and colorfully illustrated Web page; the underlying HTML code is not shown to the user unless a request is made to display it. HTML is gradually giving way to the upward compatible **XML** (eXtended Markup Language) defined by W3C, the *World Wide Web Consortium* headed by Berners-Lee.

Additional reading: *Computing Encyclopedia*, vol. 2, p. 152.

Huffman encoding   CODING THEORY

Huffman encoding, a form of **data compression**, was invented by David Huffman in 1952 in order to minimize the space to store and the time to transmit large amounts of text. As with **Morse code**, Huffman assigns the shortest (binary) **code** groups to the most frequently occurring symbols, but he devised an **algorithm** for doing this mathematically rather than assigning codes intuitively. The algorithm consists of inserting symbols to be encoded into a binary **tree** in which nodes contain probabilities of occurrence, leaves correspond to symbols, and edges are labeled either 0 or 1, depending on whether they lead to a left or a right subtree. The binary sequence recovered from a path from the root to a leaf becomes the code for the symbol associated with the leaf. In the reference, Dewdney gives an example where the letters A through G occur with the respective probabilities 0.25, 0.21, 0.18, 0.14, 0.09, 0.07, and 0.06, which, as they must, sum to 1. Then the codes assigned become A = 00, B = 10, C = 010, D = 011, E = 111, F = 1100, and G = 1101. The average length of a code group is 2.67, which can be shown to be minimum.

Additional reading: Dewdney, *The Turing Omnibus*, pp. 320–325.

Hybrid computer COMPUTER TYPES

A *hybrid computer* is a combination **analog computer** and **digital computer**. The earliest hybrid computers date back to the 1950s, but very few hybrids were manufactured until the advent of the **digital signal processor (DSP)** in 1974. In this form the hybrid computer has reached its ascendancy; a DSP is now embedded in virtually all commercial electronic products including **cellular telephones** and **digital cameras**.

Hypercube. *See* Cosmic Cube.

Hypertext DATA STRUCTURES

A *hypertext* system allows an author to create links among disconnected sections of related text and allow a reader to *traverse* or *navigate* from one node to another using these links. Considered as a **data structure**, the hypertext itself is a **linked list** whose variable-length nodes contain textual and possibly graphical and auditory information. Conceptually, hypertext originated in 1945 when Vannevar Bush proposed a **memex** system that would maintain links and annotations over locally stored printed materials, microfiche perhaps. But it was not until the 1960s that Theodor Holmes Nelson coined the terms *hypertext* and *hypermedia*. Nelson envisioned a worldwide integrated document base of hypertext-linked information that all people would be able to access and work within. Essentially, he envisioned the combination of the **Internet** and **World Wide Web (WWW)** and laid out the specifications for his *Xanadu* hypermedia system, although it was not implemented for another 30 years. Douglas Engelbart demonstrated NLS/Augment, the first distributed, shared-screen, collaborative hypertext system at the 1968 Fall Joint Computer Conference. Nelson and Andries van Dam of Brown University created an early hypertext system called HES in 1967, and van Dam and his colleagues made a successor system called FRESS in 1969. In 1987, **Apple** introduced *HyperCard*, a hypertext system designed by William Atkinson for the **Apple Macintosh**. With the appearance of the World Wide Web in 1994, **browsers** whose nodes (pages) were encoded in the HyperText Markup Language (**HTML**) became common, and the term *hypertext* entered the popular lexicon.

Additional reading: *Encyclopedia of Computer Science*, pp. 799–805.

IAS computer COMPUTERS

The *IAS computer* was a **stored program computer** built at and named for the *Institute for Advanced Study* (IAS) in Princeton, New Jersey. As the **EDVAC** was nearing completion in early 1946, John von Neumann left the Moore School to join the Institute. He wanted very much to design and build a new **digital computer** and was not daunted by the fact that IAS had no engineering component or machine shop. His reputation, persuasiveness, and ability to attract grant money overcame this hurdle, and work on the IAS computer began in late 1946 with von Neumann as project director, Herman Goldstine as his assistant, and Julian Himely Bigelow and James H. Pomerene as successive chief engineers. Others involved in the project were Ralph Slutz, Willis Ware, Arthur Burks, John Tukey, and Jule Charney. The computer was completed in 1952 and operated successfully until 1957. Since von Neumann had such influence over the **computer architecture** of the IAS computer, it is not surprising that it became the quintessential **von Neumann machine**. IAS was a one-address 40-bit-**word** machine with 1,024 words of **electrostatic memory** in which **instructions** were stored 2 per word. Among 15 copies that were close to being clones were the **MANIAC** at Los Alamos, the **ILLIAC** at the University of Illinois, **MISTIC** at Michigan State University, **Cyclone** at Iowa State University, the **AVIDAC**

at Argonne National Laboratory, the **JOHNNIAC** at the **RAND Corporation**, the **ORACLE** at Oak Ridge National Laboratory, the **ORDVAC** at the Aberdeen, Maryland, Proving Grounds, the **SILLIAC** at the University of Sydney, Australia, and the **WEIZAC** at the Weizmann Institute in Israel.

Additional reading: Goldstine, *The Computer from Pascal to von Neumann*, pp. 252–270.

Figure 21. Four giants of computer science and mathematical physics pose in front of the IAS computer in 1942. Left to right are Julian Bigelow, Herman Goldstine, J. Robert Oppenheimer, and John von Neumann. (Courtesy of the Archives of the Institute for Advanced Study)

IBM INDUSTRY

IBM was incorporated in the state of New York on 15 June 1911 as the Computing-Tabulating-

Recording Company, but its origins can be traced back to 1890 when the U.S. Census Bureau solicited competitive proposals to find a more efficient way to tabulate census data. The winner was Herman Hollerith whose **Hollerith machine** used an electric current to sense holes in **punched cards** and kept a running total of processed data. To capitalize on his success, Hollerith formed the Tabulating Machine Company in 1896. In 1911, Charles R. Flint brokered the merger of Hollerith's company with two others, the Computing Scale Company of America and the International Time Recording Company, to form the Computing-Tabulating-Recording Company (C-T-R) with George W. Fairchild as president. In 1914, Flint recruited as general manager a former executive of the **National Cash Register Company (NCR)**, the 40-year-old Thomas J. Watson. The next year, Watson was named president and began the widespread posting of his famous "THINK" signs. In 1924, C-T-R changed its name to the International Business Machines Corporation with Watson continuing as president and Fairchild elevated to chairman of the board. The company's initials were written I.B.M. for quite some time (and still are by holdout *New York Times*) but eventually elided to just IBM, now the formal name of the corporation. In 1937, Thomas J. Watson, Jr., joined his father at IBM as general manager, the initial title of his father. Ten years later, a second Watson son, Arthur K. "Dick" Watson, was hired and eventually became a vice president of the IBM World Trade Corporation in 1949. Thomas Jr., of course, later became president of IBM. Dick Watson left the company in 1970 to become ambassador to France but died in 1974 at age 55.

In 1933, IBM acquired the Electromatic Typewriter Corporation and manufactured electric **typewriters** for 57 years. By 1958, IBM had sold its one millionth machine and was deriving 8% of its revenue from the sales of typewriters. In 1961, IBM announced the *Selectric* typewriter, developed by H. S. Beattie, who was also the inventor of a number of high-performance printers. The machine was noted for proportional print, extremely high print quality, and the ability to change type fonts through use of replaceable "golf ball" typing elements. This made the Selectric a very desirable I/O device for **digital computers** such as the **IBM 1620** that supported the **APL** language. Another successful **data processing** product was the IBM Type 805 International **Test Scoring Machine** of 1937.

As of 1950, IBM had under development the Magnetic Drum Calculator (MDC), the forerunner of the **IBM 650**; the decimal Tape Processing Machine (TPM), which would later become the IBM 702; and the binary Defense Calculator, which would become its scientific machine, the IBM 701 (*see* **IBM 700 series**). IBM considered 1956 to be a turning point in its history. By that year, Univac held only a slight edge in the number of installed machines, 30 versus 24. But by a year later, the score stood at 66 702s and 46 Univacs, and the order position was even better: 193 for IBM and 65 for **Sperry Rand**, formed in 1955 through merger of **Remington Rand** and Sperry Gyroscope. And in that same year, IBM was able to announce higher-performance replacements for its two leading machines, the 704 for the 701 and the 705 for the 702, the principal advance being use of **magnetic core** memory. In May 1956, with an important antitrust suit settled and at the height of IBM's power and influence, Thomas J. Watson, Sr., passed the title of CEO to his son, Thomas J. Watson, Jr. One month later, the senior Watson died.

By the late 1950s, the only major U.S. corporations other than IBM that were able to commit sufficient resources to make computers and stay the course for at least a few years were **Radio Corporation of America (RCA)**, **General Electric (GE)**, Philco, **Burroughs**, **NCR**, Sperry Rand, and **Honeywell**, which had enjoyed the modest success described above. But eventually, RCA left the computer business and was later absorbed by GE; GE sold its computer division to Honeywell; Philco sold its division to Ford, which left the business shortly thereafter; and Burroughs was absorbed by Sperry Rand, renamed **Unisys**. Of the many computer start-up companies of that era, only the **Control Data Corporation** survived to the end of the century, but as a "solutions company" that no longer sells computers of its own manufacture. The **Digital Equipment Corporation (DEC)** almost made it, but that company was bought by **Compaq** in 1998. The shakeout was fierce. By 1964, when the **IBM 360 Series** was announced, IBM and its competitors were called "Snow White and the Seven Dwarfs," the latter consisting of Burroughs, CDC, GE, Honeywell, NCR, RCA, and Sperry Rand/Univac. IBM held 75% of the market, and none of the dwarfs more than 3%, nor did any of the leading European firms: **International Computers Limited (ICL)** in the United Kingdom, **Olivetti** in Italy, **Siemens-Nixdorf** in Germany, and **Machines Bull** in France. A few years later, major **hardware** competition was reduced to the BUNCH: Burroughs, Univac, NCR, CDC, and Honeywell. But IBM's **unbundling** decision of 1969 spawned the burgeoning **software** industry that we know today. IBM brought out the **IBM PC**, its first **personal computer**, rather late—not until 1981. But the combination of its **open architecture** and software encouraged many other companies to build compatible machines, and now, although IBM continues to market improved versions, it is not a sales leader in this category. This, combined with the gradual falloff of **mainframe** computer sales, brought to IBM a sales and earnings recession in 1993 that led to the first layoffs in company history. But the company weathered the storm and has recovered to the point where, though no longer the relative behemoth that it once was, it is nonetheless one of the most highly respected and admired companies in the world. IBM now concentrates on being a solutions-oriented "enterprise computing" company with emphasis on support for **electronic commerce** and sale of its Enterprise series mainframes, the successor line to the IBM 360/370/390 Series computers of the 1960s through the 1990s.

Additional reading: Pugh, *Building IBM*.

IBM 305. *See* RAMAC.

IBM 360 series

COMPUTERS

The rationale for the *IBM 360 series* was clear. By the early 1960s, **IBM** was writing and maintaining **software** for seven incompatible computer systems. Something had to be done. IBM began planning its New Product Line (NPL) in 1962. The idea of introducing a compatible family of the same **computer architecture**, a term coined at IBM in that year, was neither new nor unique. **General Electric (GE)** and Philco had done it first (*see* **Philco Transac S-2000**) and Univac, **Control Data**, and **Burroughs** did so at about the same time, but none of these competitive families included models that spanned a factor of 50 in performance.

The "architects" of the NPL (later S/360) were Gene Amdahl, one of the technical leaders of the **IBM 700 series**, and Fred Brooks, each reporting to Data Systems Division manager Bob Evans. Brooks loyally began work on the project even though he had urged that IBM go in a different direction, opting for maximum performance at a number of different levels of machine, compatible or not. When it became clear that concomitant development of a complex **operating system** was crucial to the success of NPL, Brooks was assigned the task of directing the efforts of almost 2,000 programmers to produce what was initially introduced as OS/360. Based on his experience with this effort, Brooks wrote the now-classic *Mythical Man Month*. The initially delivered OS/360 was disappointingly sluggish and prone to crash, but the product eventually matured into the more successful OS/MFT (**Multiprogramming** with a Fixed number of Tasks) and OS/MVT (Multiprogramming with a Variable number of Tasks).

As announced in April 1964, the original S/360 series encompassed models 30 through 70 in steps of 10, plus a model 62, but the 60, 62, and 70 were replaced by models 65 and 75 before any were shipped. A model 20 was announced in November. The next year, a specialized model 67 was added in deference to the academic community, and a model 44 was targeted for small-scale scientific applications. Over the next five years, models 85, 95, and 195 were announced in largely unsuccessful attempts to compete with **supercomputers** built by CDC and, later, **Cray Research**.

The System/360 machines used what IBM called *Solid Logic Technology* (SLT), a conservative choice midway between discrete components and true **integrated circuits**. To a good approximation, each model offered twice the performance of the one below it at a cost of only 40% more, either in validation of **Grosch's Law** or because IBM used the "law" as a guide to setting prices.

System 360 sales up through model 65 were so strong that competitors soon arose, both for the system itself and for its **peripherals**. For a while, Control Data successfully sold tapes and disk drives that were "plug to plug" compatible with their IBM counterparts and less expensive. Taking advantage of the lack of patent protection for the S/360 architecture, **Radio Corporation of America** brought out a compatible **RCA Spectra Series** that siphoned off a few percent of S/360 sales, but that company remained in the computer business for only a few years. More distressing to IBM was that its own Gene Amdahl left to found a company bearing his name, a company far more successful than RCA had been. The **Amdahl Corporation** is now a subsidiary of **Hitachi**, which continues to market IBM-compatible **mainframes**, as does the other major Japanese computer vendor, **Fujitsu, Ltd**. At one time, even the USSR brought out an S/360 compatible Syad series. But none of these efforts seriously affected IBM sales.

In 1970, IBM moved to a fully transistorized System/370 product line. Its extended S/360 architecture included *dynamic address modification* that supported efficient **time sharing**. The series began with models 155 and 165 (quickly upgraded to models 158 and 168), each rated at three to five times faster than the corresponding S/360 models 50 and 65. Over the next six years, additional models were introduced in the order 145, 195, 135, 125, and 115.

By late 1976, some of IBM's technical leaders came to believe that they had taken the S/370 as far as was feasible to stay ahead of competition and received approval to plan for FS, its Future System. Other influential IBM leaders opposed the idea. After a fierce and protracted argument, the FS was abandoned at a loss of $100 million, a significant sum even to IBM, and the System/370 was nursed along with occasional modest improvements for another 15 years. The line was split into a series of large-scale computers whose models were designated 30×× and a midrange series designated 43××. Over the system's life span, 17 different models were introduced, one fewer than the number of S/360 models. (The only 360 model number reused on the 370 was 195, the two versions being virtually identical.) A new but still upward-compatible S/390 line was brought out in the late 1980s. Machines announced in May and October of 1998 are called *Enterprise G series* computers, though they still use S/390 architecture.

Because customers are reluctant to replace **software** that they have come to trust, the S/360 architecture has proved amazingly durable. It is now 35 years since introduction of that architecture. A supposedly all-purpose language, **PL / I**, that IBM introduced along with it has long since foundered, but the architecture—extended to be sure—endures. The **DEC VAX** was a far cleaner **Complex Instruction Set Computer (CISC)** than the S/360, and the mode among newer computers is a **Reduced Instruction Set Computer (RISC)** architecture. But in the firmament of rapidly changing computer concepts, three fixed stars remain: **Fortran**, **Cobol**, and the S/360 architecture.

Additional reading: Pugh, Johnson, and Palmer, *IBM's 360 and Early 370 Systems*.

IBM 402/604 (CPC). *See* Card Programmed Calculator (CPC).

IBM 650 COMPUTERS

The *IBM 650 Magnetic Drum Calculator* of 1954 was available in two models. Models 1 and 2 had a **memory** capacity of 1,000 and 2,000 10-digit **words**, respectively, where a word contained either a signed 10-digit number or a two-address **instruction** in which the second address specified the location of the next instruction. The only significant **software** provided was the SOAP **assembler**, which optimized its **object code** by spacing instructions with regard to the *latency* (rotational delay) of the **magnetic drum** and the time to execute the prior instruction. Over 2,000 650s were installed, making it the first computer to earn a significant profit for **IBM**. Educational discounts made the computer popular in colleges and universities; Donald Knuth dedicated the first volume of the **Knuth textbooks** to the 650 he had used as a student at Case Institute of Technology.

Additional reading: Bashe et al., *IBM's Early Computers*, pp. 165–172.

IBM 700 series   COMPUTERS

The *series* in the title is plural; there were two disjoint *IBM 700 series*, a family of binary **computers** intended for scientific and engineering use and a family of decimal computers optimized for business **data processing**. The initial computer of the scientific line was the IBM 701 of 1953, a 36-bit-word one-address computer that was the commercial version of the *Defense Calculator* built in support of the Korean War effort under project leader Morton Astrahan. Jerrier A. Haddad received the 1984 Pioneer Award from the **IEEE Computer Society** for "his part in the lead IBM 701 design team." Only 19 701s were installed, perhaps because the machine used somewhat unreliable **electrostatic memory** and lacked **index registers** and **floating-point arithmetic**. A **software** system called *Speedcoding* was used to simulate the missing features, and both were added to the 1956 successor to the 701, the **magnetic core** memory IBM 704, for which **Fortran** became available a year later. The 704 was followed by the 709 in 1957, IBM's last **vacuum tube** computer, but only 38 of these were ordered because of the closely following transistorized version, the 7090 in 1959. The final machine in the Series was the IBM 7094 of 1960, which, despite the higher performance of the poorly marketed **Philco Transac S-2000** Model 212, established IBM's sales supremacy in scientific computing for a decade or more.

The first member of the IBM 700 series data processors was the IBM 702, of which 14 were installed beginning in 1955. The machine, designed by Nathaniel Rochester, used variable-length decimal **words**. A faster version with more **memory**, the IBM 705, followed shortly thereafter and proved to be the most popular member of the series with about 300 installations. The series was completed with the IBM 7080 in 1960, but only 160 were installed as the highly successful and less expensive **IBM 1400 series** began to take hold.

Additional reading: Bashe et al., *IBM's Early Computers*, pp. 158–164, 176–178, 325–332.

IBM 1400 series   COMPUTERS

The *IBM 1400 series*, introduced in 1959, began with the IBM 1401 and eventually consisted of models 1401, 1410, 1440, 1460, and 7010. The 1401, whose rental began at $2,500 per month, was a fully transistorized second-generation machine having original **memory** capacity options ranging from 1.4K to 16K 6-bit **characters**. As with the 702, arithmetic was decimal. Much of its success was due to its associated IBM 1403 chain printer that provided excellent print quality at 600 lines per minute, four times faster than had previously been available. Another factor was the introduction of the simple **RPG** (Report Program Generator), a language that mimicked what previously had to be done with plugboards. The language still has adherents to this day. Shattering the sales projection of 1,000 machines, over 14,000 IBM 1400 systems were installed until transition to **IBM 360 series** machines in the mid-1960s. By 1961, the 1401 accounted for 25% of all **digital computer** models ever installed in the world.

It is not clear that **IBM** or any other computer vendor ever sought to patent the **instruction set** of a computer, most probably because they had no reasonable expectation that the Patent Office would grant such patents. But since, years later, that office did begin to grant patents for cryptographic **algorithms** implemented in **software**, IBM may have missed an opportunity in this regard, and if they had succeeded it might have earned them an even greater share of the computer market than it did. In 1963, **Honeywell** pounced on the loophole and built a computer, the H-200, whose architecture was very similar to that of the IBM 1401. To gloss over the minor differences they wrote a **program**, wryly called the "Liberator," which could convert 1400 series programs to run on their new machine. The industry was amused, but IBM certainly was not. Within two months, its salesmen reported 196 losses to the H-200, which was carefully priced to provide a better cost–performance ratio than 1400 series machines. Rather than re-price its line or introduce new models, IBM, with what was to become the System 360 series in development, chose instead to make sure that upward compatibility to the new series could be attained through **emulation**. Despite the losses to Honeywell, the 1400 series made its mark in history as the most successful line of compatible computers prior to System 360.

Additional reading: Bashe et al., *IBM's Early Computers*, pp. 465–474.

IBM 1620   COMPUTERS

In 1962, **IBM** began delivery of a remarkable little machine originally called the *Cadet* but marketed as the *IBM 1620*. The project director for the machine was Wayne Winger. Up to that time, **digital computers** were sufficiently complex that they were tended by professional operators; not even a programmer was allowed to touch one. The operators mounted relevant tapes, fed **punched card** decks to the machine, ran the **program** that caused the machine to devour them, and then collected all input and output media for return to the program sponsor. It was the age of **batch processing**. Programmers were lucky to get four vicarious cracks a day at the machine, and the slightest error of a comma or decimal point would cause them great consternation. But the 1620, an $85,000 decimal computer whose **Fortran** compiler made excellent use of a meager 20,000-digit **memory** was a programmer's delight. Precious digits could be conserved through the judicious placement of "flag bits" that delimited **words** of variable length. These words consisted of 6-**bit** "digits" that, when used to represent a number rather than a character, consisted of a 4-bit **binary-coded decimal (BCD)** field, a sign bit, and a **parity** bit. The variable length of such numbers meant that integer arithmetic could be done to arbitrary **precision** up to the limit sustainable by the 20,000-digit memory. Multiplication speed was expedited through a form of **table lookup**. There was no **hardware** division instruction, and since the initially delivered division **subroutine** was excessively long, Francis Federighi of the **General Electric** Knolls Atomic Power Laboratory rewrote it to consume only one-fifth as much memory.

A modified version of the Selectric type-writer, which IBM had introduced just the year before, provided excellent output print quality. The machine's already relatively low cost was deeply discounted to universities, where it became a great favorite with students, providing their first "hands-on" computer experience. Their allegiance has lasted for decades, to the point where there is a Website that keeps track of 1620 lore. The 1620 was succeeded by the IBM 1130, introduced in 1965. With a main memory of 8K 16-bit words and a megabyte disk storage unit, the machine could be rented for just $895 per month. Nonetheless, the machine was less successful than the 1620, probably because many customers chose to order an **IBM 360 series** model 20 or 30 instead.

Additional reading: Bashe et al., *IBM's Early Computers*, pp. 508–513.

IBM 7030. *See* **Stretch**.

IBM AN/FSQ-7. *See* **SAGE**.

IBM AS/400   COMPUTERS

The IBM AS/400 is **IBM**'s most successful **minicomputer**. In July 1969, rather than bringing out a System/360 Model 10 (*see* **IBM 360 Series**), **IBM** introduced the System/3, a small business system that could be rented for less than $1,000 per month, half the cost of a typical S/360 model 20. The S/3 used a unique **computer architecture** and small 96-column **punched cards**. Sales of S/3 models 32, 34, and 36 were strong for nine years, but as the series began to show its age in 1978, IBM announced a successor, the S/38. Strangely, the S/38 was not compatible with the S/3 series, but it nonetheless sold very well. After another ten years, when sales of the S/38 began to sag, some elements of IBM wanted to discontinue it, but it was saved by the *Silverlake* project, whose history is documented in the reference. As the 1990s began, sales of the resulting upwardly compatible AS/400 exceeded the total annual revenue of the **Digital Equipment Corporation**, which had the second-highest revenue in the industry, and sales of the AS/400 remained strong into the new millennium.

Additional reading: Bauer, Collar, and Tang, *The Silverlake Project: Transformation at IBM*.

IBM ASCC. *See* **Harvard Mark I**.

IBM card. *See* **Punched card**.

IBM PC

COMPUTERS

By 1980, several vendors were marketing **personal computers (PCs)** based on the **CP / M** operating system and the **S-100 bus**, but not **IBM**. Urged on by IBM executive William Lowe—later considered "the father of the *IBM PC*"—a group headed by Phillip "Don" Estridge was chartered to develop a PC and bring it to market within a year. Nearly all of the **hardware**—the disk drives, **monitors**, **memory**, **microprocessor** chips, printer, and so on—was purchased from outside vendors, as was system **software**. An Intel 8088 was chosen because its **computer architecture** was close to that of the **Intel 8080** used in CP/M-based machines. An **open architecture** strategy was followed whereby the minimum configuration had only 16KB (kilobytes) of memory but could be expanded to 64KB by plugging three additional 16KB memory chips into sockets on the system board. The first BIOS (Basic Input-Output System) was written by David Bradley. Disk drives were optional on the first PCs, which came with an interface for a cassette recorder. The original **floppy disk** drives used only one side of a 5.25-inch disk and had a capacity of 160KB. IBM did not initially offer a **hard disk**. There were two display adapters, the monochrome display adapter (MDA) and a color graphics adapter (CGA). The MDA was a **character** display with a monochrome monitor. Resolution and character quality were excellent by the standards of the time, and the MDA was targeted to character-oriented applications like **word processing**, **spreadsheets**, and software development. The CGA delivered color and graphics on a more expensive RGB (red-green-blue) monitor. The **operating system** was more problematic than the hardware. CP/M could not be used directly because the 8088 architecture was not identical to that of the 8080. Initially, IBM offered a choice of three operating systems: PC-DOS, CP/M-86, and the **Pascal** *P-System*, but PC-DOS, a version of **MS-DOS**, quickly became the standard.

The IBM PC was announced in April 1981 with a sales forecast of 250,000 units over five years. But in only two years, the IBM PC had an installed base of a million machines. After its introduction, the PC underwent three major upgrades. The first was the PC-XT. The XT was a response to the needs of business and professional customers, who had clearly emerged as the major market. With the XT, IBM added a hard disk and expanded the amount of memory on the system board. The next major upgrade was the PC-AT. The AT used Intel's 80286 CPU, as had Tandy in marketing the excellent but short-lived Tandy 2000. The 80286 was faster than the 8088, and had a 16-bit data **bus**. The AT was followed by the PS/2 line in 1987. The original line consisted of six models, using the 8086, 80286, and 80386 chips (*see* **Intel 80×86**). Models using the 80386 SX and the 80486 microprocessors were announced subsequently. The PS/2s introduced 3.5-inch floppy disk drives with 1.44MB capacity and OS/2, a new operating system. OS/2 was not compatible with PC-DOS but added features including **multitasking** and a **graphical user interface (GUI)**. But then **Microsoft** began to market a competitive product, **Microsoft Windows**, that included a GUI and limited multitasking. That operating system was only moderately successful until the Windows 3.1 (and its 95, 98, and 2000 successors) became very popular in the 1990s.

By 1982, several companies were marketing PC-compatible computers. One of these, **Compaq**, combined compatibility with transportability (along the lines of the Osborne CP/M-based computers), a high-resolution display, and quality workmanship. Compaq, Zenith, and Tandy emerged as major challengers to IBM, but Zenith has withdrawn from the PC market. By the 1990s, there had ceased to be an "IBM PC" as such; there were merely PCs made by IBM. By performance, IBM's latest PC series, the *Aptiva*, is quite competitive with PCs made by other manufacturers, but IBM has dropped to fourth in U.S. sales, behind **Dell**, Compaq, and **Hewlett-Packard (*hp*)**, although, somewhat meaninglessly, it will move up to third when the latter two complete their merger.

Additional reading: *Encyclopedia of Computer Science*, pp. 832–835.

IBM SSEC. *See* **SSEC.**

ICL. *See* **International Computers Limited (ICL).**

Icon (language) LANGUAGES

In 1980, before it became the widely used term of the next article, Ralph Griswold chose *Icon* as the name of a language that he and David Hanson designed with the intent that it be a successor to **Snobol**. Icon is much more structured than Snobol but is not as much fun to use. The Snobol mechanism for **pattern matching**, its forte, was completely changed and, though more general, inhibited the move to the newer language. Nonetheless, Icon is now available for most **personal computers** and has become the most widely used **string processing** language.

Additional reading: *Encyclopedia of Computer Science*, pp. 1698–1700.

Icon (symbol) DEFINITIONS

An *icon* is a small emblem that appears on the "desktop" of a **personal computer** or **workstation** whose **operating system** is presented to the user as a **graphical user interface (GUI)**. Clicking the button on a **mouse** or **trackball** that has positioned the **cursor** of a **monitor** over an icon will launch the **program** that the icon represents.

Identifier DEFINITIONS

An *identifier* is the formal term for the textual **character** sequence used to name a **variable, constant, function**, or **procedure** of a **high-level language**. Such languages vary as to the length of a legal identifier, the characters that may be used to form one, and whether an uppercase identifier is or is not uniquely different from its lowercase counterpart.

IEEE Computer Society GENERAL

The *IEEE Computer Society* (IEEE-CS) is a direct descendant of the Committee on Large Scale Computing Devices of the American Institute of Electrical Engineers (AIEE), established in 1946 and chaired by Charles Concordia. This was followed in 1951 by the formation of the Computer Group of the Insti-

tute of Radio Engineers (IRE). In 1963, the IRE merged with the AIEE to create the Institute of Electrical and Electronics Engineers (IEEE), with "Computer Group" as the name of the merged computer subgroups. IEEE-CS assumed its current name in 1972 and now, with a membership well over 90,000, is the largest of many IEEE societies. IEEE-CS was formed to advance the theory and practice of computer and information processing technology and promote cooperation and exchange of technical information among its members. The leading IEEE-CS periodicals are the monthly *Computer* and *Transactions on Computers* and the quarterly *Annals of the History of Computing*. The IEEE-CS headquarters is at 1730 Massachusetts Avenue NW, Washington, D.C. 20036-1992.

Additional reading: *Encyclopedia of Computer Science*, pp. 881–882.

IFIP. *See* **International Federation for Information Processing (IFIP).**

ILLIAC COMPUTERS

The *ILLIAC* (ILLInois Automatic Computer), a copy of the **IAS computer**, was built at the University of Illinois under a contract that called for an identical computer, the **ORDVAC**, to be shipped to the U.S. Aberdeen Proving Ground in Maryland. The ORDVAC was the first to be completed, passing its acceptance tests in February of 1952, and the ILLIAC did so seven months later on 22 September. Like all IAS machines, the ILLIAC used a 40-bit **word**, two one-address **instructions** per word, 2's **complement** for negative numbers, and an **arithmetic-logic unit (ALU)** that comprised three **registers** and an **adder**. **Machine language** programs were encoded in the **hexadecimal number system**, five hex digits per instruction, ten to a word. Addition and subtraction required 72 μs (microseconds), multiplication 642 to 822 μs, division 772 μs, and shifting 16 μs per binary place shifted. The machine weighed five tons, used 2,800 **vacuum tubes**, and measured ten feet high, two feet wide, and eight and a half feet tall. Overlapping in time with ILLIAC I, the university also built incompatible 52-bit computers labeled ILLIAC II and III, so the original IAS class machine is now referred to as ILLIAC I. The machine was in service for ten years and dismantled in 1962.

Additional reading: James E. Robertson, "The ORDVAC and the ILLIAC," in Metropolis, Howlett, and Rota, *A History of Computing in the Twentieth Century*, pp. 347–364.

ILLIAC IV   COMPUTERS

The *ILLIAC IV* was the fourth and largest **digital computer** in the line of incompatible **IL-LIAC** computers built at or for the University of Illinois. The machine was designed by Daniel Slotnick in the mid-1960s, built under contract to the **Burroughs Corporation**, and became operational in 1975 at the Institute for Advanced Computation in Moffett, California. The ILLIAC IV, a **parallel computer** heavily subsidized by the U.S. government, was the most expensive single computer ever built. Its $30 million cost in 1965 would be equivalent to at least $100 million in 2003. Reminiscent of the **Stretch** experience, the computer achieved a speed of only 15 MIPS (millions of **instructions** per second), a fraction of its design speed. But ILLIAC IV did have a gigabit **input / output (I/O)** transfer rate that would be excellent even by current standards. The computer used a unique combination of pipelined **computer architecture** and 64 processing elements (PEs) arranged in an 8×8 **array**. As in the later **Connection Machine**, the same instruction was broadcast in parallel to all PEs. Thus the computer was equivalent to a **cellular automaton** and was a *single-instruction, multiple-data* (SIMD) computer in the **Flynn taxonomy**. This combination of *array processor* and **pipelined computer** was ideal for applications that involved **matrix** operations and partial differential equations such as those that govern weather prediction. The ILLIAC IV remained in operation for almost ten years until retired in the mid-1980s.

Additional reading: Daniel L. Slotnick. "The Fastest Computer," *Scientific American*, **224**, 2 (February 1971), pp. 76–87.

Image processing   APPLICATIONS

Digital *image processing* is the systematic modification of an input image to produce an output image that is better suited for viewing or anal-ysis. The processed images are either examined by a human observer, such as a radiologist viewing an X-ray, or they form input to a **pattern recognition** system. A digital image is represented as a discrete two-dimensional (2D) **array** of numbers. Each element in the array is called a *pixel* (picture element). Pixels from black and white (B/W) images are assigned values called *gray levels* that correspond to the relative brightness of the tiny portions of the image that they depict. Pixels from color images are assigned numeric values that encode the hue and intensity of the color at their location. Image processing deals with the systematic manipulation of these pixels. Computerized image processing began in the 1960s, principally for use in medical applications and analysis of photographs related to astronomy and space missions. The U.S. National Aeronautics and Space Administration (NASA) Jet Propulsion Laboratory (JPL) at the California Institute of Technology has been a center of research in image processing for almost 40 years.

Reducing images by **data compression** is a form of image processing. Other principal branches of the field deal with image enhancement and image restoration. The goal of *image enhancement* is to highlight or enhance a particular type of image feature, or to suppress some of its unwanted detail. These techniques encompass object/background contrast stretching, modification of dynamic range, removal of false contours introduced through inadequate quantization levels, reduction of "salt-and-pepper" noise, edge sharpening, image smoothing ("blurring"), and enhancement of detail through use of "false color." *Image restoration* deals with the problem of reconstructing a digital image given a set of image projections. There are several important application areas where the only practical way to acquire 2D or 3D images of an object is by using a set of image projections. Image reconstruction is an important requirement in **computerized tomography**, geophysical exploration, underwater exploration, and radio astronomy. Efficient processing of images requires specialized **hardware** for image acquisition, storage, manipulation, and display. Advances in image processing hardware result from advances made in fields such as electro-optics, electronics, **VLSI**, materials science, **semiconductors**, and **digital signal processing** (*see* **Digital Signal Processor [DSP]**).

Additional reading: *Encyclopedia of Computer Science*, pp. 840–847.

IMP. *See* Interface Message Processor (IMP).

IMSAI 8080   COMPUTERS

The *IMSAI 8080*, developed by IMS Associates, was designed to use the same **bus** structure as the **MITS Altair 8800** with interchangeable circuit boards. The IMSAI 8080, however, was much better built, had a more robust power supply, and a front panel. It supplanted the Altair as the standard **S-100 bus** computer. The IMSAI 8080 was the first of a complete line of micros built by this company.

Additional reading: Kidwell and Ceruzzi, *Landmarks in Digital Computing*, p. 95.

Index register   COMPUTER ARCHITECTURE

In a **digital computer**, an *index register* is a special-purpose **register** that automates the processing of successive regularly spaced **data** values in **memory**. Without use of an index register, the **program** that processes those elements would have to modify one or more addresses in the **instructions** of the **loop** used for processing the data. But an instruction in a computer having one or more index registers can include specification of such a register, whereupon the address in the instruction is augmented by the value in the index register to form an *effective address*. Then, after use of that address, the index register is incremented by the separation of the data values being accessed. An index register can also be used as a simple **loop**-control counter. In 1949, the **Manchester Mark I** became the first computer to have an index register. Its inventors, Tom Kilburn and F. C. Williams, called it the *B-line* because, on that computer, the accumulator was called the *A-line*. Its commercial successor, the *Ferranti Mark I*, had eight B-lines, but the later IBM 704 had only three and called them index registers. Modern **Reduced Instruction Set Computers (RISCs)** have many more than that.

Additional reading: *Encyclopedia of Computer Science*, pp. 436, 847–848.

Information hiding   PROGRAMMING

Information hiding is the general term that describes the central concepts embodied in **object-oriented programming (OOP)**. The information hiding principle states that certain kinds of **program** information should be hidden (*encapsulated*) in revisable modules. Information that is hidden in a module, known as its *secret*, typically includes decisions such as **data structures, hardware** characteristics, and behavioral requirements that the creator of the module and those who maintain it need to know but its users do not. In one **encapsulation** of a **pushdown stack** as a module, for example, the module's secret is the data structure used to represent the stack and the **algorithms** used to manipulate it. That data structure might be an **array** or a **linked list**. The stack module conceals the decision, and changing its implementation from one form to the other can be made without needing to change any of the programs that use the stack module. The earliest published description of the principle is that of David Parnas in 1972, but a more detailed explanation of its systematic use in the design of complex systems was given by Paul Clements, Parnas, and David Weiss in 1985. **Programming languages** that provide support for creating and using information hiding modules and their interfaces include **classes** in languages such as **Simula 67, C++**, and **Java**; *packages* in **Ada**; *modules* in **Modula-2**; and *type* classes in **Haskell**.

Additional reading: *Encyclopedia of Computer Science*, pp. 854–856.

Information retrieval
INFORMATION PROCESSING

Information retrieval as a self-descriptive term antedates computing by several millennia, although its use in conjunction with computing is attributed to Calvin Mooers. The first tool invented that relates to information retrieval was the decision to standardize the order of the letters of an alphabet. This must have happened even further back into antiquity, but the first use of alphabetization to facilitate information retrieval is believed to have been that of the Greek scholars of the third century B.C. at the library of Alexandria in Egypt. For that matter, the foundation and operation of a **library** was the first mode of information storage and retrieval. In the first century A.D., Pliny the Elder wrote a massive work called *The Natural History* in 37 volumes, a kind of encyclopedia that comprised information on a wide range of subjects. To help his readers, the first book was a massive table of contents, and each of the following books contained a list of the authors whose works he had used in compiling the information for that book. Pliny credited this practice in Latin literature to Valerius Soranus, who lived during the transitional period from the second to the first century B.C., implying that it had previously been used by Greek writers. The end-of-the-book *index* is surprisingly modern, dating back only to the onset of printed rather than handwritten books. According to Hilary Calvert, the *Gerardes Herbal* (John Gerard's *General Historie of Plants*) of 1597 had several fascinating indexes. Even more complete than an index is a *concordance*, which lists the frequencies of occurrence and the locations of *all* words in a literary work such as the Bible or the plays of Shakespeare. The once-tedious manual compilation of concordances is now done quite easily by **digital computer**.

Information retrieval (IR) has now come to mean the retrieval of information by mechanized or computerized means. More particularly, the "information" sought is stored in narrative form in a **natural language**; obtaining an answer to a query posed to an organized **database** is trivial by comparison. The simple *Rolodex*™ invented by Alfred Neustadter and first marketed in 1950 is an important contribution. More ingenious is the edge-notched-card IR method patented in the United States in 1925 and earlier in England by Alfred Perkins. Henry P. Stamford had obtained a patent for a similar system as early as 1896, but since he used only holes at card edges rather than notches, he had to devise a cumbersome way to find only one relevant search term at a time.

But it was the Perkins system, whose patent rights were sold to Royal McBee in 1932, that proved generally useful because it could cope with multiple search terms. In it, each card contains the text of some nugget of information that may or may not mention certain keywords, each of which is associated with a particular position along the top and side edges of the card. When the keyword is present, a small hole is punched at its assigned location; otherwise a notch is placed there. After all cards are prepared, a knitting needle passed through the deck at the location associated with a keyword will, when raised or pulled sideward, extract only the relevant cards; needles pass through holes but fail to engage a notch. If two or more needles are used simultaneously, the result will be the extraction of just those cards that contain the conjunction (terms connected by the AND **operator** of **Boolean algebra**) of the desired keywords. OR operations must be done with separate passes of the needles. Thus the edge-notched-card system is a mechanical realization of what is now done with a **search engine** on the **World Wide Web**, and the card file itself is a form of **associative memory**.

Computerized information retrieval dates to the 1950s when limited **memory** and processing speed made the task quite challenging. In 1953, the German engineer Hans Peter Luhn, who later headed the IR Department at **IBM** in the United States, invented an **algorithm** for the randomized generation of **file** addresses for the **RAMAC**. This method greatly increased the efficiency of information searches and started Luhn on such a productive course in IR research that he is now known as "the father of information retrieval." In February 1958, he and his group devised a program that generated automatic abstracts of documents presented to it by extracting from it the sentences that contained the keywords of highest rank. In May of the same year, Luhn proposed the *Selective Dissemination of Information* (SDI) whereby those who listed their fields of interests through a list of keywords were automatically sent relevant new reports and publications as they appeared. Luhn also invented the clever KWIC (KeyWord in Context) index (permuted index) for listing article titles. Other pioneers in the field were

Gerald Salton of Cornell who published a textbook on the mathematics of IR in 1968, Robert Korfhage of the University of Pittsburgh, and Sharon Anne Hogan, a librarian at the University of Chicago and later vice provost for information management at the University of Illinois. Hogan, the founding editor of the journal *Research Strategies*, led a national effort to teach librarians the opportunities afforded by computers in general and the **Internet** in particular.

By now, the proliferation of fast **personal computers** and the storage of the full text of so many books, papers, and magazine articles on the Web has led to maturation of the field. On 24 April 2002, a search of Amazon.com for all book titles containing the term "information retrieval" yielded references to 2,009 entries. And, of course, the very search itself exemplified an act of information retrieval.

Additional reading: Gary Forsythe at http://www.asindexing.org/site/history.shtml#toc

Information Technology (IT)   GENERAL

Information Technology (IT) is a term being used increasingly frequently by trade magazines to describe a broad range of activities associated with the use of **computers** and communication. IT generally implies the application of computers to the storage, retrieval, processing, and dissemination of **data**, particularly in the field of **electronic commerce**. But the term is sufficiently broad to encompass the activities of those who design or even merely use any device used to collect or process digital information—digital satellite and cable **television**, DVD players, digital telephony, **digital cameras**, and even photocopiers. The *Oxford English Dictionary* (*OED*) attributes the first recorded use of the term *information technology* to a 1958 article by Harold Leavitt and Thomas Whisler in the *Harvard Business Review*. By the end of the twentieth century, the usage "information technology" was widespread; it appears in the title of this book. Some universities have established departments of IT, and Peter Denning, the highly respected computer scientist and past president of the **Association for Computing Machinery** (**ACM**), writes a monthly column on the profession for *Com-*

munications of the ACM. Michael Dertouzos, the director of the MIT Laboratory for Computer Science from 1974 until 2001, was internationally known for his enthusiastic promotion of IT through his membership in the U.S. National Academy of Engineering and the U.S. Council on Foreign Relations. He was the author of eight highly influential books, including *What Will Be* in 1997 and *The Unfinished Revolution* in 2001, the year of his death. Tim Berners-Lee believes there might never have been a *World Wide Web Consortium* (W3C) without the backing and leadership of Michael Dertouzos.

Additional reading: Denning, *The Invisible Future*, chap. 23.

Information theory   THEORY

The foundation of *information theory* is the paper "A Mathematical Theory of Communication" published by Claude Shannon in the *Bell System Technical Journal* in 1948 and republished in a book of the same name by Shannon and Warren Weaver the following year. According to Shannon, the purpose of communication is to resolve uncertainty. If we toss an honest coin, communicating the outcome takes one **bit** of information, *heads* (0) or *tails* (1). But if the coin is biased and comes up heads more often than tails, the sequence of *heads* and *tails* is no longer random and can be encoded in less than one bit per toss. Shannon describes a message source as being *stochastic*, that is, *probabilistic*, and calls the degree of its unpredictability its *entropy*. Entropy can be expressed in terms of bits per symbol or bits per message. Information from a *message source* is transmitted over a *communication channel*. There is always a certain amount of noise associated with any actual **channel**. Despite errors in transmission, communication channels have a *channel capacity* measured in bits per character or bits per second. Information theory yields formulas for the capacities of various channels in terms of either probabilities of errors in transmission, or in terms of signal power, noise, and **bandwidth**. It was once believed that a highly noisy channel could not possibly be used for the transmission of error-free messages. But a theorem of Shannon states that if the rate of transmission of a source is less than the channel capacity of the noisy channel used, messages can be transmitted over the channel with an error rate as low as desired through the use of **error-correcting codes**. There are few more significant theorems in all of theoretical information science, and none is more practical.

Additional reading: *Encyclopedia of Computer Science*, pp. 869–870.

Initial orders
PROGRAMMING

Operation of the **EDSAC** of 1949 could not start until a short standard sequence of **instructions** called *initial orders* was read into the **computer's ultrasonic memory** from a mechanical **read-only memory (ROM)** made from a set of rotary **telephone** switches. The initial orders, a rite of initiation, determined the format of the additional instructions that followed on **punched paper tape**. Thus initial orders, written by Maurice Wilkes and David Wheeler, constituted the first instance of what is now called, more simply, a **loader**.

Additional reading: *Encyclopedia of Computer Science*, p. 614.

Input / Output (I/O)
DEFINITIONS

Input / output (I/O) used as a noun is the combination of input and output operations used in conjunction with a **digital computer** and is often used as an adjective in full or abbreviated form in such contexts as "input-output processing" and "I/O device." *Input* and *output*, in turn, are nouns backformed from the corresponding verb phrases "to put in" and "to put out." Thus when we "put in the plug" in the electrical sense, the plug is input to the socket, and when we "put out the cat," the cat is output to the outside world. **Digital computer** input consists either of measured or recorded **data** or of **parameters** needed to complete the specification of something to be calculated. Such input can take a wide variety of forms—**punched cards, punched paper tape**, magnetic tape, **floppy disks**, wireless **radio** signals, sounds or speech sensed by a *transducer*, **optical storage** media such as compact discs and DVD discs, **mouse**, **trackball**, or joystick motion, or digitized images produced by an **optical scanner**. The last

example illustrates how the output of one process or device can become the input to another. The most common output devices in current use are the ink-jet or **laser printer**, the **monitor**, and acoustic speakers for output that is to be viewed or heard immediately, or any of a number of magnetic or optical storage devices for output that is to be recorded for later playback or use as input.

Instruction
DEFINITIONS

An *instruction* is the fundamental unit of logic by which a **digital computer** is directed to advance the progress of a running **program**. The context is usually that of **machine language**, in which case an instruction is numeric, or **assembly language**, in which an instruction is written symbolically. In either case, an instruction consists of an **operation code (opcode)**, anywhere from 0 to 4 **operand** addresses depending on the **computer architecture** (*see also* **Multiple-address computer**) and possibly additional fields such as references to addressing modes or **index registers**. **High-level language** programs are written in terms of **statements** that are compiled into the one or more instructions needed to implement them.

Instruction lookahead
COMPUTER ARCHITECTURE

Instruction lookahead is one of the techniques first used on the IBM 7030 in 1962 to **Stretch** its performance to the limit. The basic idea of instruction lookahead is to overlap the decoding and execution of one **instruction** with the **operand** fetches needed for one or more of the following ones. The **CDC 6600** looked ahead as many as seven instructions, and since it had multiple specialized **arithmetic-logic units**, it could execute certain instructions in parallel as well as fetch operands in advance. To at least a minor degree, lookahead makes a computer a **pipelined computer**. The MC6502 **microprocessor** made the **Apple II**e the first **personal computer** to use instruction lookahead, but the technique is now common. For an alternative approach to instruction-level parallelism, *see* **Very Long Instruction Word (VLIW)**.

Additional reading: *Encyclopedia of Computer Science*, pp. 564, 1350.

Instruction set   COMPUTER ARCHITECTURE

The *instruction set* of a **digital computer**, more properly called its *command set*, consists of the repertoire of **machine language** commands (**operators**) that can be applied to one or more **operands**. There was little need for a formal instruction set before there were **stored program computers**, so it is not surprising that credit for designing the first one, for the **EDVAC** in 1945, is generally given to John von Neumann. A typical **Complex Instruction Set Computer (CISC)** may have of the order of 50 or more different commands, whereas a **Reduced Instruction Set Computer (RISC)** may have 20 or fewer. In theory, a digital computer needs only one: Subtract one number from another and jump if the result is less than zero. If a machine can subtract, it can add by subtracting the negation of a number, and a number can be negated (have its sign reversed) by subtracting it from **zero**. And a machine that can add and subtract can multiply and divide by repeated addition and subtraction. But this hypothetical one-instruction computer is just a curiosity; no designer would dare RISC everything.

Additional reading: *Encyclopedia of Computer Science*, pp. 887–891.

Integrated Circuit (IC)   HARDWARE

An *integrated circuit (IC)* is a **logic circuit** all of whose components and interconnections are etched on a wafer, or "chip," of **semiconductor** material such as germanium or silicon. The technology represents a significant advance over earlier **printed circuits** for which conduction paths but not components such as **transistors**, resistors, and capacitors could be integrated into the product. The concept of an integrated circuit is credited to British radar expert G. W. A. Dummer who, in speaking at the annual electronic components symposium in Washington, D.C., in May 1952, said:

With the advent of the transistor and the work in semiconductors generally, it seems now possible to envisage electronic equipment in a solid block with no connecting wires. The block may consist of layers of insulating, conducting, rectifying and amplifying materials, the electrical functions being connected directly by cutting out areas of the various layers.

But it would be seven more years before the vision became a reality. The first IC contained only a capacitor and a single transistor; current IC chips contain millions of transistors. The IC was independently invented by two American engineers within a few months of one another. On 24 July 1958, Jack St. Clair Kilby of **Texas Instruments (TI)** described his idea for an IC in his lab notebook, and in early 1959, Robert Noyce of **Fairchild Semiconductor** did the same. TI publicly unveiled Kilby's discovery at the Institute of Radio Engineers Show in early 1959. This accelerated the efforts at Fairchild, which were now focused on making the connections between the transistors and other components an integral part of the manufacturing process. Jean Hoerni, one of Fairchild's original founders, solved this problem when he developed a "planar" process that uses oxidation and heat diffusion to form a smooth insulating layer on the surface of a silicon chip, allowing the embedding of insulated layers of transistors and other elements in silicon. This eliminated the need to cut apart the layers and wire them back together. Fairchild filed a patent for a semiconductor integrated circuit based on the planar process on 30 July 1959, touching off a decade-long legal battle between Fairchild and TI, which previously had filed a similar patent based on Kilby's technology. Eventually, the U.S. Court of Customs and Patent Appeals upheld Noyce's claims on interconnection techniques but gave Kilby and Texas Instruments credit for building the first working integrated circuit. Kurt Lehovec, a Czech-born physicist working at Sprague Electric and a poet of recognition, also deserves credit for having patented a form of integrated circuit before Kilby did. Invention of the IC is a most significant scientific milestone, as was recognized when the 2000 Nobel Prize in Physics was awarded to Kilby. Noyce almost certainly would have shared in the prize had he not died ten years earlier.

Additional reading: *Understanding Computers: The Chipmakers.*

Intel   INDUSTRY

Intel (INTegrated ELectronics) was founded in 1968 by Robert Noyce, co-inventor of the **in-**

tegrated circuit (IC) while at **Fairchild Semiconductor**, and Gordon Moore, the expositor of **Moore's law**, in order to produce and market **semiconductor** products. In 1970, Intel made two significant advances a month apart, the *Erasable Programmable Read-Only Memory* (EPROM), invented by Dov Frohman, and the first **DRAM** memory chip. In 1971, Intel scientists developed the **Intel 4004**, the first **microprocessor**, and in 1974, the **Intel 8080**, the first microprocessor to be used in a **personal computer**. The 8088 and 8086 followed in 1978, the latter being the first of a long line of **Intel 80×86** products that extends through the current Pentium IV. Intel also produces an extensive line of specialized chips for **embedded system** applications. In 1993, Intel made a very powerful **parallel computer** called the *Paragon* whose architecture took the form of a 20 × 20 node **cellular automaton** with two Intel i860xp chips per node. But, as happened with so many other companies, **supercomputers** did not prove profitable to Intel, and the company has reverted to its strength, integrated circuits.

Through the longest stretch of its history, 1968 to 1998, Intel was led by the energetic Andrew Grove, born András Gróf in Hungary, who served first as director of operations and then later as president and chief operations officer in 1979, CEO in 1987, and chairman of the board in 1997, the year that he was named Time magazine's *Man of the Year*. Intel continues to hold the major share of the market for 80×86 chips, its only serious competitor being **Advanced Micro Devices (AMD)**.

Additional reading: Jackson, *Inside Intel*.

Intel 4004 COMPUTERS

The *Intel 4004* was the first **microprocessor**. When the Italian Federico Faggin joined **Intel** in April 1970, he was assigned to work with Ted Hoff and Stan Mazor on a contract to design a set of eight different **integrated circuit** chips for use in the *Busicom* **calculator** of the Nippon Calculating Machine Corporation (NCMC). Rather than doing precisely that, Hoff proposed that Intel develop a set of four chips that together would form a complete microprocessor: a **central processing unit (CPU)**, a **memory**, a **read-only memory (ROM)** to hold

the eight Busicom arithmetic **functions**, and an **input / output (I/O)** unit. NCMC was skeptical but ultimately agreed. The chips were designed and implemented in record time, and the November 1971 issue of *Electronic News* billed the development of the Intel 4004 as "A New Era in Integrated Electronics." The 4004 had only a 4-bit **data path**, sufficient to fetch the **binary-coded decimal (BCD)** digits from memory needed for decimal Busicom operations, but the 4004 was very slow—about the speed of the **ENIAC** of a quarter-century earlier—but it nonetheless proved to be the spark that ignited the rapid growth of the young Intel company.

Additional reading: Petzold, *CODE*, pp. 260–285.

Intel 8008 COMPUTERS

Intel announced its *Intel 8008* chip, an 8-bit **data path** extension of the **Intel 4004**, in August 1972. The Vietnamese André Thi Truong built the *Micral* industrial **microcomputer** of 1973 based on the chip and sold 500 machines in France. In the July 1974 *Radio Electronics*, Jonathan Titus attracted the attention of hobbyists by describing how to built what he called the Mark-8 "**personal computer**" based on the 8008. But the sale of fully assembled readily available personal computers did not take off until the appearance of an improved 8-bit processor, the **Intel 8080** in 1974.

Additional reading: Petzold, *CODE*, pp. 260–285.

Intel 8080 COMPUTERS

In late 1972, Federico Faggin proposed to his management at **Intel** that the performance of the **Intel 8008** could be substantially improved if it were redesigned to use Negative Metal Oxide Semiconductor (NMOS) technology based on electron movement in place of the 8008's MOS technology that was based on movement of positive charges ("holes" in a sea of electrons). He obtained approval only with difficulty, but once authorized, he and Masatoshi Shima worked on **logic circuit** design for what became the *Intel 8080*, and Ted Hoff and Stan Mazor worked on its **instruction set**. There was room in the chip's **computer architecture** to assign functional meanings to 256 **operation codes (opcodes)**, but Shima declared the design frozen when 246 were defined. This proved fateful later when Faggin and his friend Ralph Ungermann left Intel to found the *Zilog* corporation, whose first product, the **Zilog Z-80**, was a faster upward compatible 8080 that used the previously unassigned opcodes to good advantage. The Z-80, rather than the 8080, was then adopted by the Tandy Corporation for its **Radio Shack TRS-80**. But the 8080, released in April 1974, was itself a great success in that several early **personal computers** adopted it, among them the historic **MITS Altair 8800** and the **IMSAI 8080**.

Additional reading: Jackson, *Inside Intel*, pp. 108–120.

Intel 80×86 series COMPUTERS

The *Intel 80×86 series* is the successor to the **Intel 8080** of 1974. The × of the first 80×86 was null; that is, the 16-bit **data path** 8086 of June 1978 was essentially the 80[0]86. Since then, × has ranged from 1 in March 1962 through 8 in 2002. The 80286 was a 6 MHz chip containing 134,000 **transistors** and was first used in an excellent **personal computer**, the *Tandy 2000*, and then later in the IBM AT, the successor to the first **IBM PC**. The 16 MHz 80386 was introduced in 1985 and the 25 MHz 80486 in 1989. The 60 MHz 80586 of 1993 through 12 improved versions including the 300 MHz 80686 of 1997 and the 1.5 GHz 80686 of 2000 were all called *Pentium*, and the 80786 of 2001 is called the *Itanium*. Four versions of an 80886 are planned. According to **Moore's law**, the 18-year period (nine 2-year periods) should have resulted in the number of components on a single chip rising by a factor of $2^9 = 512$, which would bring the 134 million transistors of the 80286 to about 65 million on the most capable 80686, which is very close to the actual number of 60 million. Either Moore's Law really is something of a "law" or, perhaps more likely, **Intel** and its major competitor, **Advanced Micro Devices (AMD)**, have taken the adage as a script for how much of an advance they should aspire to make over each two-year span.

Additional reading: Jackson, *Inside Intel*.

Interactive computer COMPUTER TYPES

An *interactive computer* is one that responds to a query or request for minor computation quickly enough to hold the user's attention. To qualify, a computer must respond to simple requests such as the display of a **file directory** almost instantaneously, even though many users may be **time sharing** the computer simultaneously. Response to requests for compilation of a **program** within a second or two will satisfy most users, but any longer will make the user feel that he or she is not being adequately serviced. One of the earliest interactive computer systems, the *Culler-Fried System* for interactive mathematics, was implemented in 1963 at the University of California at Santa Barbara by Glen Culler and Burton Fried. **Personal computers (PCs)** are inherently interactive since, unless its user allows others to borrow some of its cycles over a **telephone** or cable line, the computer will respond as fast as its intrinsic speed permits.

Interface Message Processor (IMP)

NETWORKS

The *Interface Message Processor (IMP)* was a modified **Honeywell** DDP-516 **minicomputer** retrofitted at **Bolt, Beranek, and Newman (BBN)** in 1969 for use with the **ARPAnet**. The IMP was essentially an early **router** used as an interface between the disparate nodes in the **network**. Modification involved substantial changes to the mini's **input / output (I/O)** system and the preparation of elaborate specialized **software**. The IMP team was headed by Frank Heart and included Robert Kahn, Severo Ornstein, David Walden, Bernard Cosell, Hawley Rising, Will Crowther, and Ben Barker. BBN's initial contract was for only four IMPs, but eventually 15 were built. The ARPAnet could not have been successful without a reliable IMP. The original IMPs were about three feet wide and six feet tall, but due to the maturation of **integrated circuit** technology over the following 30 years, their function on the **Internet** is now done with **Cisco** routers that are only a few square inches in size.

Additional reading: Hafner and Lyon, *Where Wizards Stay Up Late*, pp. 75–136.

Interleaved memory

MEMORY

An *interleaved memory* is one in which its sequential addresses are alternately allocated to two or more physically distinct units of storage. Doing so decreases **memory** *access time* because interleaving creates greater opportunity to overlap a second access with the computation that the **central processing unit (CPU)** needs to process the first. Two-way and four-way interleaving are the most common since eight-way interleaving reaches a point of diminishing return. Werner Buchholz devised the first interleaved memory in 1961 in order to **stretch** the performance of the IBM 7030.

Additional reading: *Encyclopedia of Computer Science*, pp. 908–909.

Intermediate language

LANGUAGE TYPES

An *intermediate language* is both the **object code** of certain **compilers** and then the **source code** for a translator that converts that source code into executable **machine language** object code. Most current compilers generate executable object code without going through an intermediate stage. Some early compilers generated **assembly language**, which then had to be processed by an **assembler**. In 1977, Kenneth Bowles and his students at the University of California at San Diego produced what became a very popular version of **Pascal** called *UCSD Pascal*. With the object of maximizing portability to the greatest number of **digital computers**, UCSD Pascal generated an intermediate language called *P-code* (for pseudocode) that was essentially the machine language of a simple but hypothetical stack-oriented computer called a *virtual machine* (*see* **Pushdown stack**). Then, to implement a complete Pascal system, all one had to do, once for each computer of interest, was to write a P-code **interpreter**. This same technique had been used earlier by Niklaus Wirth, and it has now been resurrected to implement **Java**.

Additional reading: *Encyclopedia of Computer Science*, pp. 910–913.

International Computers Limited (ICL)

INDUSTRY

International Computers Limited (ICL) was England's "national champion" computer company for over three decades, a company once thought to be capable of worldwide competition with **IBM**. The company's origin was the 1959 merger of the British Tabulating Machine Company (BTM) and Powers Samas, a licensee of **Remington Rand**, to form the International Computers and Tabulating Company (ICT) with Sir Cecil Weir as chairman and Cecil Mead as managing director. ICT purchased RCA 301 computers and marketed them as the ICT 1500. In 1962, the company produced its own ICT 1301. In 1963, after negotiations for a possible merger with **Machines Bull** of France failed, ICT absorbed the Ferranti Packard company of Canada and decided to drop its own product and sell that company's FP6000 instead. In September 1964, ICT launched its ICT 1900 series to compete with some models of the **IBM 360 series** announced the prior April. In 1968, at the instigation of Prime Minister Harold Wilson's Labour government, ICT merged with English Electric to form ICL. In 1984, ICL was acquired by Standard Telephones and Cables (STC), but the company continued to operate under its own name. In 1990, STC sold 80% of ICL to **Fujitsu, Ltd**. and the other 20% to Nortel and began to sell IBM-compatible **personal computers**. Six years later, ICL divested its last manufacturing capability. In 1998, Fujitsu bought Nortel's share of the company, and in 2001, ICL announced that henceforth its products would carry the Fujitsu name.

Additional reading: Chandler, *Inventing the Electronic Century*, pp. 178–181.

International Federation for Information Processing (IFIP)

GENERAL

The *International Federation for Information Processing (IFIP)*, founded in 1960 at the suggestion and forceful leadership of Isaac L. Auerbach, is a federation of societies concerned with information processing. As of 2002 IFIP had 49 national members as well as 15 affiliate, associate, and corresponding members. IFIP was founded under the auspices of UNESCO (United Nation Educational, Scientific, and Cultural Organization) with whom it still maintains an official relationship. IFIP fosters international cooperation in the fields of research and development, applications, education, and information exchange pertaining to all aspects of digital computing and communication. IFIP's biannual World Computer Congress is held in a different member nation each time and attracts upwards of 5,000 participants. The technical work of IFIP is carried out by 12 Technical Committees and, under these committees, 81 Working Groups. IFIP headquarters are in Laxenburg, Austria.

Additional reading: *Encyclopedia of Computer Science*, pp. 913–915.

Internet   NETWORKS

The *Internet* was born, and **cyberspace** with it, in 1991 when the U.S. Defense Advanced Research Projects Agency (DARPA), a division of the Department of Defense (DoD), allowed commercial companies that operated other **client-server computing** networks to connect them to its NSFnet, the renamed **ARPAnet**. Thus the Internet is a **network** of networks. Connected companies, who became known as *Internet Service Providers* (ISPs), then sold connection time to their *clients*, home and office users owning **personal computers (PCs)** or **workstations** equipped with a **modem** and access to a **telephone** line. One of the earliest and still the largest ISP is **America Online (AOL)**. The Internet was made possible by the invention of **packet switching** by Paul Baran of **Bolt, Beranek, and Newman (BBN)** in the United States and Donald Davies in England, by the definition of the **TCP/IP** communications **protocol** by Vinton Cerf and Robert Kahn, by the implementation of the **Interface Message Processor (IMP)** by Frank Heart of BBN, and by related work by Leonard Kleinrock of the University of California at Los Angeles (UCLA). The Domain Name System (DNS) that maps alphanumeric Uniform Resource Locators (URLs) and **electronic mail** addresses to numeric Internet Protocol (IP) equivalents was devised in 1984 by Keith Uncapher and Paul Mockapetris, who wrote the first implementation of the *Simple Mail Transfer Protocol* (SMTP). Internet growth from 1973 to 1994 was already quite rapid but accelerated from 1994 to the present due to the introduction of graphics-based **browsers** for the

World Wide Web in that year. The Internet, the "First Wonder of the Modern World," now consists of hundreds of thousands of servers and many millions of users. Because no central computer or authority controls the Internet, it cannot be destroyed. And no one owns it.

Additional reading: Hafner and Lyon, *Where Wizards Stay Up Late*.

Interpreter   SOFTWARE

In a general context, an *interpreter* is an agent or person who can discern the meaning of utterances in one language and translate them into another. In computing, *interpreter* has had three meanings, all of which are consistent with the general definition. In the **punched card** era, most humans could not readily ascertain the meaning of the pattern of holes on a card. But the cards could be run through a special-purpose **IBM** machine that "interpreted" the holes, column by column, and placed the corresponding letter or digit at the top of the card. These (**hardware**) interpreters were produced by IBM from the mid-1920s through 1970. Another kind of hardware interpreter is an integral part of any **stored program computer**. As the computer sequences through instruction after instruction, there must be a part of its control unit that examines each **instruction**, interprets its intent by extracting its command and its **operands**, and orchestrates the tasks needed to implement it. Carrying the concept of interpretation up one higher level, a (**software**) interpreter is a translation **program** that converts **procedure-oriented language (POL)** statements one by one into **machine language** at run time. This is in contrast to a **compiler**, which translates all POL **statements** into machine language **object code** before beginning execution. Compilation produces programs that run faster, but interpretation is more flexible in that it allows dynamic modification of **high-level language** statements during execution. The first software interpreter was written under the direction of John W. Mauchly in 1949 for a mathematical language called **Short Order Code**, implemented first on the **BINAC** and then later on the **UNIVAC**. In 1953, John Backus developed an interpretive system called *Speedcoding* for the IBM 701. Perhaps the most famous interpreters were those written by Bill Gates and Paul Allen to implement **Basic** on early **personal computers**. These interpreters were placed in **read-only memory (ROM)** so that they would be instantly available when the computers were turned on.

Additional reading: *Understanding Computers: Computer Languages*, pp. 11, 14.

Interrupt   HARDWARE

An *interrupt* is a signal that enables a **digital computer** to respond quickly to external or internal events that occur at unpredictable times. Some external events of this type are input arriving from a **keyboard**, **modem**, or **network**; a signal that an I/O device has completed its task; or a signal generated by an instrument or sensor monitoring some industrial, military, or laboratory process. Internal events include invalid **memory** references, division by **zero**, or an attempt to execute an illegal **instruction**. The response to an interrupt is temporary suspension of the **program** currently in control and the invocation of a particular **subroutine** that will deal with the type of event that caused it. An internal interrupt caused by a currently executing instruction is called a *synchronous interrupt*, or *trap*, while external interrupts are clearly *asynchronous*. The **UNIVAC** did not have an external interrupt, but it did have a trap, which was triggered by arithmetic overflow or an attempt to divide by zero. The first computer with an external interrupt was the experimental computer TX-0 designed by Wesley Clark of the MIT Lincoln Laboratory in 1953 (*see* **TX-0 / TX-2**). The **Univac 1103**A of 1955 was the first commercial computer equipped with an interrupt facility. Although basically a hardware concept, the logical power of interrupt handling is supported by **programming languages** that permit **event-driven programming**, such as **Java** and some other **object-oriented programming** languages. Interrupt handling can be considered as an instance of **concurrent programming**.

Additional reading: *Encyclopedia of Computer Science*, pp. 928–931.

Intranet   NETWORKS

An *intranet* is a **network** that has all the technical characteristics of the **Internet** but is ac-

cessible only to the employees of a business or institution that owns and maintains its components. An intranet of highest security has no communications link to any computer other than its own remote computers and **servers**. Intranets existed for decades, but no particular term was used or needed to describe them until the emergence of the Internet created the need for contrasting terminology. An intranet is generally understood to be a *wide-area network* whose domain encompasses business branches in several cities, or at least several reasonably well-separated buildings. An intranet that serves at most one building or large room is more appropriately called a **Local Area Network (LAN)**.

I/O. *See* **Input / Output (I/O)**.

IPL-V   LANGUAGES

IPL-V was the fifth and most widely used version of IPL (Information Processing Language) originally written by Allen Newell, Herbert Simon, and Cliff Shaw in 1955 for research in **artificial intelligence (AI)**, a field not named until the following year. IPL was the first **list processing** language, and IPL-V was the first to build lists from noncontiguous **memory** cells connected by **pointers**. IPL was written at the **RAND Corporation**, as was the *Logic Theorist* of 1956, the next creation of Newell, Simon, and Shaw, who later moved to Carnegie Mellon University in Pittsburgh, Pennsylvania.

Additional reading: *Encyclopedia of Computer Science*, p. 1000.

Iteration   PROGRAMMING

Iteration and **recursion** are the two principal methods of calculation. To *iterate* is to repeat an action either a specified number of times or until successive estimates of a desired result *converge*, that is, differ by less than some prescribed tolerance. The simplest example of the first situation is the **Fortran** *do-loop*, DO 17 I=1,100, which will execute all **statements** from the first one after the DO up through and including the statement numbered 17 one hundred times. An example of the second situation is iteration to obtain successively better estimates of the square root of a number through

J

Jacquard loom

In 1725 in France, Basile Bouchon invented a loom partially controlled by binary encoded **punched paper tape**, but the loom required human attention and hence was not fully automatic. In 1745, Jacques de Vaucanson, who had built many **automatons**, found a way to eliminate need for a human operator, but his system was too complex to be marketed widely. Finally, in 1801, Joseph Marie Jacquard built an automatic **punched card** driven machine that proved so successful that by 1812 more than 11,000 were being used in France alone. The Jacquard loom is believed to have been the inspiration for Charles Babbage's decision to use punched cards as input to his **Analytical Engine**.

Additional reading: Kidwell and Ceruzzi, *Landmarks in Digital Computing*, pp. 26–27.

Java

Java is a **high-level language** for use on **client-server computing** networks, particularly the **Internet**. Java, developed in 1995 by James Gosling and others at **Sun Microsystems**, is an **object-oriented programming** language in which all **data** and **functions** must be in **classes**. Java is especially useful for implementing **applets** that are invoked by a **browser** running on the client computer, usually a **personal computer (PC)**, to provide **computer animation** and other features that require local com-

Figure 22. The [Joseph Marie] Jacquard loom of 1725, which was essentially a special-purpose computer whose input medium was a chained sequence of punched cards and whose output was an intricately woven tapestry. (Deutsches Museum, Munich)

putation. Java source programs are usually compiled to *byte code* that is run on an **interpreter** called the *Java Virtual Machine* (*see* **Byte**; **Code**; **Intermediate language**). Some **compilers** for **Ada** also provide the option of producing byte code. Java **character strings** are encoded in **Unicode**, and the language does not have **pointers**. There are specialized versions of Java such as *JavaCard* for **smart cards** and *EmbeddedJava* for **embedded systems**.

JavaScript is a *scripting language* developed by **Netscape Communications**, not Sun, for its *Netscape Navigator* browser and successors; it is not syntactically compatible with Java and in fact produces **Common Gateway Interface (CGI)** programs for server-side rather than client-side applications, as the latest versions of Java can also do. **Microsoft** announced its own Java-like language called *C#* (*C sharp*) in June 2000. The language, designed and implemented by Anders Hejlsberg and Scott Wiltamuth, is intended for use with a new Microsoft **distributed system** framework called .Net ("dot Net"). Like Java, C# looks much like C++ and compiles to an intermediate language.

Additional reading: Lohr, *Go To: The Programmers Who Created the Software Revolution*, pp. 181–202.

Jevons logic machine
HISTORICAL DEVICES

The Englishman W. Stanley Jevons was a pioneer in symbolic logic and a great admirer of Boole's work in **Boolean algebra**. In 1869, Jevons built a logic machine that, because of its appearance, he called a "logical piano." The device contained 16 letters, each visible in a small window, that represented all possible combinations of four Boolean **variables**. The terms of a logical equation could be keyed into the bottom of the machine, an action that automatically eliminated from the top windows all combinations of terms (letters) that were inconsistent with the input. Sadly, Jevons drowned before his fiftieth birthday.

Additional reading: Lee, *Computer Pioneers*, pp. 400–401.

JOHNNIAC
COMPUTERS

JOHNNIAC (JOHN [von] Neumann) Integrator and Automatic Computer), a copy of the **IAS**

computer, was built at the **RAND Corporation** in Santa Monica, California, in the early 1950s. Willis H. Ware received the 1993 **IEEE Computer Society** *Pioneer Award* for its design and that of the IAS. JOHNNIAC, whose construction team was led by Bill Gunning, first ran successfully in March 1953 and continued in operation for 13 years. In addition to its long useful lifetime, the machine is noted for its being the host for **JOSS**, an early **time sharing** system.

Additional reading: http://ed-thelen.org/comphist/johnniac.html

JOSS
LANGUAGES

JOSS (Johnniac Open-Shop System) was an interactive **high-level language** created by J. C. Shaw of the **RAND Corporation** in the early 1960s. Programmers Leola Cutler and Mary Lind helped with the implementation. JOSS was also a rudimentary **operating system** that supported 12 terminals in a **time sharing** mode. To facilitate interactivity, JOSS was implemented as an **interpreter**, first for the JOHNNIAC and later for a DEC PDP-6 (*see* **DEC PDP series**). JOSS itself did not spread beyond the RAND community, but its principles migrated to many successor languages.

Additional reading: Wexelblat, *History of Programming Languages*, pp. 495–513.

Jovial
LANGUAGES

Jovial (Jules' Own Version of the International Algebraic Language) was an implementation of Algol 58 written by a team led by Jules Schwartz of the Systems Development Corporation (SDC) for the IBM AN/FSQ-7 (*see* **SAGE**) and IBM 709 (*see* **IBM 700 series**).

Additional reading: Wexelblat, *History of Programming Languages*, pp. 369–401.

K

Kalman filter
ALGORITHMS

The *Kalman filter* is an **algorithm**, not a device, that is indispensable in the **automatic control** of **real-time systems**, particularly in the field known as *time series analysis*. The algorithm was invented in 1960 by the Hungarian mathematical systems theorist Rudolf E. Kalman for the analysis of sequences of data values that track a moving object or financial trend in the presence of noise that must be filtered out (hence the name "filter"). The algorithm uses **recursion** to obtain an update of the best estimate of the current value of a time-dependent variable in terms of the prior one and known **parameters** of the system. The results produced are comparable to the dynamic production of a smooth **least squares** curve that best approximates the trend of fluctuating data values. Because of its widespread application, the Kalman filter rivals the **Fast Fourier Transform (FFT)** as the most innovative result in **numerical analysis** in the last 50 years. And just as there is an FFT, there is an FKF (Fast Kalman Filter), a computational improvement developed by the Finn Antti Lange in 1989. The reference is not frivolous; it contains a sidebar that presents the two coupled Kalman filter equations with text that describes how to apply them.

Additional reading: André Guéziec, "Tracking [Baseball] Pitches for Broadcast Television," *IEEE Computer*, **35**, *3* (March 2002), pp. 38–43.

Kelvin tide machine
HISTORICAL DEVICES

In 1876, William Thomson, Lord Kelvin, developed a mechanical **analog computer** that he called a *harmonic synthesizer*. The device, more commonly known as the *Kelvin tide machine*, used an integrator invented by Kelvin's brother James in order to synthesize and predict tidal movements based on the amplitude, phase, and periods of their constituent sine functions (*see* Figure 23). In essence, the machine used wires and pulleys to compute and print a Fourier analysis of tidal data and record it on a roll of punched **paper tape**; *see* **Fast Fourier Transform (FFT)**. Once the machine was completed, Kelvin realized that its operation was more generally applicable to the solution of linear second-order differential equations.

Additional reading: Aspray, *Computing Before Computers*, pp. 172–177.

Keyboard
I/O DEVICES AND MEDIA

A computer *keyboard* is much like that used with a **typewriter**, right down to the choice of key layout devised by Christopher Latham Sholes in 1863. The layout, plus the addition of a numeric keypad and special function keys, has persisted despite the advantages of the more ergonomic layout perfected by August Dvorak over a 12-year period in the 1930s and 1940s. Desk **calculators** need numeric keyboards, and the modern 10-key design arranged in the 3-row sequence [7, 8, 9] [4, 5, 6] [1, 2, 3] plus

Figure 23. Lord Kelvin's tide machine of 1976, an early analog computer, could synthesize tidal movements based on up to ten constituent harmonics. This image is of the United States Tide Predicting Machine No. 2 of 1912 that was built on the same principles as Kelvin's, but it was a double machine, with one side summing the heights of up to 37 tidal constituents and the other side summing their derivatives. Thus it could give both the height of the tide and the exact times of high and low tides. This machine, used until 1966, still resides in the lobby of the National Oceanic and Atmospheric Administration (NOAA) in Silver Springs, Maryland. (CO-OPS/NOAA)

a **zero** key was introduced in 1914 by Oscar J. Sundstrand of Rockford, Illinois. Later, he founded the *Sundstrand Adding Machine Company* with his brother David. Early **digital computers** did not have keyboards because they were not *interactive*; that is, they ran their **programs** from start to finish without human intervention. If an error occurred that stopped the machine, its operator would reset it through toggles or switches. **Personal computers (PCs)** and **workstations** need keyboards because they are highly interactive and used heavily for **word processing**. The first commercial digital computer to use a keyboard was the PDP-1 **minicomputer** of 1960 (*see* **DEC PDP series**). The first PCs that came with a keyboard were the Processor Technology **SOL** and the **Commodore PET** in 1977. Dvorak keyboards and the **software** needed to support them can be procured as an option for use with most current PCs.

    Additional reading: White, *How Computers Work*, pp. 221–223.

Knowledge engineering. *See* Expert system.

Knuth textbooks   GENERAL

Over the period 1968–1973, Stanford University professor Donald Knuth published three volumes of an originally planned seven textbooks on **computer science**. These books are the only ones that in and of themselves are a true milestone of the field; their impact has been compared to what the publication of Euclid's *Elements* did for plane geometry. Now translated into several languages, the books were mentioned in the citations supporting Knuth's *Turing Award* that he received from the **Association for Computing Machinery (ACM)** in 1974 and for the National Medal of Science awarded to him by President Jimmy Carter in 1979. Despite the erudition they show, the books are not without a touch of humor. Volume 1 is "affectionately dedicated" to the **IBM 650** that Knuth used early in his career at Case Institute of Technology. And the index contains both "Circular definition" and "Definition, circular," which point to each other. The series is entitled *The Art of Computer Programming*, but the titles of the individual volumes better convey their breadth: *Fundamental Algorithms*, *Seminumerical Algorithms*, and *Searching and Sorting*. Rather than proceeding on to volumes 4 through 7, Knuth revised the first three volumes to meet his exacting standards, not only of mathematical precision but also of typography. Because of concerns about the latter, Knuth took nine years off to perfect T_EX, a notational language for preparing mathematical equations for typesetting, and *Metafont*, a language for designing typefonts, and persuaded his own publisher, Addison-Wesley, and many others to use them for printing mathematically oriented material. A decade after initial publication, each of the Knuth textbooks was still selling at the rate of 2,000 copies per month and have continued to do so to this date.

Additional reading: Slater, *Portraits in Silicon*, pp. 343–351.

L

Lambda calculus

The *lambda calculus* (λ-calculus) was invented by American logician Alonzo Church in 1934 to model the mathematical notion of substitution of values for **variables**. The λ-calculus is equivalent in computational power to the **Turing machine** in that any computable **function** may be represented as a λ-expression, but its computational mechanism is much closer to that of a **programming language** than Turing machines are. This has led to attempts to model languages such as **Algol 60** in terms of the λ-calculus. Such models capture certain concepts, such as nested block structure, binding of variables to values, and the order of evaluation of **operators** in **expressions**, but have difficulty in capturing other concepts, such as the assignment **statement**, variables shared among **subroutines**, side effects, and unconditional branching. Thus the λ-calculus is better suited to modeling **functional languages**, which may be regarded as notational variations of the λ-calculus. This is precisely why John McCarthy and Marvin Minsky used it as the basis for the syntax of **Lisp**, their **list processing** language.

Additional reading: *Encyclopedia of Computer Science*, pp. 953–955.

LAN. *See* Local Area Network (LAN).

Language translation. *See* Machine translation.

Laptop computer

A *laptop computer*, as the name implies, is a **personal computer** that is small enough and light enough to hold in one's lap but is nonetheless fully functional. That is, a laptop version of an **Apple** iMac, for example, can run all the **software** that a desktop iMac can. Laptops use a flat-panel display housed on the inside of a cover that can be closed to cover the **keyboard** and the underlying case that contains a battery, a **floppy disk** drive, a CD- or DVD-ROM, and **input / output (I/O)** ports. This "clamshell" design was pioneered by **Toshiba**. The principal attribute of a laptop is not that it is normally held in the lap but rather that it is portable. Some of the latest laptops are called *notebooks* to emphasize their low weight, but there is no standard guideline as to the weight that defines the borderline between the two. All major vendors that sell desktop computers—**Dell**, **Compaq**, **IBM**, and **Gateway**, for example—also sell laptops, whatever the terminology. What may have been the first laptop was designed in 1979 by a Briton, William Moggridge, for Grid Systems Corporation. The Grid *Compass* was one-fifth the weight of any **personal computer** of equivalent performance and was used by the U.S. National Aeronautics and Space Administration (NASA) in its space shuttle program of the early 1980s. The Compass had a 340KB (kilobyte) **bubble memory** with a diecast magnesium case and a folding electroluminescent graphics display screen. A rival claim is that of Manny Fernandez, who started the *Gavilan Computer Corporation* and in 1984 promoted his machine as the first "laptop" computer. This early **Intel 8008**–based laptop had an eight-line **liquid crystal display (LCD)** screen, an innovative touchpad, and an optional printer that attached to the back. The first **IBM PC**–compatible laptops did not appear until the early 1990s.

Additional reading: *Computing Encyclopedia*, vol. 3, p. 88.

LARC
COMPUTERS

The *LARC* (Livermore Automatic Research Computer), like its contemporary **Stretch**, was one of the first **supercomputers**. The project was initiated in 1954 by a request for proposals by the University of California Radiation Laboratory (UCRL) whose scientists wanted a computer a hundred times faster than its installed **UNIVAC**. Sperry Univac won the competition, and the machine was built during 1959–1960 at the Sperry Univac engineering facilities in Philadelphia with Herman Lukoff as project director. LARC architecture was designed by Arthur Gehring and Albert Tonik and its circuitry by Josh Gray assisted by Lukoff, Bill Winter, and Lloyd Stone. Specifications were negotiated by a team from Sperry and one from UCRL led by Sydney Fernbach, director of its computing laboratory. Only two LARCs were manufactured, one delivered to UCRL and the other to the former David Taylor Model Basin (now the Naval Ships Research and Development Center) near Washington, D.C.

The LARC comprised two units, an I/O processor and a parallel **central processing unit (CPU)** capable of both fixed and floating-point **binary-coded decimal (BCD)** arithmetic. Addition and multiplication times were 4 and 8 microseconds (μs), respectively. **Magnetic core** main **memory** was shared by the I/O processor and either one or two CPUs. Each of 8 up to 39 memory units contained 2,500 **words** of 11 decimal digits plus a sign digit, a maximum of 97,500 words. Eight units were used in the basic system on a high-speed **bus** to provide an effective access time of 0.5 μs per word. Up to 24 **magnetic drums** could be included in the system, each capable of storing 250,000 words.

The drums could maintain a continuous data transfer rate of 2,500 words every 83 milliseconds between the drums and CPU.

LARC was a technological success, but the quality of its **software**, particularly its **Fortran** compiler, inhibited its usefulness. Nonetheless, both the UCRL and David Taylor computers continued to operate until phased out over the 1968–1969 period. LARC was the largest *decimal* computer ever built, a distinction it is likely to retain forever. Its story is colorfully and lovingly told by its project director in the reference.

Additional reading: Lukoff, *From Dits to Bits*, pp. 145–177.

Laser
HARDWARE

The *laser*, an acronym for *light amplification by stimulated emission of radiation* coined by Gordon Gould, is the basis for the **laser printer** and for **optical storage** memory devices such as the CD-ROM and DVD-ROM. Albert Einstein predicted that stimulated emission of radiation could occur in 1916. The laser is an extension of the *maser* (*microwave amplification by stimulated emission of radiation*), invented in 1954 at Columbia University by Charles Townes and Arthur Schawlow to create coherent beams of microwaves. Once the maser was demonstrated, several researchers sought to extend the principle for use in the visible portion of the electromagnetic spectrum. The first working laser, built by Theodore Maiman at Hughes Research Laboratories in Malibu, California, became operational on 16 May 1960, the same year in which Townes and Schawlow received a patent for their laser, but a year after Gould, a doctoral student working under Townes, had filed his own patent application. Gould's patent, later granted legal priority, pertained to the optically pumped and discharge-excited laser amplifiers now used in most industrial, commercial, and medical applications of lasers. In 1962, Robert Hall of the GE Research & Development Center in Niskayuna, New York, built the first semiconductor injection laser. Hall's device, for which he obtained a patent in 1967, was based on a specially designed highly efficient p-n junction semiconductor that generated coherent light from a very compact source; the semiconductor crystal was a cube a third of a millimeter on a side. **Semi-conductor** lasers based on Hall's original design are used in all CDs and CD-ROMs, all laser printers, some TV remote controls, and most **fiber optic** communications systems. In 1964, Townes was awarded a Nobel Prize in Physics shared with Russian physicists Aleksandr Prokhorov and Nikolai Basov. In 1981, Schawlow and Nicolaas Bloembergen of Harvard University received the Nobel Prize, also in physics, for their contributions to the development of laser spectroscopy. In 1991, Gordon Gould was inducted into the National Inventors Hall of Fame, as was Robert Hall in 1994.

Additional reading: *Computing Encyclopedia*, vol. 1, pp. 200–201.

Laser printer I/O DEVICES AND MEDIA

The *laser printer* was invented in 1969 at a Xerox research facility in Webster, New York, by Gary Starkweather, who did so by combining a **laser**, **xerography**, and something called a Research Character Generator (RCG) developed by Butler Lampson and Ron Rider. The laser "draws" the printable image on a special light-sensitive drum, which then collects powdered ink ("toner") using an electrostatic charge and rolls it onto the page. A hot roller melts the toner to the paper to fix the image. By 1971, the first perfected version of Starkweather's invention could print two pages per second at a resolution of 300 dpi (dots per inch). In 1976, **IBM** introduced its IBM 3800 laser printer, capable of printing 20,000 lines per minute. Xerox's first commercial laser printer, the 9700, did not come until 1977. In 1984, **Hewlett-Packard** introduced its LaserJet printer featuring 300 dpi resolution for $3,600, and a year later **Apple** released its Apple LaserWriter. Laser printer technology continued to advance over the years until 1,200 dpi black and white printers with a speed of ten pages per second for less than $1,000 became typical.

Additional reading: Hiltzik, *Dealers of Lightning*, pp. 127–144.

Least squares method ALGORITHMS

In 1809, Carl Friedrich Gauss claimed to have used the statistical method of *least squares* as early as 1795, but he credited Adrien Marie Legendre with its first publication. Legendre did

so in 1805 in the form of an appendix entitled "Sur la méthode des moindres quarrés" to his "Nouvelles méthodes pour la determination des orbites des comètes." Given data that appears to follow a linear trend, the least squares **algorithm** computes the slope m and the intercept b of the straight line $y = mx + b$ that best approximates the data in the sense that the sum of the deviations of measured values from their corresponding points on the approximating line is minimized. The method is an important milestone in **data reduction** and can be generalized to apply to **data** that is nonlinear with respect to an independent **variable**.

Additional reading: *Encyclopedia of Computer Science*, pp. 963–964.

Leibniz calculator CALCULATORS

The *Leibniz calculator*, the first major improvement over the **Pascal calculator**, was invented in 1671 by the German mathematician Gottfried Wilhelm Leibniz, the codiscoverer of **calculus** and the first person to describe the **binary number system**. By adding a "stepped drum" and wheels based on an active/inactive pin principle and a delayed carry mechanism, Leibniz modified Pascal's machine so that it could multiply and divide as well as add and subtract (*see* Figure 24). The stepped drum became the basis for virtually all **calculators** until late into the nineteenth century. Knowledge of the precise workings of the Leibniz calculator was almost lost. Leibniz gave his machine to the University of Göttingen in the late 1670s, and it lay dormant in an attic until discovered in 1879 by workers trying to fix a leaky roof.

Additional reading: Aspray, *Computing before Computers*, pp. 42–49.

Figure 24. In 1671, Gottfried Wilhelm Leibniz, co-discoverer of the calculus, invented the "stepped wheel" calculator, which, since it could perform all four basic arithmetic functions, was a significant advance over the Pascal calculator. (Deutsches Museum, Munich)

Figure 25. Joe Formoso's conception of one of the cells that comprise Jorge Luis Borges's "The Library of Babel," which begins: "The universe (which others call the Library) is composed of an indefinite and perhaps infinite number of hexagonal galleries, with vast air shafts between, surrounded by very low railings. From any of the hexagons one can see, interminably, the upper and lower floors." (Courtesy of Joe Formoso)

LEO COMPUTERS

The *LEO* (Lyons Electronic Office), the first business computer, was built in London in 1951 by J. Lyons & Co., then Britain's leading caterer. The initial LEO was similar to the **EDSAC**. **Hardware** extensions to the EDSAC design included a doubling of the store to 2,048 17-bit **words** and the incorporation of enhanced **input / output (I/O)** facilities. Buffered input and output permitted **data** for one transaction to be read while the previous transaction was being processed and the results for the one before that were being printed. LEO supported three parallel input streams, one from **punched paper tape** and two from **punched card** readers. The two output streams were to a line printer and a card punch. This system started running the Lyons payroll as an integrated application from clock card to pay envelope in February 1954. The project was led by Raymond Thompson with John Pinkerton as head of engineering and David Caminer as head of application design and programming. The LEO II of 1957 was used as a testbed for **magnetic tapes**, **magnetic drums**, and **magnetic core memory**. In all, ten systems were delivered to industrial and government users. The LEO III of 1961 used **semiconductor** technology, **microprogramming** to provide a 93-operation **instruction set**, and support for **real-time systems**. Sixty LEO IIIs were delivered over the period 1962–1966. LEO Computers merged with English Electric in 1964 and became part of **International Computers Limited (ICL)** in 1968.

Additional reading: *Encyclopedia of Computer Science*, pp. 981–982.

Library GENERAL

Just as the concept of **algorithm** is the foundation of **computer science**, the concept of a *library* is central to **information technology**

(IT). Reading a book at a library or borrowing one therefrom was the earliest form of **information retrieval**. The first libraries were collections of scrolls accumulated by the ancient Greeks, and the most famous library of antiquity was the library of the Museum of Alexandria in Egypt founded in about 290 B.C. by Ptolemy I and gradually expanded by his successors Ptolemy II and III in accord with organizational plans formulated by Demetrius of Phaleron. The first recorded librarian was Zenodotus of Ephesus, holding that post from the end of Ptolemy's reign until 245 B.C. The third librarian, Eratosthenes (275–194 B.C.), famous for devising the algorithm known as the **Sieve of Eratosthenes** to compute **prime numbers**, calculated the circumference of the earth to within 1% based on the measured distance from Aswan to Alexandria and the fraction of the whole arc determined by differing shadow-lengths at noon in those two locations. Whether accidentally or deliberately, the Alexandrian library was destroyed in the civil war that occurred under the Roman emperor Aurelian in the late third century A.D. But the concept lived on (*see* Figure 25).

The first lending library in the United States was the *Library Company of Philadelphia*, founded in 1731 by Benjamin Franklin and a group of his friends as a subscription library. It was the nation's first lending library and is now its oldest cultural institution. Currently, the world's largest libraries are the U.S. *Library of Congress*, the *British Library*, and the *Bibliotheque nationale de France*. The Library of Congress holds 25 million volumes, the Harvard University Library, 14 million, and the New York (City) Public Library, about 11 million. Although **electronic books (e-books)** are becoming popular, most people find no substitute for the real thing. In this age of **automation**, only a few computing centers contain books, but all libraries of any size contain multiple **Internet**-ready **personal computers**.

Additional reading: Flower and Zahran, *The Shores of Wisdom: The Story of the Ancient Library of Alexandria*.

Librascope LGP-30   COMPUTERS

The *Librascope LGP-30*, manufactured by the Royal Precision Electronic Computer Company, was an early binary **minicomputer** produced in competition with the **ALWAC III-E**. "LGP" stood for Librascope General Precision. This desk-sized **digital computer** contained only 113 **vacuum tubes** and 1,450 diodes but consumed 1,500 watts of power. **Memory** was a 4,096-word **magnetic drum**. I/O consisted of a small **cathode ray tube (CRT)** and a Flexowriter as a printer. **Word length** was 32 **bits**. The LGP-30's **instruction set** consisted of only 16 **instructions**. Seven machines were produced in 1956–1957 and sold for about $40,000 each.

Additional reading: Ceruzzi, *A History of Modern Computing*, p. 42.

Light Emitting Diode (LED)   HARDWARE

The first *light emitting diode (LED)* of any kind was invented by J. W. Allen (United Kingdom) and P. E. Gibbons (United States) in 1960. The first practical visible-spectrum LED was developed in 1962 by Nick Holonyak, Jr., of **General Electric (GE)**. Holonyak's LED emitted red light. His colleague and former student George Craford developed one that emitted yellow light, and Shuji Nakamura, now at the University of California at Santa Barbara, made the first LEDs that emit blue or green light. LEDs were commonly used in digital watches and in experimental flat-panel displays until 1977, after which **liquid crystal displays** (LCDs) became the preferred choice for such applications. But the availability of red, green, and blue (RGB) LEDs now make large RGB flat-panel displays possible. The NASDAQ billboard in New York City's Times Square, currently the world's largest video screen, uses 18,677,760 LEDs that cover 10,736 square feet. It may also be possible to combine RGB LEDs to form bright and extremely efficient white-light fixtures that ultimately replace incandescent bulbs.

Additional reading: George Craford et al., "In Pursuit of the Ultimate Lamp," *Scientific American*, **284**, 2 (February 2001), pp. 62–67.

Light pen   I/O DEVICES AND MEDIA

A *light pen* consists of a light-sensitive stylus wired to a video **monitor**. The device is used to draw pictures on the screen or select from displayed menu options. The user brings the pen to the desired point on the screen and presses a button on the pen to make contact. The pen senses light from the screen, not the reverse. During screen refresh, the pen senses the particular **pixel** being pointed at, and its coordinates are stored for whatever use the running **computer graphics** program wants to make of them. The first light pen—more of a light gun than a pen—was invented at MIT in 1952 for use on the **Whirlwind**.

Additional reading: http://www.epanorama.net/documents/pc/lightpen.html

LINC
COMPUTERS

The *LINC* (Laboratory INstrument Computer), designed by MIT physicist Wes Clark in 1962, has a strong claim to having been the first **personal computer**. Prior to conceiving the LINC, Professor Clark had worked on two historically important **transistor** computers, the **TX-0 / TX-2** of 1956 and 1959. In 1963, about 60 LINCs were built at an MIT "Summer Camp" where researchers from across the country met to assemble computers that they could use in their laboratories rather than having to submit to the tyranny of **batch processing** at a central computing center. The LINC was built by Clark and Charles Molnar with **Digital Equipment Corporation (DEC)** modules that used diode-transistor and diode-capacitor **logic circuits** before that company began making its own **digital computers** (*see* **DEC PDP series**). LINC had 1,024 12-**bit words** of **magnetic core** memory, later expanded to 2,048. The LINC used an early **graphical user interface (GUI)** with a 256 × 256 **cathode ray tube (CRT)** display and four pre-**mouse** "knobs" used to enter variable **parameters**. The screen editor of the LAP-6 (LINC Assembly Program) was integrated with the **assembler** itself and the computer's **file** system. Each of two **magnetic tape** drives held a spool that held 512 blocks of 256 12-bit words (512 6-bit characters).

Additional reading: Hiltzik, *Dealers of Lightning*, pp. 40–43.

Linear Bounded Automaton (LBA)
THEORY

When a **Turing machine** solves a problem of size n, it is allowed to use an amount of tape (**memory**) that is an arbitrary function of n,

perhaps kn^2 or kn^3 or even ka^n squares of tape (where k and a are constants). In contrast, a *Linear Bounded Automaton (LBA)*, though it operates logically as does a Turing machine, is limited to using an amount of tape that is at most kn, a linear function of n. An LBA is the second highest of the four levels of the **Chomsky hierarchy**; it is able to recognize Type 1 (context-sensitive) languages. The LBA is also the subject of the second most important open problem in **computer science**, namely, whether a nondeterministic LBA can recognize languages that are any more complex than can be recognized by an ordinary (deterministic) LBA. (For the most important open problem, *see* **NP-complete problem**.) A language is *recognizable* by an **automaton** if it can parse an arbitrary symbol **string** and determine whether the string is or is not syntactically correct with regard to the rules of a particular grammar.

Additional reading: Dewdney, *The Turing Omnibus*, pp. 46–47.

Linked list
DATA STRUCTURES

A *linked list* is a dynamic **data structure** that is useful for representing symbolic information characteristic of **artificial intelligence**, **computer algebra**, and **natural language** processing. Like the **graph**, a linked list consists of interconnected *nodes*, but with the restriction that nodes are connected only to their successors, or only to their predecessor and successors, so as to form a singly- or doubly-linked chain. A list that has a specific last node is said to be *linear*, and a list that loops back to its "first" node is said to be *circular*. This gives rise to four kinds of linked list: the singly-linked linear list (SLLL), the singly-linked circular list (SLCL), the doubly-linked linear list (DLLL), and the doubly-linked circular list (DLCL). All nodes of a particular linked list are of uniform size and structure; a node consists of a data field and either one or two **pointers** that serve as the links for singly- and doubly-linked lists, respectively. In contrast to the **array**, it is relatively easy to insert a new element into a linked list or to delete an element from one (by rearranging pointers). **Programs** for the early one-and-a-half-address **magnetic drum** computers such as the **IBM 650** were essentially SLLLs, and when debugging revealed that a program was missing a key **instruction**, it was easy to place it in an unused **memory** location and splice it into the flow of control. John von Neumann wrote of linear linked lists implemented in **machine language** as early as 1947, and SLLLs were the foundation data structure for the first three widely used list processing languages, IPL (Information Processing Language) of Allen Newell, Herbert Simon, and Cliff Shaw in 1956 (*see* **IPL-V**), John McCarthy and Marvin Minsky's 1958 **Lisp** (and its many variants to this day), and Joseph Weizenbaum's SLIP (Symmetric List Processor) of 1963.

Additional reading: *Encyclopedia of Computer Science*, pp. 992–1000.

Linux   OPERATING SYSTEMS

Linux, pronounced "Lynn-ucks," is one of three major **free software** versions of **Unix**, the others being *GNU Unix* and *FreeBSD*. Linux is named for the Finnish programmer Linus Torvalds who created it in 1994 while still a student at Helsinki University. Linux, along with the **Apache** program for Web **servers**, is one of the two most widely used **open source software** products. Linux brought Unix, originally a **mainframe** operating system, to the **personal computer (PC)** and is now available for all commonly used PCs. Its **source code** and a host of related ancillary programs, all written in **C** or **C++**, may be downloaded from the **Internet** without charge or may be purchased for a nominal fee from companies such as Red Hat and Caldera who ship it on CD-ROM bundled with a choice of **graphical user interface (GUI)** programs and supporting manuals. Linux is now second only to **Microsoft Windows** as an **operating system** for **Intel 80×86**–compatible PCs, and at least one **handheld computer**, the Sharp Zaurus, uses Linux.

Additional reading: *Computing Encyclopedia*, vol. 2, pp. 213–214.

Liquid Crystal Display (LCD)

HARDWARE

The *liquid crystal display (LCD)*, invented by James Fergason in 1979, is the most common type of flat-panel display (FPD). LCDs are based on an unusual form of matter discovered by Austrian botanist Friedrich Reinitzer in 1888 and called a *liquid crystal* by German physicist Otto Lehmann a year later. A liquid crystal, a seemingly contradictory term, behaves mechanically like a liquid but optically like a solid. Lehmann found that liquid crystals are so sensitive that they respond to heat, light, sound, mechanical pressure, electromagnetic fields, radiation, and certain chemical vapors. In physics terminology, a liquid crystal is optically *active*; that is, its natural nematic (twisted) structure can be used to turn the polarization of light by up to 90°. Two crossed polarizers normally do not transmit any light, but if a 90°-twisted liquid crystal is inserted between them, light will be transmitted. Applying an electric field will unwind the helical structure, the liquid crystal loses its ability to affect polarization, and the display turns dark. In the 1930s, the American scientist John Dreyer used liquid crystals to make polarizing lenses for use in sunglasses and glasses for viewing three-dimensional (3D) movies. The LCD invented by Fergason consists of an array of **pixels** that can be individually addressed to create a desired image. Thin-film transistor LCDs (TFTLCDs) are now widely used in the lightweight **laptop computers** called *notebooks*. The flat-panel LCD has begun to displace the **cathode ray tube (CRT)** display **monitor** typically used with desktop **personal computers (PCs)**, and HDTV (high-definition **television**) will accelerate the ascendancy of the LCD over use of CRTs in that medium as well.

Additional reading: Linda Hamilton, "Liquid Crystals," *Information & Technology*, **17**, 4 (Spring 2002), pp. 20–29.

Lisa. *See* **Apple Lisa**.

Lisp   LANGUAGES

Work that led to creation of the **list processing** language *Lisp* (LISt Processor) was initiated in late 1958 by professors John McCarthy and Marvin Minsky in conjunction with the MIT Artificial Intelligence Project. Lisp was designed for use in **artificial intelligence (AI)** in general and **computer algebra** in particular. The major features of Lisp, which is modeled after the **lambda calculus**, are **recursion**; **Polish notation** syntax in fully parenthesized prefix form; conditional **expressions**; representation of symbolic information externally by lists and internally by **linked lists**; and the representation of **programs** and **data** in exactly the same way, a property usually associated only with **machine language**. Lisp is a **functional language**, but it also has the imperative features common to **procedure-oriented languages**. The heart of Lisp is McCarthy's *eval*, a theoretically motivated universal **function** that is more compact than a universal **Turing machine**. Lisp 1.5 was stable, well documented, and executed through an **interpreter**. Later implementations include a **compiler** for fast execution and compile-time debugging. Compilers for Lisp were written in Lisp as far back as 1959, an early example of language *bootstrapping*.

Attempts to improve Lisp 1.5 led to versions for the DEC PDP-6 and PDP-10 (*see* **DEC PDP series**). Major systems came from MIT, Stanford, **Bolt, Beranek, and Newman (BBN)**, and **Xerox PARC**. Next came MIT's MACLisp for the **time-sharing** system **Multics** on the **GE 600 Series**, and then to special single-user **hardware** implementations called "Lisp machines." In the early 1970s, special hardware/**software** systems for InterLisp were produced at Xerox and Carnegie Mellon University, but neither version survived. The popular pedagogical *Scheme* dialect of Lisp created by Guy Steele and Gerald Sussman in the 1970s is used at MIT to support the first programming course. By 1984 the proliferation of Lisp dialects became sufficiently alarming that a grassroots effort was formed to produce a new unified version called *Common Lisp*, ANSI standard X3J13 of 1994.

Additional reading: *Encyclopedia of Computer Science*, pp. 991–992.

List. *See* **Linked list**.

List processing

INFORMATION PROCESSING

List processing refers to the creation, storage, and modification of **data** stored in the form of a **linked list**. List processing may be done in any **programming language** that supports **pointers**, such as **Ada**, **Pascal**, **PL / I**, and **C++**, but is most easily done in special-purpose languages such as **Lisp**. The operations most often performed during list processing are those that fetch the first element of a list, the rest of the list that follows that element, and operations that insert or delete nodes of a list.

Additional reading: *Encyclopedia of Computer Science*, pp. 992–1000.

Literate programming

PROGRAMMING METHODOLOGY

Literate programming is a system proposed by Donald Knuth in the late 1970s for combining the writing of a **program** and its documentation, a method used in developing his typesetting language TEX. The system converts documentation and **code** into beautifully typeset material with no additional effort by the programmer. Close correspondence between a program and its documentation is enforced, but program code can be extracted from a literate program and compiled or processed in the usual way. Stephen Wolfram's **computer algebra** system *Mathematica* uses a form of literate programming in that its "notebooks" allow mathematical articles to be written mixing text with executable mathematical expressions.

Additional reading: *Encyclopedia of Computer Science*, pp. 1000–1002.

Loader

SOFTWARE

A *loader* is a short sequence of **instructions** that must be loaded into the main **memory** of a **digital computer** before a full **program** can be read and executed. In the days of **batch processing**, each **punched card** deck constituting a program in the batch had to be preceded by one or more cards that defined the loading sequence. Creation of a "one-card loader" was considered to be a work of programming craftsmanship. A loader necessarily has to specify the **memory** address of the starting point of the program to follow, plus an input **loop** whose execution begins upon completion of the loading stage. All early **digital computers** needed a loader, the first of which may have been one called **initial orders** on the **EDSAC** of 1949. In the current era of

operating systems that, barring serious error, never halt, the kind of "bootstrap" loading described herein is seldom needed because a permanently stored part of such systems is constantly ready to accept new input.

Additional reading: *Encyclopedia of Computer Science*, pp. 988–991.

Local Area Network (LAN)   NETWORKS

A *local area network (LAN)*, a concept of the 1970s, is an inexpensive **network** suitable for use in a classroom, laboratory, office suite, or an entire building to enable **personal computers (PCs)** and **workstations** to share **software** and **peripherals** such as **laser printers** and large **hard disk** drives. Standardization in the early 1980s was followed by worldwide deployment of 10Mb/s (megabits per second) LANs. By the end of the decade, LANs were common, and by the late 1990s, transfer rates had reached 100Mb/s. In its most basic form, a LAN is realized by attaching all devices directly to the transmission medium. Each data packet (or *frame*) transmitted by any station is seen by all the others, each of which examines the destination address field of the packet to determine whether it should receive it (*see* **Packet switching**). As a consequence, such a network does not have to use a **router**. LANs can be connected as a "star," like spokes in a wheel, with the most powerful computer at the hub. But the two most common LAN topologies are the **bus** and the *ring*. The two principal designs associated with these topologies are **Ethernet** and the *token ring*. As the name of the latter implies, access to the transmission medium is based on the use of a *token*. A station can transmit only while it holds the token. When it finishes transmitting, it passes it to the next station, typically the next (active) station in a physical ring. The token ring network was patented by W. D. Farmer and E. E. Newhall in 1971, but the Swedish engineer Olaf Soderblom also obtained a token ring patent in 1981 and claimed that the invention dated back to 1967.

Additional reading: *Encyclopedia of Computer Science*, pp. 1008–1114.

Logarithm   MATHEMATICS

The *logarithm* of a number a is the exponential power of a chosen base b that is equal to that number. For example, the logarithm of 64 to the base 2 is 6 because $2^6 = 64$. But to base 8, the logarithm of 64 is 2 because $8^2 = 64$. Logarithms are not integers except in the rare cases where a is an integral power of the base b. For example, the logarithm of 25 to base 10 is approximately 1.39794 because $10^{1.39794}$ is very close to 25. The most common bases used to generate tables of logarithms are 10 and e (2.718281828459045 . . .), a fundamental **constant** like **pi** called "the base of the natural logarithms." The latter choice is called "natural" because if the function e^x is plotted versus x, the slope of the function (its *derivative* in the language of **calculus**) is equal to itself, and this is the only function with that property. The notation $\log(x)$ usually implies use of base 10, and the notation $\ln(x)$ implies use of natural logarithms. Computer scientists often use $\lg(x)$ to indicate base 2, which is useful in the **analysis of algorithms** that involve halving or doubling of problem size.

Logarithms were invented in 1614 by the Scott John Napier, Eighth Laird of Merchiston, in order to expedite calculation. The Swedish mathematician Jobst Bürgi also developed a system of logarithms at about the same time, but this was not known until much later. Natural logarithms are also called *Napierian logarithms*, but Napier did not have the concept of number base, and what he did was equivalent to using a base of 1/e, not e itself. In 1618, Edward Wright produced an English translation of Napier's description of logarithms with a 16-page appendix, believed to have been compiled by William Oughtred, that included the first table of natural logarithms. Napier's work was also popularized by Henry Briggs, who, in his classic *Arithmetica Logarithmica*, proposed use of base 10, or *common logarithms*. He showed how to use his own precomputed tables of common logarithms, which are believed to have been computed by **difference calculation**, to transform the difficult labor of multiplication and division of a sequence of numbers into the simpler process of addition and subtraction. For example, $a \times b / c$ can be computed by adding the logarithms of a and b, subtracting the logarithm of c, and then looking up the inverse, or *antilogarithm*, of that result to obtain the desired answer. This **algorithm**, now taught in high school trigonometry, is not practical for use with **digital computers** because of the storage that would be needed to compute to, say, seven-digit **precision**, but it is suitable for use with analog devices such as the **slide rule** whose precision is limited to three or four significant digits. In 1688 John Pell devised an efficient interpolation formula for obtaining antilogarithms based on the three ordinates and corresponding logarithms closest to the one desired, but the method is now called *Lagrangian interpolation* after Joseph-Louis Lagrange, who rediscovered it a century and a half later. And even Pell was apparently anticipated by Fermat in 1640 and, though used in a different context, by the Indian mathematician Brahmagupta in A.D. 628.

Additional reading: Aspray, *Computing Before Computers*, pp. 16–27.

Logic circuit
HARDWARE

A *logic circuit* consists of interconnected *gates*, each of which implements one of the operations of **Boolean algebra**. The study of how to design and interconnect these gates to achieve a particular result is called **switching theory**. The input to a logic circuit consists of one or more **bits**, and its output is a bit that represents the result of the function implemented by it. The earliest logic circuits were made from electromagnetic relays. Nikola Tesla invented an AND gate in about 1892 and in 1898 successfully defended his patent application against the claim of the Canadian Reginald Fessenden, the first person to transmit wireless voice messages, that his similar invention had precedence. The first electronic logic circuit was the *Eccles-Jordan trigger circuit* of 1919, which implemented a **flip-flop**. Another early logic circuit was the 1-bit **adder** devised by Stibitz in 1937. An important **algorithm** for minimizing the number of gates needed to implement a particular Boolean **expression** was devised in 1956 by Willard Van Orman Quine and Edward J.

McCluskey. Current logic circuits are packaged as **integrated circuits**

Additional reading: Petzold, *CODE*, pp. 86–142.

Logic programming
PROGRAMMING METHODOLOGY

Logic programming is the creation of computer **programs** that pertain to **automatic theorem proving** and the manipulation of **expressions** in propositional calculus. Most logic programming is now done in **Prolog** by mathematicians and theoretical computer scientists who are interested in reducing computation to fundamental operations in pure logic.

Additional reading: *Encyclopedia of Computer Science*, pp. 1017–1031.

LOGO
LANGUAGES

Logo, a name derived from the Greek λογος, meaning "word," is a dialect of **Lisp** designed for educational use. Logo is a general-purpose **functional language** with special emphasis on symbolic computing. The first version of Logo was developed in 1967 at **Bolt, Beranek, and Newman (BBN)** by Wallace Feurzeig, Seymour Papert, Daniel Bobrow, and Cynthia Solomon. Inspired by psychologist Jean Piaget's argument that students learn mainly by *construction*—comparing new ideas to those already understood—their goal was to provide an environment for mathematical thinking in terms of concrete examples. For example, Logo uses *turtle graphics* rather than traditional Cartesian graphics. The pupil imagines that segments are being drawn by a pen controlled by a robot turtle. At any given moment the turtle is at a known position and facing in a known direction. A new line segment is then drawn by either moving the turtle a given distance in the same direction or by first turning the turtle relative to its original heading and then moving it a given distance. Thus a single procedure can draw a given shape anywhere on the screen with any desired orientation. The way the turtle creates a new figure is identical to the way that a person would do so by following the outline of that shape on a bare floor and marking progress with a crayon. Logo teachers consider procedure definition as "teaching the computer" to do something new. The keyword that initiates a definition is TO, rather than DEFINE, to suggest the sentence, "I'm going to teach you how TO SORT" (or TO DIFFERENTIATE, or TO AVERAGE). The keyword TO suggests that the procedure name is a verb, as befits an action being taught to a computer.

Additional reading: Maddux and Johnson, *Logo: A Retrospective*.

Lookahead. *See* **Instruction lookahead**.

Loop   DEFINITIONS

As soon as there were **stored program computers**, it became possible to write a "jump" **instruction** that transferred control back to a lower-numbered **memory** address. If the jump is "unconditional"—one that always jumps—and if none of the jumps, if any, in the repeated sequence of instructions ever jumps, what results is an *infinite loop* that will not terminate (until, perhaps, a clock sends a "time expired" **interrupt** to the **operating system**). Creation of such a *loop* (other than to deliberately cause a time delay) is always an error—a **bug** in the program. But a properly controlled loop—one that iterates a finite number of times—is a very valuable programming tool (*see* **Iteration**). Finite loops are controlled by a test of some kind that will terminate the loop when, for example, a *counter* counts down to 0 or a logical condition among **variables** is met.

Lull logic machine   HISTORICAL DEVICES

In about 1275, the Catalan mystic Raymon Lull (Raymundus Lullus) wrote a tract called *Ars Combinatoria* in which he described his invention of what was probably the very first "logic machine." His paper machine—literally made of stiff paper—used connected geometrical figures that followed a precisely defined framework of rules in an attempt to produce all the possible declarations that the human mind could possibly imagine. These declarations or statements were represented only by a series of signs and letter sequences (**strings**). Run long enough, the machine would thus produce the totality of human wisdom by mechanically combining a strictly limited number of signs. The device was quite useless, of course, because it would produce many more false propositions than accidentally correct ones.

Additional reading: http://www.c3.hu/scca/ butterfly/Kunzel/synopsis.html

M

M (Mumps)

M, originally named *MUMPS*—an acronym for *Massachusetts* (General Hospital) *Utility MultiProgramming System*—is a **high-level language** created in 1962 at the institution cited in the acronym. Mumps was chartered by Otto Barnett and programmed by Neil Pappalardo, Robert Greenes, and Curtis Marble for a DEC PDP-7 (*see* **DEC PDP series**). The language was general purpose but had especially powerful **string processing** operations. By 1969, Pappalardo and Marble had left to form Meditech, Greenes founded Automated Health Systems, and their companies made MUMPS commercially available. A Mumps **User Group** (MUG) was formed, and the language became an American National Standards Institute (ANSI) standard in 1977. Renamed M, the language has been ported to many **digital computers** and remains viable to this day.

Additional reading: http://www.mcenter.com/mtrc/whatism.html

Machine language

Machine language is that language usable for direct **program** creation without use of intervening **software**. It is also the language of the **object code** produced by an **assembler** or **compiler**. Although there may be some telltale signs, it would be very difficult for the uninformed person who examines object code to as-

certain whether it was written from scratch by another person or whether it was generated as the output of assembly or compilation of a higher-level language program. Machine language is the lowest-level language of a **digital computer** unless its **instruction set** has been implemented in **microinstructions** made accessible to the programmer rather than as hardwired **logic circuits** (*see* **Microprogramming**). Machine language has not been standardized; there are as many different dialects as there are variant **computer architectures**. But neither are instruction sets patentable, so any manufacturer may choose to build computers having the same architecture, and hence the same machine language, as that of a competitor. The historic example is the **RCA Spectra series**, which, though it contained some embellishments, could run machine language programs written for the **IBM 360 series**.

Additional reading: *Encyclopedia of Computer Science*, pp. 1043–1045.

Machine learning

Machine learning is the study of methods for designing intelligent **software** by analyzing examples of its desired behavior. The goal of machine learning is the production of an **expert system** for use in a particular domain of knowledge. Machine learning is appropriate in settings where prospective users of a system are unable to provide precise specifications for its behavior but where examples of the behavior are available. Examples include **optical character recognition (OCR)**, handwriting recognition, **speech recognition**, and steering an automobile. People who perform these tasks quite easily usually cannot explain exactly *how* they do them in sufficient detail to model them algorithmically. Machine learning methods have been used to assess credit risk, to filter news articles or **electronic mail (e-mail)** spam, to refine **information retrieval** queries, to predict user behavior in browsing the **World Wide Web**, and to find interesting patterns in **databases**, a task called **data mining**.

Additional reading: *Encyclopedia of Computer Science*, pp. 1056–1059.

Machine-readable form   DEFINITIONS

Since **data** in *machine-readable form* is information that can be read by a machine, more particularly by a computing machine, the definition of the term keeps expanding with the invention of each new device that can "read" something. When the term arose in the 1950s, the machine-readable media were **punched cards**, **punched paper tape**, and prerecorded **magnetic tape**, certainly not printed documents and music recordings. But now, with the development of online compact disc drives and the combination of **optical scanners** and **software** capable of **optical character recognition (OCR)**, typed or typeset documents and musical CDs are certainly in machine-readable form, and the term is falling into disuse.

Machine translation   APPLICATIONS

Machine translation (MT) is the use of a computer **program** whose input is a document written in one **natural language** and whose output is its translation into another. The American Warren Weaver was the first to mention the possibility of using a computer to translate, both in a letter to Norbert Wiener (*see* **Cybernetics**) of March 1947 and in a conversation with Andrew Booth, a British X-ray crystallographer who was visiting the United States. Booth's memorandum entitled "Translation," written in July 1949 at Carlsbad, New Mexico, is considered the principal milestone in the field. Some early work with **punched card** machines was published in England by Booth and Richard Richens, but not until 1955. The earliest work in the United States was done by Harry Huskey at the University of California at Los Angeles (UCLA) and Anthony Oettinger at Harvard.

In 1954 **IBM** and Georgetown University sponsored the first public demonstration of the feasibility of MT. The earliest systems consisted primarily of large bilingual dictionaries where entries for words of the source language gave one or more equivalents in the target language and some rules for producing the correct word order in the output. The quality of the output was disappointing. Early workers in **artificial intelligence (AI)** thought that 20 years or so would bring great progress, but the task has proved exceedingly difficult. Relative to MT, translating (compiling) **high-level language** statements into (**machine language**) **object code** is straightforward because the source language is context free—each word or phrase means the same thing regardless of the surrounding **characters**. But a natural language like English is context sensitive, and hence notoriously ambiguous, as is exemplified by the truck-stop sign "Eat here and get gas."

In 1964 in the United States, the *Automatic Language Processing Advisory Committee* (ALPAC) was formed, and in a famous 1966 report the Committee concluded that MT was slower, less accurate, and twice as expensive as human translation. The report was so damaging that it brought MT research in the United States to a virtual standstill, and it also impacted efforts in the Soviet Union and in Europe. MT research did continue, however, in Canada, France, and Germany. Within a few years, two reasonably successful projects were completed. Peter Toma, one of the members of the Georgetown University project, developed *Systran* for operational use by the United States Air Force (USAF) and National Aeronautics and Space Administration (NASA), and shortly afterward *Systran* was installed by the Commission of the European Communities for translating from English into French and later between other Community languages. At about the same time, the METEO system for translating weather reports was developed at Montreal University. Perhaps the greatest current use of MT is by the **search engine** *Google*, which, at user request, will translate a Web site expressed in French, German, Spanish, or Italian into reasonably useful English. The result is never great literature, but one can usually discern the intent.

Additional reading: *Encyclopedia of Computer Science*, pp. 1059–1066.

Machines Bull INDUSTRY

In 1922, the Norwegian Fredrik Bull formed a company to make **punched card** machines. Through merger, the firm became *Egli-Bull* in 1929. This company, in turn, merged with the French group *Caillies* to form the *Compagnie des Machines Bull* S.A. in 1932. The first Machines Bull **digital computers**, the *Gamma 3* and the *Gamma 60*, were built in 1958 and 1960, respectively. But in 1964, financial difficulties forced Bull to sell a controlling two-thirds interest to **General Electric (GE)** coincident with GE's acquisition of the computer division of **Olivetti**. The De Gaulle government then tried to keep France competitive (to **IBM**, principally) by forming a new company, the *Compagnie International de l'Informatique* (Cii). Cii then began producing minicomputers based on licenses from Max Palevsky's *Scientific Data Systems*. In 1970, GE withdrew from the computer business by selling its computer division to **Honeywell**, which became *Honeywell-Bull*. In 1975, the French government gained control of Honeywell-Bull through the formation of *Cii-Honeywell-Bull* (Cii-HB). In 1981, François Mitterrand's government nationalized Cii-HB and returned it to its original name, Compagnie des Machines Bull. In 1989, Bull acquired the **personal computer** division of Zenith and in 1993, a 20% interest in the PC line of **Packard Bell**. By the late 1990s, Bull was marketing under its own label a **mainframe** computer produced by **Nippon Electric Corporation (NEC)**.

Additional reading: Chandler, *Inventing the Electronic Century*, pp. 181–183.

Macintosh. *See* Apple Macintosh.

Macro instruction PROGRAMMING

A *macro instruction* (or just *macro*) is an **assembly language** or **higher-level language** statement that generates an *open subroutine*, a sequence of **instructions** or **statements** that is copied into a main **program**'s flow of control wherever needed. Even the earliest **assemblers** had at least a rudimentary macro facility, but their use with **compilers** was first proposed by Douglas McIlroy in 1960. A macro is defined through a **declaration** and invoked by placing its name and any associated actual **parameters** in the flow of program control as often as desired. Each invocation written causes the assembler or compiler being used to *expand* the macro into a customized copy of the instructional sequence specified in its declaration. Macros are a programming convenience; their expansion does not usually account for a large percentage of **object code**. But in one interesting application, a main program could consist of nothing but macro declarations followed by a sequence of macro invocations, each of which corresponds to a foreign **instruction set**, that is, the instruction set of an otherwise incompatible *target computer*. In such a case, the collection of macro declarations would comprise a simulator for the target computer, which is then programmed by writing any desired sequence of macro invocations (*see* **Simulation**). Programs written with such a simulator would run faster than one that consists of closed subroutines (because of the absence of jumps to and out of the subroutines) but would obviously require much more **memory**.

Additional reading: *Encyclopedia of Computer Science*, pp. 1066–1068.

MADDIDA COMPUTERS

MADDIDA (MAgnetic Drum DIgital Differential Analyzer), now generally rendered as *Maddida*, was built at Northrop Aircraft in 1949. The project began with the intention of producing the first digital **differential analyzer**, a DIDA, in contrast to the specialized **analog computer** that Vannevar Bush had invented. But when the decisiion was made to use a **magnetic drum** as **memory**, "MAD" was prefixed to "DIDA," and affectionately pronounced "Mad Ida." The director of the project, Glenn

Hagen, was later a cofounder of Logistics Research, Inc., which produced the **Alwac III-E** in the United States. But the conceptual leader of the Maddida design group was Floyd Steele, who had been inspired not only by the work of Bush but also by that of Lord Kelvin (*see* **Kelvin tide machine**). Steele hired Donald Eckdahl and Hrant (Harold) Sarkissian, two of the coauthors of the reference, and Richard Sprague to work on magnetic recording and on the Maddida's germanium diode **logic circuits**. In contrast to **ENIAC** and **UNIVAC**, which used electrical pulses to represent **bits**, Maddida was the first computer to use voltage levels, and was the first computer whose entire logic was specified in **Boolean algebra**, either of which makes development of Maddida a significant milestone. The original Maddida is now part of the collection of the Computer History Museum in Mountain View, California.

Maddida was essentially a special-purpose **digital computer** for solving a system of ordinary differential equations. By the end of 1952, six had been delivered and installed, but only four UNIVACs and one IBM 701 (*see* **IBM 700 series**). Gradually, it was realized that any reasonably powerful general-purpose digital computer could be used to behave as a differential analyzer through use of an appropriate **problem-oriented language (POL)** such as **Dynamo**. Sensing this, Steele, Eckdahl, Sarkissian, Sprague, and Irving S. Reed left Northrop to form the Computer Research Corporation (CRC) on 16 July 1950, a year after the first Madidda was demonstrated, in order to produce general-purpose digital computers. After developing the *Cadac*, which would have been called a **minicomputer** had the term been extant, CRC was sold to **National Cash Register (NCR)** in February 1953, thus launching NCR into the thick of the digital computer business.

Additional reading: Donald E. Eckdahl, Irving S. Reed, and Hrant H. (Harold) Sarkissian, "West Coast Contribution to the Development of the General-Purpose Computer: Building Maddida and the Founding of Computer Research Corporation," *IEEE Annals of the History of Computing*, **25**, *1* (January–March 2003), pp. 4–33.

Magnetic core MEMORY

Magnetic core **memory** consists of a two-dimensional **array** of tiny ferrite tori ("cores") interconnected with setting and sensing wires that enable storage and readout of one **bit** per torus. Cores are usually said to be "doughnut-shaped," but "washer-shaped" would be more accurate. The physical basis of magnetic core operation is the law of *hysteresis* discovered by Charles Proteus Steinmetz of **General Electric (GE)** in 1891. In 1947, Frederick Viehe, an inspector of streets and sidewalks for the Los Angeles, California, Department of Public Works patented a magnetic core memory that he developed in his home laboratory. **IBM** eventually purchased rights to his patent for a substantial sum in 1956, four years before he died. The Viehe patent was apparently not known to, or not considered pertinent by, three other inventors, all of whom independently invented a magnetic core memory in 1949: Chinese-American An Wang of Wang Laboratories, Polish American Jan Rajchman of the **Radio Corporation of America (RCA)**, and American Jay Wright Forrester of MIT. Rajchman made cores by compressing ferrite material with a converted aspirin press. The Forrester magnetic core memory, based on research by Dudley Buck and William Papian, was the first actually installed in a **digital computer**, the **Whirlwind**, in 1953. IBM's magnetic core effort was led by Michael Haynes, who had begun experimentation with magnetic core logic circuits while a student at the University of Illinois. Kenneth Olsen, founder of the **Digital Equipment Corporation (DEC)**, and Richard Best invented an improved form of magnetic core memory in 1960 and received a patent for it in 1964, the first awarded to DEC. The company built its own magnetic core manufacturing business and by the mid-1970s was producing 30 billion magnetic cores per year. Early magnetic core memories could not be entirely mass-produced; women at IBM using stereo microscopes worked hour after hour stringing exceedingly thin wires through individual cores by hand. Magnetic core remained the technology of choice for **digital computer** main memory from about 1955 to 1972 when it was displaced by **solid state memory**. *See also* **DRAM**.

Additional reading: Williams, *A History of Computing Technology*, pp. 322–323.

Figure 26. Jay Wright Forrester inspecting a plane of magnetic core memory for use with the Whirlwind in 1953. (MIT Museum)

Magnetic disk. *See* **Floppy disk; Hard disk**.

Magnetic drum   MEMORY

A *magnetic drum* is a **memory** device consisting of a rotating cylinder whose surface is coated with magnetizable material. In 1939, John Vincent Atanasoff used a drum memory for his **ABC**, but its surface storage elements were capacitors rather than magnetized spots. Perry Crawford described the magnetic drum concept in his 1942 master's thesis at MIT but did not build one, although he successfully advocated their use for **Whirlwind**. In 1946, engineers at **Engineering Research Associates (ERA)** used parts of captured German Magnetophones to build magnetic drums and disks for cryptographic devices for the U.S. Navy. The

magnetic drum designed by ERA was used in several **digital computers** built in the United States over the next few years. Heinz Billing of West Germany made a magnetic drum in 1947 by wrapping strips of magnetic tape around the drum, but later drums were surface coated. Andrew Donald Booth in England made a magnetic drum in 1948 that was just two inches long and two inches in diameter and capable of holding only 10 **bits** per inch. Eventually, drum technology reached the point where a drum can hold several megabytes. **Data** is recorded circumferentially in parallel tracks. As the drum rotates at high speed, the data can be read or written only as it passes under a read/write head. There may be only one such head per track, but often two to eight per track are arrayed around the circumference so as to reduce access time to a few milliseconds. Magnetic drums were once used as main memory for computers such as the **IBM 650** and **Alwac III-E**, but once **random access memory (RAM)** became predominant, drums were relegated to use as **auxiliary storage** and are seldom used today.

Additional reading: Bashe et al., *IBM's Early Computers*, pp. 90–91, 169–170.

Magnetic Ink Character Recognition (MICR)   APPLICATIONS

Magnetic Ink Character Recognition (MICR), in contrast to **optical character recognition (OCR)**, is a restricted form of **data recognition** in which the **characters** to be identified must be members of a stylized typefont that is recognizable by sensing the patterns defined by the magnetically impregnated ink used to print them. MICR is most commonly used to encode paper checks and other documents used in banking transactions. MICR characters can also be read with OCR devices, but the reverse, of course, is not true. The basic MICR patent, numbered 3,000,000 by the U.S. Patent Office because of its perceived significance, was granted to Kenneth R. Eldredge on 12 September 1961. But it is Alfred Zipf of the Bank of America who is called "the father of MICR" because of the encouragement he gave to those who perfected it for use with **General Electric**'s ERMA computer: Jerre Noe, Bill Kautz, and Phil Merritt.

Additional reading: *Computing Encyclopedia*, vol. 3, p. 9.

Magnetic tape   I/O DEVICES AND MEDIA

In 1900 Danish engineer Valdemar Poulsen invented a way to record sound on steel wire and called his device the Telegraphone. Poulsen had also invented an arc light wireless **telegraph** system. Steel wire or tape was used until 1935 when quarter-inch cellulose acetate plastic tape coated with iron oxide began to be used in Germany in a device called the Magnetophone. In the United States, Marvin Camras argued forcefully for the use of *magnetic tape* rather than magnetic wire. Camras was credited with over 400 patents related to continually improved means of magnetic recording. In the late 1940s, Charles Ginsburg led the research team at Ampex Corporation that developed the first practical videotape recorder. **Engineering Research Associates (ERA)** used plastic tape **auxiliary storage** for its **BINAC** but metallic tape for their *Univservo* tape drives on the **UNIVAC**. After the mid-1950s, magnetic tape was invariably plastic mylar tape. Eight-track audio magnetic tape was invented by William Lear in 1965. Through the late 1980s, **minicomputer** and **mainframe** configurations typically included at least four magnetic tape drives to facilitate **sorting** data, but once other mass storage devices in the multi-megabyte range became common, most installations kept at most one tape drive, principally for portability and archival backup of **data**.

Additional reading: Bashe et al., *IBM's Early Computers*, pp. 187–230.

Mainframe   COMPUTER TYPES

Originally, the *mainframe* of a **digital computer** system was the cabinet that housed its **central processing unit (CPU)** and **memory**, as distinguished from the cabinets that housed *peripheral* devices (disks, printers, tape drives, etc.). Typically, it was the largest component in size and cost, but modern electronics has allowed great reductions in both. The central processor and main memory were housed to-

gether to increase processing speeds through proximity and to improve reliability through maintenance of a common environment. The term *mainframe* came from the use of a rack or *frame* as a device to hold electronics, and the frame holding the electronics that does the computing might reasonably be called the *main frame*. In this sense, even a **personal computer (PC)** has a main frame, the tower or case that holds its **microprocessor** and memory. But at some point after the term *minicomputer* came into vogue, *mainframe* as a single word came to be used to designate a computer that was more powerful than a minicomputer but less powerful than a **supercomputer**. Thus we now speak of a mainframe as a type of computer in the sequence **handheld computer**, personal computer, **workstation**, minicomputer, mainframe, and supercomputer. So the term *mainframe* is not quite a milestone because neither its original or current meaning can be precisely dated or attributed to a particular person.

Management Information System (MIS)   INFORMATION PROCESSING

A *Management Information System (MIS)* is an organized assembly of resources and procedures required to collect, process, and distribute **data** for use in decision making. The term subsumes **data processing** and **database** but is much more broad in that it includes promulgated procedures and business plans that, together with computerized **information retrieval**, help the leadership of an organization make sound decisions. A search of the literature reveals many descriptions of particular MI Systems but none that stakes a claim to originality. The term dates back to at least the 1950s and possibly before. There is an increasing tendency to drop the "Management" and write just IS. In those cases, the significance depends on what the meaning of IS is.

Additional reading: *Encyclopedia of Computer Science*, pp. 1070–1077.

Manchester Mark I   COMPUTERS

In 1947, Tom Kilburn, Fred Williams, and Geoff Tootill of Manchester University in England began to design a **digital computer** to use the **electrostatic memory** storage tube invented by Williams. A prototype with 32 32-bit **words** called the "Baby Machine" was built and, on 21 June 1948, ran a 52-minute factoring **program** to became the world's first operational **stored program computer**. Two research students were added to the team, D. B. G. Edwards and G. E. Thomas. Professors M. H. A. (Max) Newman and Alan Turing provided theoretical support. A full machine was built that ran its first program in early 1949, one to search for Mersenne primes (*see* **Prime numbers**). The machine, the *Manchester Mark I*, was also known as MADM (Manchester Automatic Digital Machine) and MUC (Manchester University Computer). Main **memory** consisted of 128 40-bit words of Williams tube storage, backed by **auxiliary storage** in the form of a 1,024-word **magnetic drum**. Mark I was the first computer with an **index register**, one called a *B-box* to distinguish it from the accumulator, or *A-box*. **Logic circuits** were made from war surplus thermionic "valves," which is what the British called **vacuum tubes**. Addition took 1.8 milliseconds. **Input / output (I/O)** was via a five-channel **punched paper tape** unit designed by Turing, Edwards, and Thomas. In late 1948, Sir Ben Lockspeiser, chief government scientist, sponsored a government contract with **Ferranti, Ltd**. to make a production version of Mark I. The first Ferranti Mark I was installed at Manchester University in February 1951. Eight more were sold, making the Ferranti Mark I the first commercially available **digital computer**. Manchester's involvement with Ferranti continued through the 1950s with the design of a Mark II computer whose 1957 production version was known as the *Ferranti Mercury* and with the **Atlas** of 1962.

Additional reading: Lavington, *Early British Computers*, pp. 36–43.

MANIAC
COMPUTERS

MANIAC (Mathematical Analyzer, Numerical Integrator, and Computer), a copy of the **IAS computer**, became operational at the U.S. Los Alamos National Laboratory in New Mexico on 15 March 1952. The project leader was Nicholas Metropolis. The machine was used by such luminaries as John von Neumann and Stanislaw Ulam for calculations based on the **Monte**

Figure 27. The Manchester Mark I. On 21 June 1948, the "Baby machine," the prototype of the Manchester Mark I, ran a 52-minute factoring program to became the world's first operational stored-program computer. The full Mark I, shown here, became operational a year later. (Reprinted with permission of the Department of Computer Science, The University of Manchester)

Carlo method, Enrico Fermi for problems in nuclear and particle physics, George Gamow for early work on the genetic code, and Edward Teller for classified work relative to nuclear weapons. *See also* **Computer chess**.

Additional reading: Metropolis, Howlett, and Rota, *A History of Computing in the Twentieth Century*, pp. 457–464.

Marchant calculator. *See* Baldwin / Odhner calculator.

Mark I (Harvard). *See* Harvard Mark I (ASCC).

Mark I (Manchester University). *See* Manchester Mark I.

Mark-sense test scoring. *See* Test-scoring machine.

Matrix
MATHEMATICS

A *matrix* is a two-dimensional **array** of numbers—integer, real, or complex—subject to the operations of *transposition* (rotation about a diagonal) and *inversion*. Since this meets the definition of **data structure**, the matrix is a milestone of **computer science** as well as of mathematics. Matrices of the same size may be added or subtracted, element by element. Under restrictive conditions, a pair of matrices A and B may be multiplied in accord with a certain **algorithm**, but the operation is not necessarily *commutative*; that is, $A \times B$ may differ from $B \times A$. Other operations commonly applied to a matrix are the calculation of its *determinant* and its *eigenvalues*. Many of the most effective algorithms of **numerical analysis** apply to numbers arranged in matrix form. Interestingly, the physicist Werner Heisenberg's version of quantum mechanics called *matrix mechanics* could not have been so named until Max Born pointed out to him that certain numeric structures he had used in his system were already known to mathematicians as *matrices*. The term *matrix*, Latin for "womb," was not coined until 1848 by J. J. Sylvester, although Leibniz had studied determinants as early as 1693, and Cramer's rule for solving systems of linear equations was invented in 1750. But the full flowering of matrix algebra awaited the work of Arthur Cayley in 1855.

Additional reading: Weisstein, *Concise Encyclopedia of Mathematics*, pp. 1143–1147.

Matsushita Electronics INDUSTRY

Matsushita Electronics was founded in Osaka, Japan, by Konosuke Matsushita in 1918 to make and sell electrical plugs and battery-powered bicycle lamps. The company now sells **television** sets and DVD players under the Panasonic brand name, Technics audio products, and televisions to **General Electric (GE)** and other companies for sale under their own names. The company's many subsidiaries employ nearly 45,000 people, and collective sales are about $60 billion annually.

Additional reading: Chandler, *Inventing the Electronic Century*, pp. 51–54.

Memex HISTORICAL DEVICES

Vannevar Bush, later the developer of the **differential analyzer**, first wrote of the device he called a *memex* in the early 1930s, but his now-famous article "As We May Think" was not published in *Atlantic Monthly* until 1945. The article is quoted again and again by those such as Douglas Engelbart and Ted Nelson who have acknowledged its influence on the eventual development of **personal computers** in general and **hypertext** in particular. In his 1945 article, Bush wrote that the memex is "a device in which an individual stores all his [*sic*] books, records, and communications, and which is mechanized so that it may be consulted with exceeding speed and flexibility." In his vision, described in current terminology, a memex would resemble a desk with **touch screen** display **monitors** and an **optical scanner**. Within it would be several billion **bytes** (gigabytes) of **memory** filled with textual and graphic information, all indexed according to a universal scheme equivalent to present-day hypertext. Bush considered the ability of the memex user to navigate through its enormous **database** as far more important than the yet-to-be-invented **hardware** needed to implement it. The one thing he did not envision was that instead of storing relevant information locally, it could be stored remotely as a **distributed system** on something like an **Internet**. Bush described the building of paths to connect related information thus:

When the user is building a trail, he names it, inserts the name in his **code** book, and taps it out on his **keyboard**. Before him are the two items to be joined, projected onto adjacent viewing positions. At the bottom of each there are a number of blank code spaces, and a **pointer** is set to indicate one of these on each item. The user taps a single key, and the items are permanently joined. . . . Thereafter, at any time, when one of these items is in view, the other can be instantly recalled merely by tapping a button below the corresponding code space. Moreover, when numerous items have been thus joined together to form a trail, they can be reviewed in turn, rapidly or slowly, by deflecting a lever like that used for turning the pages of a book. It is exactly as though the physical items had been gathered together from widely separated sources and bound together to form a new book.

Despite its vintage, there can be no more apt description of how to form a link between nodes in current hypertext systems such as is done in constructing pages for the **World Wide Web**. Though only a concept, memex ranks very high as a milestone of **computer science**.

Additional reading: Nyce and Kahn, *From Memex to Hypertext*.

Memory
HARDWARE

Memory is one of the two principal parts of a **digital computer**, the other being its **central processing unit (CPU)**. Less anthropomorphically, Charles Babbage called the memory of his **Analytical Engine** of 1834 its *store*, and British computer scientists tended to use that term, or *storage*, in connection with their early computers of the late 1940s and early 1950s, and some still do. Thus they spoke of a computer's main memory as its *local store* and supplemental memory as **auxiliary storage**. The earliest form of memory that has survived to this day consists of etchings on recently found ochre stones dating back 77,000 years. Once engraved, these stones were thereafter **read-only memory (ROM)**, whereas at least some of the memory of a computing device needs to be rewritable. Mechanical **calculators** needed only enough memory units (**registers**) to hold three or so **operands**, but **stored program computers** need enough main memory to hold a **program** of reasonable size. Early main memory was either **ultrasonic memory**, **electrostatic memory**, or **magnetic drum** memory. Second-generation computers used **magnetic core** memory, and now main memory consists of various forms of solid state components. Magnetic drums and **magnetic tape** drives were used as early forms of auxiliary storage, but the major current forms of such storage are **floppy disks** (diskettes), **hard disks**, and various kinds of **optical storage** media, especially CD-ROM and DVD-ROM and newer forms of compact disc and DVD storage that are rewritable. Some early computers held **data** in decimal form—the last large decimal computer was **Sperry Rand**'s **LARC**—but now computers invariably use the **binary number system**. A digit of a binary number is called a **bit**, and eight bits comprise a **byte**. The capacity of main memory is typically smaller than that of an auxiliary memory, perhaps 128 to 256 megabytes (millions of bytes) of **random access memory (RAM)** versus 20 to 80 gigabytes (billions of bytes) of hard disk on **personal computers**, but main memories have faster access times. Main memory is usually supplemented by a relatively small number of even faster access storage units called *registers*, some of which serve as **index registers**. Small specialized storage might be **bubble memory** or **flash memory**, which are nonvolatile—their content persists even after power is withdrawn. Auxiliary memory in general is nonvolatile, as are some types of main memory, but the latter is often volatile. Volatility of main memory is of no concern to computer users because it is routine to *back up* valuable data and **programs** by storing copies of them on removable auxiliary storage media.

Additional reading: *Encyclopedia of Computer Science*, pp. 1130–1144.

Mercury delay line. *See* **Ultrasonic memory**.

Metcalfe's law
GENERAL

Metcalfe's law is the claim that the value of a **network** increases as the square of the number of its interconnections. The observation follows quite naturally from noting that in a network of N nodes there are $1/2N(N-1)$ possible 2-way interconnections, and thus if all such interconnections are equally valuable when N is large, overall value increases as N^2. The law was named (by George Gilder) after network pioneer Robert M. Metcalfe, who proposed the **Ethernet** in his 1973 Ph.D. dissertation at Harvard University, continued his research at **Xerox PARC** and Stanford University, and co-founded the **3Com** Corporation in 1979.

Additional reading: *Encyclopedia of Computer Science*, pp. 960–963.

MICR. *See* **Magnetic Ink Character Recognition (MICR)**.

Microchip. *See* **Microprocessor**.

Microcomputer
DEFINITIONS

A *microcomputer* is a small but complete **digital computer** consisting of a **microprocessor**, **memory**, and **input/output (I/O)** devices. Rendered as "micro computer," the term was first used by Intel in 1972 to describe its MCS-4 chip set. As one word, the term has become virtually synonymous with **personal computer (PC)**.

Microinstruction

COMPUTER ARCHITECTURE

A *microinstruction* is a very low level **instruction** stored in the **read-only memory (ROM)** of a microprogrammable **digital computer**. The instructions in the native **instruction set** of such a computer are then encoded as sequences of microinstructions rather than as hardwired **logic circuits**. When the microinstruction set is made accessible to the programmer, alternative higher-level instruction sets can be formed that allow **emulation** of otherwise incompatible computers. *See also* Microprogramming.

Microprocessor

COMPUTER TYPES

A *microprocessor* is the aggregate of all **integrated circuit (IC)** semiconductor chips needed to build the **central processing unit (CPU)** of a **digital computer**. The first microprocessor was the 4-bit **Intel 4004** designed by Ted Hoff, Federico Faggin, and Stan Mazor in 1970, and **Intel** followed with the 8-bit **Intel 8008** in 1972. The first single-chip microprocessor, the "computer on a chip," was patented by **Texas Instruments** in 1973.

Additional reading: Malone, *The Microprocessor: A Biography.*

Microprocessor chip. *See* Integrated Circuit (IC)

Microprogramming

PROGRAMMING METHODOLOGY

As Maurice Wilkes of Cambridge University started work on the **EDSAC** in the 1940s, he realized that the sequencing of control signals that implemented the computer's **instruction set** was similar to the sequencing actions required in a regular **program** and that one might use a stored program to represent the sequences of control signals rather than implementing them in hardwired **logic circuits**. In 1951, he published the first paper on this technique, which he called *microprogramming*. The **microinstructions** that he proposed be kept in a *writable control store* had a simple format: the unencoded control signals were stored with a next-address field. Initial selection of the appropriate microprogram was handled by using the **operation code (opcode)** value appended with

zeros as a starting address in the control store, and normal sequencing used the contents of the next-address fields thereafter. Conditional transfers were handled by allowing conditions to modify **bits** in the next-address fields. In a 1953 publication that followed, Wilkes expanded on the idea in collaboration with his colleague, J. B. Stringer. The Cambridge University group went on to implement the first microprogrammed computer, the EDSAC 2, in 1957 using an 8 × 6 **magnetic core** matrix.

In the late 1950s, microprogramming was not yet a mainstream technology, but John Fairclough at **IBM**'s laboratory in Hursley, England, led a development effort that explored use of a **read-only memory (ROM)** for the control unit of a small computer. Fairclough's experience with microprogramming played a key role in IBM's 1961 decision to pursue a full range of compatible computers to comprise its **IBM 360 series**. All but two of the initial 360 models, the high-end models 75 and 95, were microprogrammable, and this approach not only allowed the same program to be executed on most computers in the Series, but it also allowed use of substitute ROMs to emulate the instruction set of older IBM computers so as to run legacy programs through **emulation**.

Additional reading: *Encyclopedia of Computer Science*, pp. 1169–1170.

Microsoft

INDUSTRY

Microsoft, originally Micro-Soft, was founded by Bill Gates and Paul Allen on 4 April 1975. The pair had achieved prior success by making a very compact **interpreter** for **Basic** for the **MTS Altair 8800** and other **Intel 8080** and **Zilog Z-80** computers that followed. In 1980, Microsoft scored its first major success by landing an **IBM** contract to produce PC-DOS for the **IBM PC** and marketed it to many other vendors of **personal computers** as **MS-DOS**. In 1985, Microsoft moved its offices from Seattle to Redmond, Washington, the year it introduced its *Microsoft Excel* **spreadsheet** and *Microsoft Word* **word processing** program. In 1986, Microsoft amassed substantial capital through its initial public stock offering. By 1990, Microsoft so dominated the market for personal computer **software** that it was sued three times by the U.S. Department of Justice (DoJ) for restraint of trade. The third suit centered on whether it was proper to bundle the Microsoft *Internet Explorer* browser with **Microsoft Windows** at no extra charge. At the time this was done, *Netscape Navigator* had captured two-thirds of the market for **browsers**, and that share subsequently fell far below that of *Explorer* even though both products were and are **freeware** (*see* **Netscape Communications**). Navigator became an **open source software** product in 2000, but its share of the market relative to *Explorer* has fallen dramatically. By 2002, *bundling* rather than **unbundling** had become the issue, and in January 2003, a federal judge ruled that Microsoft must henceforth package **Sun Microsystems'** version of **Java** along with **Windows XP**.

Additional reading: *Computing Encyclopedia*, vol. 3, pp. 37–39.

Microsoft Windows   OPERATING SYSTEMS

Microsoft Windows is an *operating system* based on the concept of a **graphical user interface (GUI)**. **Microsoft** released what might now be called Windows 1.0 in November 1985 in response to the widespread approval of the GUI used on the **Apple Macintosh**. Response to the first two versions of Windows was lukewarm, but sales took off with the introduction of Windows 3.1 in April 1992 and Windows NT in 1993, developed by a team led by Microsoft CTO (chief technology officer) Nathan Myhrvold and intended for use on **local area networks (LANs)**. Windows NT and Windows 95, 98, 2000, and XP in 2001 have served to bring Microsoft more than a 90% share of the market for operating systems for **IBM PC**–compatible computers, the only alternatives being BeOS and the Free BSD and **Linux** versions of **Unix**.

Additional reading: *Computing Encyclopedia*, vol. 4, pp. 201–203.

Minicomputer   COMPUTER TYPES

A *minicomputer* (or just *mini*) is a computer whose functionality and physical size is smaller than that of a **mainframe** but larger than that of a **workstation**. Except for the fact that there were no workstations in the era, the first computers that met that definition were the **Alwac III-E** and the **Librascope LP-30** of the 1950s. But the term *minicomputer* was not coined until the head of the **Digital Equipment Corporation**'s operations in England, John Leng, sent back to the United States a sales report in 1968 that started: "Here is the latest minicomputer activity in the land of miniskirts." The first computer officially called a minicomputer was built by InterData, but the first highly successful one was the DEC PDP-8 (*see* **DEC PDP series**). Because of the steadily increasing power of **personal computers (PCs)** and workstations, categories that are themselves beginning to merge, the term minicomputer is falling into disuse, although a very successful line of minis is still being produced by IBM (*see* **IBM AS/400**).

MISTIC   COMPUTERS

MISTIC (MIchigan STate Instructional Computer), a copy of **ILLIAC** and hence of the **IAS computer**, was declared operational at Michigan State University (MSU) in East Lansing, Michigan, in November 1957. Richard James Reid, who later became chairman of the Computer Science department at MSU, worked on the design and construction of MISTIC along with colleagues Julian Kateley, M. Glenn Keeney, and Jerry Weeg, all under the direction of professor Lawrence Von Tersch. Over the 18 months following MISTIC startup, Reid extended the design of MISTIC by replacing its **electrostatic memory** with **magnetic core** that quadrupled its **memory** capacity to 4,096 40-bit words.

Additional reading: http://www.msu.edu/unit/complab/mainframe.html

MITS Altair 8800   COMPUTERS

In 1974, Ed Roberts, the owner of a small company called *Micro Instrumentation and Telemetering Systems* (MITS), decided to build a small **digital computer** based on the recently developed **Intel 8080** circuit chip and sell it as a kit to electronics hobbyists for just $367. The resulting *MITS Altair 8800* **microprocessor** was designed by Roberts, William Yates, and Jim Bybee. The computer was featured on the cover of *Popular Electronics* magazine in January 1975, after which the company was flooded with orders. Although the Altair didn't have a **monitor** or a **keyboard** or any **software**, it was nonetheless the very first **personal computer** on the market. It had to be programmed in **machine language** using toggle switches. But when two Harvard students, Bill Gates and Paul Allen, called Roberts and offered to develop an **interpreter** for **Basic** for the Altair, he agreed, and the product was delivered a mere six weeks later. Roberts sold MITS in 1977, and the company foundered two years later. Allen and Gates went on to found **Microsoft**.

Additional reading: Freiberger and Swaine, *Fire in the Valley*, pp. 41–60.

Mitsubishi INDUSTRY

Mitsubishi, which means "three diamonds," was founded as a shipping firm by Yataro Iwasaki in 1870. After World War II, Mitsubishi moved strongly into the **semiconductor** and consumer electronics business. Mitsubishi is now a $35 billion sales conglomerate of many companies with hundreds of thousands of employees and operations in 34 countries. Its principal **computer**-related products are high-performance **monitors**, **touch screens**, and **liquid crystal displays (LCDs)**. Mitsubishi is also heavily engaged in support of electronic commerce and financial services.

Additional reading: *Computing Encyclopedia*, vol. 3, p. 44.

Mobile computing COMPUTER TYPES

Mobile computing is the use of a **handheld computer** equipped with a wireless connection to the **Internet** (*see* **Wireless connectivity**). The ultimate in mobile communications is a unified device that combines the features of a mobile handheld computer, a **cellular telephone**, and a **Global Positioning System (GPS)** device. Qualcomm introduced such a combination in 1998, but it proved too heavy to be practical. Lighter devices are now being marketed by Qualcomm, Palm, Kyocera, and **Motorola**. *See also* **Ubiquitous computing**.

Modem COMMUNICATIONS

A *modem* (modulator-demodulator) is an electronic device used to connect a **digital com**puter to an analog **telephone** line. The earliest modems were called *acoustic couplers* because they contained circular padded receptacles into which the headset of the telephone was inserted so that the coupler could "listen" to the signal. Such devices were inherently slow, transmitting data at a mere 100 *baud* (**bits** per second at low speed). In 1976, Dennis Hayes and Dale Heatherington invented the first all-electronic modem, laying out the prototype on Hayes's dining room table. Hayes founded a company whose first products were modem logic boards for **S-100 bus** computers and the **Apple II**. The company's most successful product was the Hayes Smartmodem introduced in June 1981. The latest modems can transmit data at 56,000 bits per second, and even faster when **data compression** is used. Modems are being slowly phased out in favor of **Ethernet** cards as the popularity of **fiber optics** cable TV connections and **Digital Subscriber Lines** (**DSLs**) increases.

Additional reading: *Computing Encyclopedia*, vol. 3, pp. 44–45.

Modula-2 LANGUAGES

After perfecting **Pascal**, Niklaus Wirth designed a modular **high-level language** called *Modula* but became dissatisfied with it until recast as *Modula-2* in 1978. Modula-2 was an early attempt to achieve a degree of **information hiding** that was not realized until the maturation of **object-oriented programming (OOP)** languages. Several good textbooks were written for Modula-2 that resulted in its use for several years in the **data structures** course in the typical undergraduate **computer science** curriculum, but that role has largely been preempted by **C++**. In conjunction with Jurg Gutknecht in 1986, Wirth went on to design *Oberon*, an object-oriented language that still has adherents in academic circles. Others made a Modula-3, but it is not widely used.

Additional reading: http://arjay.bc.ca/Modula-2/Text

Modular arithmetic NUMBER SYSTEMS

While still a college student, Carl Friedrich Gauss [1777–1855] outlined an alternative form of arithmetic that involves only the integers 0,

1, 2 . . . up to one less than some chosen *modulus*. When a modulus of 24 is chosen, counting in the resulting *modular arithmetic* proceeds such that 0 not only precedes 1, but it follows 23, just as a (military) clock cycles back to 0 at midnight. For this reason, modular arithmetic is sometimes called "clock arithmetic," and it is also called *remainder arithmetic* or *residue arithmetic* because in the system just used as an example, 53 divided by 24 is 5, not 2, because we care about the remainder, not the normal quotient. Following Gauss, this result is written $5 \equiv 53 \bmod 24$, which is called a *congruence* and read "5 is congruent to 53 modulo 24." In other words, both 5 and 53 have a remainder of 5 when divided by 24. Modular arithmetic is very important in **computer science** and is, in fact, necessarily an integral part of any **digital computer** having a finite **memory** and **registers** of finite length (which is all of them). For example, on a binary computer with 2^n addressable units of memory (**bytes** or **words**), the address that "follows" the largest address, $2^n - 1$ (all 1-**bits**) is 0. Modular arithmetic can be used to implement the **Euclidean algorithm** and together with that **algorithm** is very important in cryptography (*see* **RSA algorithm**). Many **high-level languages** have a *mod* **operator**; the **Pascal** mod operator is spelled out, so that the **expression** 53 mod 24 looks just like what was used in the example. The corresponding **C++** expression is 53 % 24; that is, the percent sign is used as a mod operator.

Additional reading: Flannery, *In Code*, pp. 113–142.

Molecular computer. *See* DNA computer.

Monitor I/O DEVICES AND MEDIA

A *monitor*, more particularly a *display monitor*, is a **cathode ray tube (CRT)** or *flat-panel display* on whose screen text, numbers, or graphic images can be displayed (*see* **Computer graphics**). Monitor screens are subdivided into thousands of tiny rectangular *pixels* (picture elements), a typical *raster* (two-dimensional array) being 1,024 x 768 or 1,280 x 1,024 pixels. The earliest monitors were either monochrome

or supported shades of gray, but now monitors are invariably color monitors whose pixels may be any one of millions of shades of color. The CRT was invented by Vladimir Zworykin in 1929 and the flat-panel display by James Fergason in 1979 (*see* **Liquid crystal display [LCD]**). Early **operating systems** were also called *monitors*, but that usage has faded, and *monitor* is also used to describe a construct in **concurrent programming**.

Additional reading: *Understanding Computers: Computer Images*, pp. 22–25.

Monroe calculator. *See* Baldwin / Odhner calculator.

Monte Carlo method ALGORITHMS

The partial differential equations that govern physical phenomena are derived from an analysis of the statistical behavior of elementary particles in a particular medium. When it is deemed impractical to solve these differential equations by the methods of conventional **numerical analysis**, resort can be made to a **simulation** based on *constrained randomness*, that is, following the scattering or absorption of millions of individual particles whose movement is random in some respect but nevertheless constrained by known probabilities. Such behavior is called stochastic, and because of its chance element, the computational technique described is called the *Monte Carlo method*. A simple example is the calculation of an approximation to **pi** as described in **Buffon's needle**. The use of **digital computers** to solve Monte Carlo problems was pioneered by John von Neumann and Stanislaw Ulam in the mid- to late 1940s. Monte Carlo calculations need a copious supply of numbers produced by **random number generation**.

Additional reading: *Encyclopedia of Computer Science*, pp. 1192–1193.

Moore's law GENERAL

In 1965, Gordon Moore noted that the number of components per chip seemed to increase at roughly a factor of two every year. In 1975, Moore revised his observation to state that the number of **transistors** on a chip doubles every two years, a prediction that has become known as *Moore's law*. Over the 30-year period from

Figure 28. Gordon Moore's law looks into the future and is stated in terms of the density of transistors per unit area of semiconductor chip as a measure of the degree of increasing microminiaturization of logic circuits. But this photograph shows that similar progress toward more compact circuitry was occurring in the **vacuum tube** era. From left to right, the women are holding comparable functional units from the **ENIAC** of 1946, the **EDVAC** of 1948, the **ORDVAC** of 1950, and the BRLESC (Ballistic Research Lab Electronic SuperComputer) of 1962. Note that the first three dates are two years apart, and the second and third units are each about half the size of their predecessor. The BRLESC unit of a decade after ORDVAC is incongruous; it is representative of the beginning of the transistor era. (Courtesy of the United States Army)

1971, when the **Intel 4004** was introduced, through the introduction of the **Intel** Iridium chip in 2001, the number of transistors on a single **microprocessor** chip did indeed double every two years (*see* **Intel 80×86**). How long this trend can continue is a matter of continuing speculation.

    Additional reading: *Encyclopedia of Computer Science*, pp. 960–963.

Morse code   CODING THEORY

In the years leading up to the filing of his **telegraph** patent in 1837, Samuel F. B. Morse needed to develop a way to relate the dots and dashes (*dits* and *dahs*) being transmitted to meaningful text. What he devised is called *Morse code*, but since he chose to encode character by character rather than word by word, his system would have been called a *cipher* by the cryptologist. The Morse code is actually a trinary (3-symbol) system, the third "symbol" being the space (time interval) that separates dots and dashes. According to the U.S. Army Technical Manual TM-11-459/TO 31-3-16 of September 1957, the following relationships exist between the elements of the code (dits and dahs), the characters (letters, digits, and three punctuation marks), and words:

The DIT is the basic unit of length.

The DAH is equal in length to three DITS.

The SPACE (time interval) between the DITS and DAHS within a character (letter) is equal to the time to transmit one DIT.

The time interval between characters (letters) in a word is equal to that of three DITS.

The time interval between words is equal to seven DITS.

It is interesting that Morse perceptively assigned the shortest code sequences, a single dot or dash, to the letters most frequently used in English, *E* and *T*, respectively. Similarly, the four combinations of a dot and a dash were assigned to the next four most frequently used letters, and so on. Legend has it that he obtained approximate letter frequencies by sending his friend and assistant Albert Vail to a newspaper to inspect the number of instances of each letter in a typesetter's tray. Thus Morse anticipated the general principle of **Huffman encoding** of a hundred years later. The current International Morse Code is a slightly modified form of Morse's original code, now called American Morse. The famous SOS distress signal is . . . _ _ _ . . . only in International Morse. SOS was not adopted to replace the previously used CQD until 1908. In 1912, the *Titanic* in its desperation alternately transmitted both CQD and SOS. The *Carpathia*, 58 miles away, responded. The *Californian*, 15 miles away, did not.

Additional reading: Petzold, *CODE*, pp. 1–14, 40–46.

MOS 6502 COMPUTERS

In 1975, Chuck Peddle, the founder of *MOS Technology*, developed a **microprocessor** chip called the *MOS 6502*. The chip was fast, powerful for its time, and inexpensive—characteristics that appealed to Steven Wozniak, who used the 6502 in a "homebrewed" computer that proved attractive to Steve Jobs, his future partner in founding **Apple** (*see* **Homebrew Computer Club**). The 6502 was also used in many other **personal computers** of the period, including the **Apple II**, the **Atari**, and the **Commodore PET**.

Motorola INDUSTRY

The company now known as *Motorola* was founded in Chicago, Illinois, on 25 September 1928 by the brothers Paul and Joseph Galvin. The company they incorporated was called the *Galvin Manufacturing Corporation*, whose initial product was a "battery eliminator" that enabled home **radios** to use house current. In 1929, Paul Galvin coined the name "Motorola" for the company's first automobile radio, invented by William Lear, who much later marketed the Learjet airplane. In 1947 the name of the company was officially changed to *Motorola*. By the early 1950s, Motorola had become one of the world's largest manufacturers of **semiconductor** products. In 1955, a germanium **transistor** for car radios became Motorola's first mass-produced **semiconductor** and the world's first high-power transistor in commercial production. In 1964, Motorola marketed the transistorized *Pageboy* pager, the first personal communications device to become a standard business tool. The company's first **microprocessor**, the 8-bit **data path** 6800, was introduced in 1974, followed by the 16-bit data path 68000 in 1979 (*see* **Motorola 68 × × × series**). Motorola's MicroTAC personal **cellular telephone**, the smallest and lightest then on the market, was introduced in 1989. In 1991, the company's one-millionth pager was supplied to Nippon Telegraph and Telephone in Japan, and a technology alliance among **Apple**, **IBM**, and Motorola began development of the *PowerPC* family of microprocessors. In 1999, the i1000plus handset using Motorola's iDEN digital communications technology was the first to integrate a digital phone, two-way radio, alphanumeric pager, **wireless connectivity** to the **Internet**, **electronic mail**, **facsimile transmission (fax)**, and two-way messaging capability. By 2000, Motorola had net sales of $37.6 billion and 147,000 employees and merged with General Instrument Corporation to provide integrated video, voice, and data networking for Internet and high-speed data services. In 2001, Motorola shipped its five billionth microcontroller, its 18 millionth digital TV set-top terminal, and its five millionth cable **modem**. Motorola and **Texas Instruments** now control almost the entire market for **digital signal processors (DSPs)**.

Additional reading: *Computing Encyclopedia*, vol. 3, pp. 53–54.

Motorola 68××× series COMPUTERS

The *Motorola 68××× series* began in 1974 with the 8-bit **data path** Motorola 6800, the first member of **Motorola**'s family of 16- and 32-bit **microprocessors**. The 68000 was introduced in 1979. The 68000 had 32-bit **registers** but only a 16-bit **arithmetic-logic unit (ALU)** and external data **bus**. It had 24-bit addressing and a linear address space, with no need for the **memory** segment registers of contemporary **Intel** processors that made programming them in **assembly language** so difficult. That meant that a single directly accessed **array** could be larger than 64KB (kilobytes) in size. The 68000 had 16 32-bit registers, eight data registers, and eight address registers. One address register is reserved for the stack **pointer**. Any register, of either type, can be used for any function except direct addressing. Only address registers can be used as the source of an address, but data registers can provide the offset from an address. Like the Zilog Z8000 (*see* **Zilog Z-80**), the 68000 featured a supervisor and user mode, each with its own stack pointer. The Zilog Z8000 and 68000 had similar capabilities, but the 68000 was faster because of its 32-bit addressing that obviated memory segmentation. The 68000 also had a 2-stage **instruction lookahead** pipeline. It was used in many **workstations**, notably early machines of **Sun Microsystems** and **personal computers** such as the **Commodore Amiga** and the first **Apple Macintosh**. The 68000 was used in the **Apple Lisa**, and Andrew Morton used it in the design of his Applix 1616 **multitasking** computer of 1985. Other members of the 680×0 series, introduced at two- to three-year intervals, included the 68010, 68020, 68030, 68040, and 68060. The 68020 was the first true 32-bit personal computer in that it had both a 32-bit ALU and a 32-bit data path.

Additional reading: *Computing Encyclopedia*, vol. 3, p. 53.

Mouse I/O DEVICES AND MEDIA

A *mouse* is a small handheld interactive input device that, when rolled over a flat surface, controls placement of the **cursor** on a **monitor**. The oval-shaped palm-sized device is usually connected to the computer by a wire that is suggestive of a tail, hence the affectionate name "mouse." The mouse (or a similar device invented earlier called a **trackball**) was essential to the development of the **Graphical User Interface (GUI)** and the **browser** used to navigate the **World Wide Web**. The mouse was invented by Douglas Engelbart at the Stanford Research Institute in 1965, but its technology has changed considerably since then. The Englebart mouse, which was so complex that six months' training was considered necessary to master its use, was marketed briefly and unsuccessfully at a price of $400 for a mouse and $300 for its interface. Steve Jobs of **Apple**, after seeing the Engelbart mouse used with the **Alto** at **Xerox PARC** in 1979, hired a Design Group founded by Dean Hovey and David Kelley to develop a far smaller and less costly mouse for use on the **Apple II** and the **Apple Macintosh**. The Hovey-Kelley team of Jim Sachs, Rickson Sun, Jim Yurchenco, and Douglas Dayton came up with a prototype within ten days, one made of a Ban deodorant ball mounted in a butter dish. The final product, the mouse as we know it today, was delivered in March 1981.

The most common mouse is electromechanical. As the mouse is moved over a flat surface, the rolling motion of a rubber-coated steel ball that protrudes from its bottom is detected by two orthogonal rollers that touch the surface of the ball. These rollers act as transducers that convert the speed and direction of the rolling ball to electrical signals that are fed to a **software** driver that moves the screen cursor accordingly. To provide good traction, an electromechanical mouse is generally used with a flat soft-cushioned *mouse pad*. The first such pad was "invented" by Jack Kelley, who went on to become a noted designer of furniture. In late 1999, **Microsoft** introduced an electro-optical mouse, the *IntelliMouse Explorer*, that has no moving parts and may be used on any surface other than glass; no mouse pad is needed. In 2001, Logitech introduced its MouseMan wireless optical mouse.

Additional reading: Alex Soojung-Kim Pang, "The Making of the Mouse," *American Heritage of Invention & Technology*, **17**, *3* (2002), pp. 48–54.

MP3   ALGORITHMS

MP3 (Moving Picture Experts Group Audio Layer 3) is a **data compression** algorithm format that shrinks audio **files** to about one-tenth their size with only a small degradation from CD sound quality. A 4MB (megabyte) MP3 file contains about three minutes of music. MP3 has made the downloading of music files from the **Internet** extremely popular, to the point where the Recording Industry Association of America took legal action to shut down Shawn Fanning's *Napster* Website that facilitated distribution of copyrighted music. One of the legal problems with Napster was that it was essentially a **client-server computing** network; files were moved from PC (**personal computer**) to PC via a central **server**. To obviate this, the Gnutella (GNU-Tella) **protocol** was developed by Justin Frankel and Tom Pepper as **open source software**. Gnutella is a so-called peer-to-peer (P2P) protocol since it allows cooperating PCs to exchange files directly, each acting as a server while uploading and a client while downloading. The protocol was perfected by Gene Kan, who founded InfraSearch and then sold his company to **Sun Microsystems** in 2001, joining that company in the process. Then, in 2002, Kan took his own life at age 25.

The MP3 compression **algorithm** was invented in the mid-1980s at the Fraunhofer Institut in Erlangen, Germany, with the help of Dieter Seitzer, a professor at the University of Erlangen. In 1989, Fraunhofer was granted a German patent for MP3, and a few years later it was submitted to the International Standards Organization (ISO) and integrated into the MPEG-1 specification. In 1997 Tomislav Uzelac, a developer at Advanced Multimedia Products, created the AMP MP3 Playback Engine, the first portable MP3 player, and a very popular portable MP3 player called iPod is now available from **Apple**. Casio markets an MP3

player, a *wearable computer* worn like a wristwatch, and Bill Gates has hinted that **Microsoft** will soon offer a similar device. When the Napster Website was posted to the Internet in 1999, it allowed anyone with a connection to find and download just about any type of music they wanted, in minutes. By connecting users to other users' **hard disk** drives, Napster created a virtual community of music lovers that grew at an astonishing pace until shut down in late 2001. But users who own music CDs can convert their tracks to MP3 format for concise storage and later playback by using **software** called *RealAudio Jukebox*. MP3 can also be used to compress voice to an even greater degree than music. Each day, the voice of a *New York Times* narrator reading the paper is broadcast in MP3 format over the Internet.

Additional reading: *Computing Encyclopedia*, vol. 3, pp. 56, 66–67.

MS-DOS   OPERATING SYSTEMS

MS-DOS (MicroSoft Disk Operating System), initially written under contract to **IBM** as PC-DOS for the **IBM PC** of 1984, is a command line **operating system** that succeeded the **CP / M** used on most earlier **personal computers**. Bob O'Rear, **Microsoft** employee number 7, led the team that developed MS-DOS. It was originally booted from a **floppy disk**, but the product went through a long succession of versions of increasing capability until it was ultimately packaged with **Microsoft Windows** and distributed on a CD-ROM (*see* **Optical storage**). For the sake of legacy **programs** that run only under MS-DOS, **Microsoft** continued to do this up through Windows 2000 but stopped doing so as of Windows XP in 2001.

Additional reading: *Computing Encyclopedia*, vol. 2, pp. 39–40.

MTAC   GENERAL

The journal *Mathematical Tables and Other Aids to Computation* (MTAC), originally published by the National Research Council (NRC) of the U.S. National Academy of Sciences, was founded by Professor Raymond Clare Archibald in 1943. For quite some time it was the only journal dealing exclusively with computation and computing devices. The landmark article "The Electronic Numerical Integrator and Computer (**ENIAC**)" by Herman H. Goldstine and Adele Goldstine was published in the July 1946 issue. In 1960 the name of the journal was changed to *Mathematics of Computation*, and in 1963 the NRC transferred publication rights to the American Mathematical Society (AMS).

Multics OPERATING SYSTEMS

Multics (MULTiplexed Information and Computing Service) was the second of two historic **time sharing** systems developed in the 1960s by MIT **Project MAC**. Since the earlier **CTSS** was based on the IBM 7094 computer (*see* **IBM 700 series**), it was assumed by many, including **IBM**, that its successor would be written for the **IBM 360 series**, more particularly for the IBM 360/67, which allegedly had the enhancement necessary for **multiprogramming**. But to the great consternation of IBM, who also lost the competition for a similar system at Bell Labs, the contract was awarded to **General Electric (GE)** in 1965, and Multics was implemented for the **GE 600 Series**. Eventually, Multics was installed at over 100 sites, but not all on GE or successor **Honeywell** equipment.

Additional reading: Waldrop, *The Dream Machine*, pp. 243–244, 311–315, 425–426.

Multiple-address computer

COMPUTER TYPES

A *multiple-address computer* is a **stored program computer** whose **instruction set** consists of a *command* and two to four *operand addresses*. Early **digital computers** that used a **magnetic drum** as main **memory** often used a 4-address structure where, for example, ADD A B C D meant "Add the contents of address A to the contents of address B, store the result at address C, and jump to address D." Because of drum *latency* (rotational delay), the worst place to put a successor **instruction** was the next location after a prior one, so optimal programming of such computers consisted of creating physically separated instructions linked through their D address fields. The **EDVAC** was the earliest *4-address computer*. Upon the advent of **random access memory (RAM)**, the

D address became dispensable, reducing operand addresses to at most three. The **NORC** was a *3-address computer*. In a *2-address computer*, the instruction ADD A B means "Add the contents of the A address to the contents of the B address" (or the reverse, depending on the design). The **Univac 1103** was a 2-address computer, as are the much later **IBM 360 series** and the **Intel 80×86 series**. As a special case, a drum-based 2-address computer in which the second address merely specified the location of the next instruction is sometimes called a "one and a half address computer." *Zero-* and *one-address computers* are described in the article **computer architecture**. As befitting its status as the epitome of a **complex instruction set computer (CISC)**, the **DEC VAX series** included some instructions that took one address, some that took two, and some that applied to three.

Additional reading: *Encyclopedia of Computer Science*, pp. 435–436.

Multiplexing DEFINITIONS

Multiplexing is the shared, quasi-simultaneous use of a single communications channel by two or more I/O devices. The **bandwidth** (capacity) of the **channel** is shared in either time (*time-division multiplexing*—TDM) or frequency (*frequency-division multiplexing*—FDM).

Multiprocessing

COMPUTER ARCHITECTURE

Multiprocessing requires multiple processors of one kind or another. In one form, synchronous identical processors operate in parallel, each performing the same operation to dispersed data values belonging to the same program, and in another, possibly different processors cooperate to work on a single stream of data supplied by the same program (*see* **Parallel computer**). In the **Flynn taxonomy**, the former are called *Single Instruction Multiple Data* (SIMD) machines, and the latter are called *Multiple Instruction Single Data* (MISD) machines. The **ILLIAC IV** was possibly the earliest SIMD machine, and the **CDC 6600** was an MISD machine. There are also MIMD (*Multiple Instruction Multiple Data*) machines.

Additional reading: *Encyclopedia of Computer Science*, pp. 1205–1207.

Multiprogramming   OPERATING SYSTEMS

Multiprogramming is the interleaved execution of multiple **programs**, initiated by possibly different users, by a single processor. Contrast with the definitions of **multiprocessing** and **multitasking**. Implementation of multiprogramming is a prerequisite to **time sharing** and depends for its success on either the swapping of whole programs between main and **auxiliary storage** or the use of **virtual memory**. Multiprogramming was first implemented in 1962 for the **Atlas** at the University of Manchester by a team led by Tom Kilburn. The **LARC** was the first **supercomputer** capable of multiprogramming.

Additional reading: *Encyclopedia of Computer Science*, pp. 1207–1210.

Multitasking   DEFINITIONS

Multitasking is the concurrent performance or interleaved execution of two or more distinct tasks initiated by the same user. The same process applied to tasks initiated by different users is called **multiprogramming**, and occasionally the terms are used as if they were synonyms. Since a **personal computer (PC)** is ordinarily used by only one person at a time, multitasking came to the fore only when PCs evolved to the point where **memory** size and processor speed made it possible to create a multitasking **operating system**, about 1995. The most common tasks performed in (apparent) parallel are the printing of output from a prior task while the user goes on to initiate the next and possibly unrelated one.

Additional reading: *Computing Encyclopedia*, vol. 3, pp. 63–64.

Multithreading   DEFINITIONS

Multithreading is a form of **concurrent programming** in which different parts of the *same* program are executed in parallel. Contrast with **multitasking**, where the tasks running in parallel are distinct.

Additional reading: *Computing Encyclopedia*, vol. 3, p. 64.

MUMPS. *See* M (Mumps).

N

Nanoelectronics HARDWARE

In 1959, the Nobel physicist Richard Feynman predicted that it would one day be possible to synthesize any chemical substance atom by atom. Twenty years later, MIT engineer Eric Drexler advocated beginning work on *nanoelectronics*, building electronic logic elements from individual molecules. The prefix "nano" means "one-billionth." Light travels only about a foot in a nanosecond. Nanoelectronics implies the harnessing of electronic action at scales of a nanometer, one one-hundred-thousandth of the thickness of a human hair and about the size of the sugar molecule. An **integrated circuit (IC)** with a thousand carbon nanotubes acting like **transistors** was devised by Phaedon Avouris of **IBM** in early 2001. *Nanotubes*, discovered by Sumio Iijima of the **Nippon Electric Corporation (NEC)** in 1991, are cylinders of nanometer radius made of hexagonal arrangements of carbon atoms that, magnified, look like rolled-up chicken wire. Nanotubes can sustain current densities hundreds of times greater than that of common metals and can be made in either metallic or **semiconductor** form.

Conventional **microcomputer** chip manufacture is a "top-down" process whereby increasingly higher transistor densities are achieved by deposition of ever smaller clumps of molecules in a race with **Moore's law** that is destined to be lost. Nanoelectronics, in contrast, proceeds from the bottom up: First build individual molecular logic gates, then interconnect them to form a complete computer. The latter feat, when realized, will be a major **computer science** milestone, but the first step—construction of molecular logic gates—was announced in back-to-back papers in the 9 November 2001 issue of *Science* magazine. In the first, Yu Huang et al. describe how a group at Harvard used nanowire building blocks to configure AND, OR, and NOR logic gates and used them for simple computations. In the second, Adrian Bachtold et al. announced that a group at the Delft University of Technology in the Netherlands had created field-effect transistors (FETs) made of single carbon molecules called *nanotubes* and assembled them into inverters, NOR gates, and a static random-access memory cell. The first *nanocomputer* may be only a decade away.

Additional reading: Huang et al., "Logic Gates"; Bachtold et al., "Logic Circuits."

Napier's bones CALCULATORS

A few years before his invention of **logarithms** in 1614, John Napier, Eighth Lord of Merchison, built a device that he called *Rabdologia* but which his friends called *Napier's bones*. Napier did not publish a good description of his invention until 1617, the year of his death. The "bones" were strips of ivory inscribed with individual decimal digits that could be aligned to compute the partial products of multiplication.

The correspondence of the German mathematician Athanasius Kircher with his fellow Jesuit Gaspard Schott helped spread knowledge of the device from Europe to China in just a few years. In 1668 Schott developed improvements to Napier's bones in which numbers were arranged on boxed rotating cylinders, making setting of the rods ("bones") for multiplication a much faster process.

Additional reading: Aspray, *Computing before Computers*, pp. 16–23.

National Cash Register (NCR)

INDUSTRY

National Cash Register (NCR) was founded by John Patterson in 1884 in order to market the product for which the company is named. In 1906 Charles F. Kettering designed the first NCR cash register powered by an electric motor. Patterson adopted the motto "Think" that one of his managers, Thomas J. Watson, Sr., brought to the Computing-Tabulating-Recording Company (C-T-R, later **IBM**) in 1914. During World War II, NCR employed Navy WAVES to produce **Bombe** machines to British specifications, machines used to emulate German *Enigma* cipher machines. In 1952, NCR entered the **digital computer** business through purchase of the Computer Research Corporation (CRC), whose staff had made the **MADDIDA** and other special-purpose computers for aviation. In 1953, NCR and **General Electric (GE)** jointly produced the NCR 304, the first fully transistorized large-scale computer. In 1968, the NCR Century 100 and the CDC 7600 became the first computers built entirely with **integrated circuit** logic. In 1974, with cash registers no longer its principal product, the company changed its name to just NCR Corporation. By 1979, NCR was second only to IBM in revenue derived from computer products. In 1991, just as NCR was moving into **workstations** and acquiring the Teradata Corporation, the company suffered an unfriendly takeover by AT&T and was renamed Global Information Systems (GIS) in 1994. But by 1995, AT&T became dissatisfied with the result and divested GIS, which reassumed the name NCR. Over the last seven years, NCR has divested its computer manufacturing division but

acquired several specialized companies that support the company's new focus on full-service **electronic commerce**. The NCR marketing division is credited with introducing two widely used computer terms, *online*, and *tower*, the latter describing a **personal computer** whose case stands vertically rather than horizontally.

Additional reading: Chandler, *Inventing the Electronic Century*, pp. 99–100, 114–117, 227.

National Semiconductor

INDUSTRY

National Semiconductor (NS) was founded in Danbury, Connecticut, on 27 May 1959 by eight engineers who left **Sperry Rand** to do so. NS began as a manufacturer of **transistors** and then began to concentrate on analog to digital and digital to analog conversion products. In 1980, NS introduced the Digitalker™, the first **speech synthesis** system. In the same year, NS formed National Advanced Systems (NAS) to sell **IBM**-compatible **mainframes**, but sold it to Hitachi in 1989 for $386 million. In 1987 NS acquired **Fairchild Semiconductor** for $122 million from Schlumberger Ltd. and then sold it for $550 million in 1997 to investors who continued to operate under the Fairchild name. In the same year Cyrix was purchased in what turned out to be an unsuccessful attempt to compete with **Intel**. Cyrix was sold to VIA Technologies, Inc. in 2000. NS still markets a wide range of electronic and communications products and has retained its strength in analog to digital conversion devices.

Additional reading: *Computing Encyclopedia*, vol. 3, p. 73.

Natural language

LANGUAGE TYPES

A *natural language* is one spoken (and usually written) by a tribe, nation, or ethnic region. The first natural language was a milestone not for computing in particular but for everything human. But its origin dates back too many thousands of years to be accurately traced. Mandarin Chinese is spoken by about 1 billion of the earth's people, and English and Hindustan by about a half-billion each. Spanish is third, with about 400 million speakers, and fifth to tenth in order are Russian, Arabic, Bengali, Portuguese, Malay, and French, ranging downward from 280 to 130 million. In the enthusiasm of the earliest days of **artificial intelligence (AI)**, it was thought that it would be routine by now to be able to write programs that could translate written works in one natural language to another, but the goal has proved elusive (*see* **Machine translation**). Similarly, it was once thought that a language such as English could be made the basis for a **high-level language** superior to, say, **Algol 60** or **Pascal**, but the richness of English (and virtually any natural language) creates such ambiguity that the likelihood is that we will continue to rely on artificial languages of rigidly controlled syntax into the indefinite future. Although some progress is being made in **speech recognition** and **speech synthesis**, these remain research areas of applied artificial intelligence. *Natural language processing* has become of great interest to researchers in the humanities. Perhaps the most interesting early result was the 1964 computer analysis of Frederick Mosteller and David Wallace in which they compared the text of the various Federalists papers to works of known authorship to determine with high probability that James Madison had written 12 papers whose authorship had been disputed. The full text of many a classic novel is now available online, and the computer generation of concordances to the Bible, Shakespeare, etc. has become routine.

Additional reading: *Encyclopedia of Computer Science*, pp. 1218–1222.

NCR. *See* **National Cash Register (NCR)**.

NEC. *See* **Nippon Electric Corporation (NEC)**.

Netscape Communications   INDUSTRY

Netscape Communications was founded in 1994 by Jim Clark, who had earlier founded **Silicon Graphics, Inc. (SGI)**, and Marc Andreesen, who had created *Mosaic*, the first **browser** for the **World Wide Web**. The new company's first product was the *Netscape Navigator* browser. After introduction in October 1994 at a nominal cost, Navigator quickly captured 85% of the market but not long thereafter was forced to offer it free of charge to match what **Microsoft** was doing with its *Explorer* browser. Netscape's revenue stream was thus reduced to whatever it could glean from advertisements embedded in its browser, and its fortunes declined to the point where the company was sold to **America Online (AOL)** in 1999. Surprisingly, AOL has continued to use *Explorer* to support its popular dial-up service, but the Netscape browser continues to hold the loyalty of many users in the face of Microsoft's dominance in this area.

Additional reading: *Computing Encyclopedia*, vol. 3, pp. 77–79.

Network   GENERAL

Topologically, a network is a **graph**. Other than the specialized **neural net** of the following article, the term *network* or *net* in **information technology (IT)** implies use of interconnected terminals or **digital computers** as nodes that can at least receive information from other nodes, or from a central source, or more capably, nodes that can send as well as receive. An example of the former is a **television** network such as CNN (Cable News Network), and the best-known example of the latter is the **Internet**. A network all of whose nodes reside in the same room or building is called a **Local Area Network (LAN)**. One whose nodes span an area the size of a city is called a *Metropolitan Area Network* (MAN). One whose reach is even further is called a *Wide Area Network* (WAN). The first significant WAN was the **ARPAnet**, the forerunner of the Internet, which began operations in 1969. The ALOHAnet of a year later, which used the ALOHA **protocol** of Norman Abramson, covered the Hawaiian Islands and was the first to use packet **radio** for network communication via satellite. In 1981 Ira Fuchs, vice chancellor of the City University of New York, promoted the organization of BITnet (Because It's Time net), a U.S. national network of universities that used large **IBM** computers (**mainframes**). BITnet operated for 15 years until the growth of the Internet rendered it irrelevant.

Additional reading: *Computing Encyclopedia*, vol. 1, pp. 25, 74–75.

Neural net NETWORKS

An artificial *neural net* is a network of physical components arranged so as to simulate the biological neural nets in the brains of living creatures. The study of neural nets that have the possibility of "learning" from their environment is one of the two principal branches of **artificial intelligence (AI)**, the other being the attempt to achieve simulated intelligence through programmed **algorithms**. The foundation of the study of neural nets is the 1943 paper "A Logical Calculus of the Ideas Immanent in Nervous Activity" by Warren McCullough and Walter Pitts in the *Bulletin of Mathematical Biophysics*. Current net researcher John Holland considers this paper as influential to the development of **computer science** as was Turing's 1937 paper on the **Turing machine**. An electromechanical neural net called the **perceptron** was built by the American Frank Rosenblatt of Cornell University in 1960. Modeling a biological neural net is a formidable task because of disparity of fan-out, a term defined in the article **graph**. A neuron in the human central nervous system is connected to 1,000 to 10,000 other neurons (its fan-out), whereas the fan-out of a typical electronic component is typically less than 10. Perhaps for this reason, perceptrons using photocells as "eyes" had very limited success in **pattern recognition**, and neural nets fell into disfavor for two decades under severe criticism from McCarthy and Minsky, founders and strong advocates of the algorithmic and **heuristic** approach to AI. In the early 1980s, however, there arose a resurgence in the study of neural nets led by researchers such as physicist John Hopfield of the California Institute of Technology, Stephen Grossberg of Boston University, Gail Carpenter of Northeastern University, and MacArthur fellow John Holland of the University of Michigan. Holland is also a fellow of the Santa Fe Institute in New Mexico where he studies the interrelation of **chaos theory**, **genetic algorithms**, **artificial life**, and the emergent potential of neural nets. In his words, "When a neural network begins to form stimulus-related reverberating assemblies, it acquires behaviors and organization not present in the initially supplied random connections. In both cases we see simple mechanisms gener-

ating behaviors that transcend the capacities of the designer."

Additional reading: Holland, *Emergence: From Chaos to Order*, pp. 18–21, 81–114.

Newton's method ALGORITHMS

A cumbersome **algorithm** for the direct extraction of square roots is (or once was) taught in grade school even though a **recurrence relation** that can be used to extract roots by **iteration** was known as long ago as 600 B.C.: Given an estimate to a square root, then the average of that estimate and the number whose root is wanted divided by that estimate is a still better estimate. As a formula, if x_n is an estimate of the square root of A for $A > 0$, then $x_{n+1} = (x_n + A/x_n)/2$ is a better estimate. Iteration is continued until some error criterion is met, such as requiring that the absolute value of the relative error $(A - x_n^2)/A$ be less than some small quantity. Thus if one guesses that the square root of 10 is 3, then successively higher iterates are $3.1666\ldots$, $3.16228\ldots$, $3.1622773\ldots$ as compared to the exact answer $3.16227766\ldots$. Note that we obtained two additional digits of accuracy for each iterate. Based on his newly invented **calculus**, Newton extended this algorithm to apply to any reasonably well behaved function. Given a function $f(x)$ whose roots are desired and a guess x_n, then

$$x_{n+1} = x_n - f(x_n)/f'(x_n)$$

where $f'(x_n = df(x)dx$, the derivative of $f(x)$ evaluated at x_n. Applying this to the function $f(x) = x^2 - A$ yields precisely the recurrence relation for the square root given earlier, and applying it to $f(x) = x^3 - A$ yields a similar relation for extracting a cube root. Newton's method is sometimes called the *Newton-Raphson method* because it was described by Joseph Raphson in his *Analysis aequationum universalis*, published in 1690. Newton gave its formula in his *Method of Fluxions*, written in 1671, and used it to show how to find the roots of $x^3 - 2x - 5 = 0$, but this book was not published until 1736, nearly 50 years after Raphson's. But Raphson was a friend of Newton's and almost certainly learned the algorithm from him.

Newton's method is also applicable to functions of a complex variable, in which domain strange things can happen. For example, the function $f(z) = z^4 - 1$ has four roots, $+1$, -1, i, and $-i$, where $i = \sqrt{-1}$. When a first guess of z_n is made, Newton's recurrence relation will converge to *one* of the four roots but not necessarily to the one closest to that guess. The approach to a root is chaotic (*see* **Chaos theory**), and graphs that indicate the so-called *basins of attraction* for each of the four roots make beautifully colored **fractal** diagrams.

Additional reading: de Carvalho, "Chaotic Newton's Sequences."

Nippon Electric Corporation (NEC)

INDUSTRY

The *Nippon Electric Corporation (NEC)*, founded in 1899 by Kunihiko Iwadare, is now one of the world's largest computer and electronics firms in the world. NEC began making **transistors** in 1950 and built the NEAC 2201 transistorized **digital computer** in 1959. The company began the manufacture of **semiconductors** in 1978 and now vies with **Toshiba** as the second-largest semiconductor firm after **Intel**. NEC began research in **nanoelectronics** in 1991 with the discovery of the nanotube by NEC employee Sumio Iijima. And while the most powerful U.S. computers are now **Beowulf** clusters, NEC and **Fujitsu, Ltd**. of Japan continue to make very powerful unified **supercomputers**.

Additional reading: *Computing Encyclopedia*, vol. 3, p. 74.

NORC

COMPUTERS

NORC (Naval Ordnance Research Calculator) was a decimal **digital computer** built by **IBM** for the U.S. Navy Bureau of Ordnance under a nonprofit contract to build the most powerful and effective **digital computer** that the state of the art would permit in 1951. The goal was a processing speed of 200 times that of the **SSEC**. The outgrowth of a research project under Byron L. Havens at IBM's Watson Scientific Computing Laboratory at Columbia University, NORC was designed for the rapid solution of very large scientific computational problems. Assembly started in late 1953, and

the machine was demonstrated and turned over to the Navy on 2 December 1954, at which time it calculated **pi** to over 3,000 decimal places. In 1955, it was installed at the Naval Proving Ground at Dahlgren, Virginia, and remained in highly productive use until replaced by an IBM **Stretch** in 1968. NORC was based on the use of a microsecond delay unit designed by Havens, diode switching, a 3,600-word CRT storage unit with 8-microsecond access, and four-channel **magnetic tape** units that transferred 71,340 decimal digits per second. Computing speed was 15,000 three-address **instructions** per second. Numbers were stored in 16-digit words as floating-point values with a 13-digit integer field, a 2-digit exponent, and a sign digit. The arithmetic unit featured fast multiplication that produced the product of two such numbers in 31 microseconds. One of the early uses of the NORC was computing the exact positions of the Moon, Earth, and planets in space at all times to the year 2000 for Project Vanguard. One of its last jobs was a large astronomical calculation for which the answers could be rigorously checked. The NORC run lasted 65 hours, performing over 75 billion operations without error.

Additional reading: Bashe et al., *IBM's Early Computers*, pp. 132–133, 181–183.

NP-complete problem

THEORY

An *NP-complete problem* is one that belongs to a large class of problems such that if a polynomial-time **algorithm** could be found for any one of them, then such an algorithm could be found for *all* of them. A *polynomial-time algorithm* is one whose running time rises no faster than n^a, where a is usually a small integer like 2 or 3, and n is some measure of problem size. (The "P" in NP stands for *polynomial*, and the "N" for *nondeterministic*, a concept to be explained momentarily.) Conversely, if any requires exponential time—time proportional to, for example, a^n—they all do. Currently, and likely forever, the only known algorithms for solving the NP-complete problems have exponential running times. Stephen Cook in the United States and Leonid Levin in Russia independently formulated the concept of NP-completeness in 1971. Cook showed that the problem of testing a Boolean **expression** to see if some combination of true-false values assigned to its **variables** will "satisfy" the formula (cause it to be true overall) is NP-complete. For example, the Boolean expression

(¬A and B) and (¬B or C) and (¬C or D)

(where ¬ means NOT) might be false for all truth-value combinations of A, B, C, and D, or there might be one or more combinations for which it is true ("satisfied"). Of the 16 combinations of values for the four variables, only one combination happens to satisfy the expression, namely, A is false and B, C, and D are all true. A human could inspect a particular formula like this one and arrive at the result pretty quickly, but the question is: Do we know how to tell a computer to find a solution to *any n*-variable expression that might be given to it other than telling it to systematically try all 2^n possibilities? We do not. Of course, if we had a highly **parallel computer**, one with n processors, for example, we could ask each processor to try a different combination and thus get the work done in time linear in n (and n^1, of course, is a *monomial*, a degenerate polynomial). Satisfiability problems for n as high as 20 have been solved by highly parallel **DNA computers**. But only a hypothetical computer that could magically reconfigure itself as n processors no matter how large n is could do this, so a time of this kind is called *nondeterministic polynomial* (NP) time. Theorists do not like to associate nondeterminism with parallelism; they prefer to imagine that a nondeterministic machine functions like an oracle that can "guess" a correct solution and then, in the context of this article, take no more than polynomial time to verify that the guess is correct. The concept of nondeterminism, formulated in 1959 by Michael Rabin and Dana Scott, is discussed further in the article **Chomsky hierarchy**.

Shortly after Cook's work, Richard Karp extended the **set** of known NP-problems to include several other problems of practical interest, and many hundreds are now known. Through an article in *Scientific American* in 1978 and through their textbooks, Harry Lewis

and Christos Papadimitriou have also made significant contributions. To show that a new problem is NP-complete, it suffices to show that it is an NP-problem and that any one problem already known to be NP-complete is reducible to it. The work by Cook and Karp constitutes a very important milestone; whether the class of NP-problems is or is not equivalent to the class of P (polynomial-time) problems (that is, whether P = NP or P ≠ NP) is the most important open question in **computer science**.

Additional reading: Dewdney, *The Turing Omnibus*, pp. 154–160, 252–257.

Number theory   MATHEMATICS
Number theory is that branch of mathematics that is concerned with relationships among integers and the behavior of functions of real and complex **variables** that are affected by integers. Originally thought to be pure mathematics, number theory has now been shown to have many practical applications; see, for example, the long title of the reference. It is difficult to date the origins of number theory as a milestone—the **Sieve of Eratosthenes** is indicative of the interest that the ancient Greeks had in **prime numbers**. But several particular facets of number theory or its application to **computer science** are the subjects of articles in their own right, among them being the **Euclidean algorithm**, **modular arithmetic**, **error-correcting code**, **public-key cryptosystem (PKC)**, **RSA algorithm**, and **Pretty Good Privacy (PGP)**. More generally, combinatorial analysis involving combinations and permutations is essential to the **analysis of algorithms**, and **Stirling's approximation** is needed to simplify the factorial functions that are inherent to them. See also the discussion of the *binomial theorem* in **discrete mathematics** and that of *Mersenne primes* in **cooperative computing**.

Additional reading: Schroeder, *Number Theory in Science and Communication With Applications in Cryptography, Physics, Biology, Digital Information, and Computing*.

Numerical analysis   THEORY
Numerical analysis is concerned with the development, analysis, and use of **algorithms** for

solving the mathematical equations that describe physical processes. Numerical analysis is so much older than **computer science** that it is not always considered a branch of the latter; it is just as likely to be taught in university departments of mathematics. Many famous mathematicians from the seventeenth through the nineteenth centuries—including Gauss, Newton, Lagrange, Jacobi, and Fourier—developed numerical algorithms that are still widely used. At the **machine language** level, a **digital computer** knows only arithmetic—addition, subtraction, multiplication, division, and AND, OR, NOT, and possibly other operations of **Boolean algebra**. Even such a simple concept as extraction of a square root must be expressed in terms of the foregoing operations (see **Newton's method**). Thus it is the task of the numerical analyst to reduce a mathematical operation of interest to a well-defined algorithmic sequence that involves only the fundamental operations cited. In doing so, the analyst seeks to devise algorithms that are computationally fast and efficient, with due regard to *error analysis*. Because **digital computers** work to finite **precision** and because conversions from one **positional number system** to another are seldom exact, *round-off errors* are inevitably introduced. Numerical analysts must be concerned about the effect of such errors on the **accuracy** of the results. Numerical analysis typically deals with root-finding methods for a single equation or for systems of equations; interpolation; approximation; **least squares methods**; numerical differentiation and integration (*see* **Simpson's rule**); solution of linear equations; **matrix** eigenvalue problems; solution of ordinary and partial differential equations; boundary value problems; **random number generation**; and Fourier analysis and the **Fast Fourier Transform**.

The towering figure of numerical analysis is Carl Friedrich Gauss [1777–1855]; many of the twentieth-century milestones in the subject were rediscoveries or improvements on methods that Gauss had invented. The many scientists who have advanced the field since the advent of digital computers include J. Wallace Givins, Richard Hamming (*see* **Error-correcting code**),

Figure 29. The name of Carl Friedrich Gauss (1777–1855), the towering figure of numerical analysis, is associated with more topics in applied mathematics than any other person. The graph on the left of this German bank note depicts the normal curve of error or *bell curve* and is often called the *Gaussian distribution* even though it was derived by Abraham de Moivre in 1733, 76 years before Gauss worked on the subject. See www.ms.uky.edu/~mai/java/stat/GaltonMachine.html for a fascinating Java applet that lets the viewer watch a normal distribution form through simulation of the fall of hundreds of small marbles through an array of regularly spaced pins. (From the collection of the author)

Alston S. Householder, Douglas Rayner Hartree, Anthony Ralston, Richard Varga, Eugene L. Wachspress, Herbert S. Wilf, James Hardy Wilkinson, and David M. Young.

High-quality **subroutine** libraries and mathematical **software** packages first appeared in the 1970s. These include libraries of algorithms and several systematized collections such as BLAS (Basic Linear Algebra Subroutines), EISPACK (matrix operations), FUNPACK (higher transcendental functions), and LINPACK (linear systems of equations). The **Association for Computing Machinery** *Transactions on Mathematical Software* regularly publishes mathematical software that is available in **machine-readable form** from the ACM Algorithms Distribution Service. From 1986 through 2002, a series of books with the generic title *Numerical Recipes* has been published by Cambridge University Press. Written by the team of William H. Press of Harvard, Brian P. Flannery of EXXON, Saul A. Teukolsky of Cornell, and William T. Vetterling of the

Polaroid Corporation, the various volumes have included the **source code** for, successively, algorithms written in **Fortran**, **C**, **Pascal**, and **C++**. All algorithms are well documented and accompanied with extensive narratives that discuss derivation, applicability, stability, and error analysis. Some editions include source code on an accompanying CD-ROM.

Additional reading: *Encyclopedia of Computer Science*, pp. 1260–1273.

O

Object code

DEFINITIONS

Object code is the executable form of a **program** that results from the assembly or compilation of symbolic **source code**. Source code is intelligible to the human reader who knows the **programming language** in which it is expressed, whereas binary object code is easily read only by the computer itself.

Object-Oriented Programming (OOP)

PROGRAMMING METHODOLOGY

Object-oriented programming (OOP) separates the internal structure of a **program** from its external, visible interactions. OOP emphasizes **information hiding**, the **encapsulation** of the inner state of entities called *objects* with the specification of interactive properties defined through *methods*, **procedures** whose details do not need to be revealed to the user. Typical objects are **abstract data types** such as lists, stacks, **sets**, **queues**, and **matrix** arrays, or simulated real-world entities such as banks, airports, restaurants, zoos, or any organization that processes information or manages physical entities in accord with well-defined rules. A **programming language** is said to be *object-based* if it supports the **declaration** of objects as a language feature and is said to be *object-oriented* if, additionally, objects are required to belong to **classes** that can be incrementally modified through *inheritance*, a property whereby a subclass can inherit the capabilities of a parent class and also extend or modify these capabilities. OOP, a rather natural extension of **structured programming**, originated with the development of **Simula** in 1967 and **Smalltalk** in 1972. Among the many later object-oriented languages are **C++**, *Eiffel*, **Ada**, and **Java**.

Additional reading: *Encyclopedia of Computer Science*, pp. 1279–1284.

Octal number system

NUMBER SYSTEMS

The octal number system is a base 8 **positional number system** that expresses numbers through use of the digits 0 through 7. Just as there is no single symbol for *ten* in decimal, there is no 8 (or 9) in octal; in any positional number system, the base is written 10. Any number base that is an exponential power of 2 makes a good shorthand notation for writing down numbers in the **binary number system**, and 8, of course, is 2^3. Thus the octal equivalent of 101 011 111 is 537—the conversion is trivial because all that needs be done is to replace each triad of **bits** with its octal equivalent, and there are only eight equivalents to memorize. Octal was widely used in the documentation of early binary computers because they tended to have **word lengths** and **instruction** layouts that were divisible by three. For example, consider a 36-bit-word computer whose **instruction set** layout is a 6-bit command and two 12-bit **operand** addresses, each of which is followed by a 3-bit reference to one of eight **index registers**. Then the octal expression 72 4037 5 2771 3 could be used as shorthand for command number 72 (111010 in binary) applied to operands octal 4037 indexed by index register 5 and octal 2771 indexed by index register 3. On more recent computers whose word lengths are more usually a multiple of four, the **hexadecimal number system** (base 16) becomes the notation of choice.

Additional reading: *Computing Encyclopedia*, vol. 3, p. 92.

Odhner calculator. *See* **Baldwin / Odhner calculator**.

Olivetti   INDUSTRY

The *Olivetti* corporation of Italy was founded as a family business by Camillo Olivetti on 20 October 1908 to produce **typewriters**. Under the leadership of Camillo's son Adriano, the company broadened its product line in the 1930s to include adding machines and by the 1950s was Europe's largest producer of office machinery. During the 1950s, Olivetti acquired the American typewriter firm of Underwood. In 1959, Olivetti began to market **digital computers** made by **Machines Bull** of France and to offer its own line of computers called the *Elea* system. Five years later, Olivetti sold its computer division to **General Electric (GE)** but continued to produce electromechanical office equipment and a highly innovative programmable desktop computer, the *Programma 101*, which Olivetti considers to have been the forerunner of the **personal computer**. By 1978, Olivetti was near bankruptcy, and the family firm was sold to Carlo De Benedetti and his brother, who returned to the large computer business by buying OEM (Original Equipment Manufacture) plug-compatible (to **IBM**) mainframes from the Japanese firm of **Hitachi** and marketing them as the Olivetti M-Series. In 1983, Olivetti began selling OEM PCs to AT&T and Xerox in the United States and **Toshiba** in Japan and two years later was third (to IBM and **Apple**) in PC revenue. But, like Machines Bull and so many other European computer companies, Olivetti's fortunes declined and on 20 January 1997, its computer division

was sold to a buyout group led by a London-based American lawyer, Edward Gottesman. By acquired right, the Gottesman group continues to sell PCs under the Olivetti name, and the original Olivetti firm's prospects have brightened through emphasis on telecommunications.

Additional reading: Chandler, *Inventing the Electronic Century*, pp. 183–185.

Open architecture   GENERAL

An *open architecture* is a **personal computer (PC)** organization that allows insertion of additional logic cards into the interior of the computer chassis beyond those used with the most primitive configuration of the system. This is done by inserting the cards into *slots* in the computer's *motherboard*, the main logic board that holds its **central processing unit (CPU)** and **memory** chips. The cards have access to the computer's expansion **bus** in order to integrate the **peripheral** devices that they support into the overall system. A computer vendor who adopts such a design places the electronic characteristics of the motherboard slots into the public domain so that third-party vendors can design and market customized logic cards. The rationale is that the greater the variety of cards marketed, the greater will be sales of the host computer itself. The logic cards provide support for such enhancements as one form of **hard disk**, enhanced **monitor** resolution, supplemental memory or **input / output (I/O)** ports, sound, an **optical scanner**, and an **Ethernet** board for high-speed **Internet** connection. The open architecture tradition, an important milestone in computing, can be dated to the first generation of fully assembled PCs of the late 1970s that were based on the **S-100 bus** and the **Intel 8080** (or **Zilog Z-80**) **microprocessors**. A reversal of position with regard to the merits of open architecture has played a significant role in the commercial history of **Apple**. The **Apple II** used an open architecture, whereas its **Apple Macintosh** line was closed. But the competitive **IBM PC** used open architecture. The combination of open architecture and a proprietary BIOS (Basic Input-Output System) that was easy to reverse-engineer led to the marketing of a plethora of IBM PC compatibles, which, perhaps more than any other factor, led to the rapid growth of personal computing (*see* **Compaq**; **Dell**; **Gateway, Inc.**). The Macintosh, which was much more difficult to clone, also enjoyed sales success, but not to the same degree as the aggregate of PC-compatible products.

Additional reading: *Computing Encyclopedia*, vol. 3, pp. 98–99.

Open source software   SOFTWARE

Open source software is a term coined by Eric S. Raymond in a paper entitled "The Cathedral and the Bazaar," first presented at **Linux** Kongress '97, to establish the concept that providing **software** that is "open" (freely available) to all implies distribution of its **source code**, not just compiled **object code**. In January 1998 when **Netscape Communications** set out to make the source code to its **browser** available, the company struggled with some parts of the **free software** model enunciated by Richard Stallman. Instead of using the Stallman General Public License (GPL), the company created the Netscape Public License (NPL) used for the distribution of Linux and of the similar license developed for the distribution of *FreeBSD*, another open source version of **Unix**. In addition to Linux, the other major open source product in widespread use is **Apache**, the software used by the great majority of **servers** used with the **Internet**. Because of differences in philosophy that others believe are slight but he considers profound, Stallman denies that he and the GNU (Gnu's Not Unix) programs with which he is associated are part of the open source software movement.

Additional reading: Lohr, *Go To*, pp. 203–320.

Open Systems Interconnection (OSI)

NETWORKS

The *Open Systems Interconnection (OSI)* is a set of standard **protocols** for intercomputer communication. The term "open" is meant to imply freedom from technical barriers to communication. Work to create the OSI standards was begun by the International Organization for Standardization in 1978. The first step was the creation of the OSI Reference Model, the communication architecture that provides a framework for the various component standards of the OSI family of protocols. Individual members of the family were published serially throughout the 1980s. The OSI Reference Model defines seven layers ranging from the highest, the *application layer* that performs the functions that are the reason for the communication, down to the lowest, the *physical layer* that is concerned with electrical compatibility. Agreement among vendors to follow the OSI standards is an important milestone because without them each network **server** would have to devise a specialized **data** conversion unit for each other server in its network that uses a proprietary protocol.

Additional reading: *Encyclopedia of Computer Science*, pp. 1288–1289.

Operand   DEFINITIONS

In a medical context, the surgeon is the *operator* and the patient is the *operand*. Similarly, in the **instruction set** of a **digital computer**, an *operand* is an item of **data** to be processed by the **operator**, or command, that (usually) precedes the **memory** address of the operand (*see* **Operation code [opcode]**). An instruction may have more than one operand address (*see* **Multiple-address computer**). **High-level languages** use operands as constituents of **expressions**. In most such languages, expressions use normal algebraic infix notation in which binary (2-operand) operators appear between their operands, as in $a + b$, but some languages (and some **calculators** such as the **HP-35**) use reverse **Polish notation** (RPN) in which that addition would be written as $ab+$, or prefix Polish notation, $+ab$, the order used by an instruction in a *two-address computer*.

Operating system   OPERATING SYSTEMS

An *operating system* (OS), sometimes called a *supervisor, executive system,* or *master control program,* is a continuously running master **program** that controls the interleaved execution of subordinate programs submitted to it for processing. Some rudimentary early operating systems were called *monitors*, an example being the *Fortran Monitor System* (FMS) written at North American Aviation in 1958, but the term now invokes the image of a display **monitor**. The 1950s operating systems supported only batch processing, but the advent of **time sharing** in the 1960s saw the birth of several now-historic systems: **CTSS** and **Multics** of **Project MAC** at MIT; one for **Atlas**, the first **virtual memory** operating system; and *OS 360/MVT* for the **IBM 360 series**. The 1970s brought forth **Unix** for several different computers, *Exec-8* for the **Univac 1100 Series**, *Scope* for the **CDC 6600**, and *VMS* for the **DEC VAX series**. In the 1980s, operating systems were developed for a plethora of **personal computers (PCs)**, principally **CP / M** for **Intel 8080** and **Zilog Z-80** computers, *Apple DOS* (Disk OS) for the **Apple II**, **MS-DOS** for the **IBM PC** and compatible computers, and *Mac OS* for the **Apple Macintosh**, which caused something of a sensation as the first commercial PC with an OS based on the concept of a **graphical user interface (GUI)**. Unix and several **open source software** versions of it, including **Linux** and *FreeBDS*, attained great popularity in the late 1990s and retained it into the new century, and **Microsoft Windows** for **Intel 80×86** machines and *Mac OS X* for **Apple** products are now the most popular GUI-based operating systems.

Additional reading: *Encyclopedia of Computer Science*, pp. 1290–1324.

Operation code (opcode)

COMPUTER ARCHITECTURE

An *operation code*, or *opcode*, corresponds to the command portion of an **instruction**. As stored in **memory**, opcodes are numeric; as expressed in **assembly language**, they are symbols such as ADD, LOAD, STORE, CLEAR, and JUMP.

Operator

DEFINITIONS

An *operator* is the mathematical and **high-level language** equivalent of the command used with an **instruction** in the **machine language** of a **digital computer**. A *unary operator* operates on only one **operand**, examples being ~A in **Boolean algebra** (where the unary operator ~ means "negate what follows") and !5 in **APL** in which the ! is the unary factorial operator. Binary (or *dyadic*) operators operate on two operands and may be used with either infix notation (as in $a+b$), prefix notation ($+ab$), or reverse **Polish notation** (postfix notation), $ab+$.

Optical Character Recognition (OCR)

APPLICATIONS

Optical character recognition (OCR) is the process of converting images of machine printed or handwritten numerals, letters, and symbols into **machine-readable form**. Commercial OCR is generally usable only with machine-printed text; the technology for recognition of handwriting is distinctly different. Online systems allow recognition of **characters** and words as they are written on a surface, as is done by **personal digital assistants (PDAs)** and some **handheld computers**. Offline systems include address-reading machines used by post offices and check-reading machines used by banks.

In 1912, Emanuel Goldberg built a device for converting scanned text into **Morse code** to be sent over a **telegraph** line. Modern OCR technology was born in 1951 with David Shepard's invention of GISMO, a **robotic** reader-writer. In 1954, Jacob Rabinow developed a prototype machine that was able to read uppercase typewritten output, but only at the unacceptably low rate of one character per minute. But by 1967, several companies, including **IBM**, Recognition Equipment, Inc., Farrington, **Control Data**, and Optical Scanning Corporation, were marketing much faster and reasonably successful OCR systems. In 1976, Raymond Kurzweil invented the *Kurzweil Reading Machine*, which combines OCR and **speech synthesis** to read to the blind. Several commercial **software** packages are now available for **personal computers (PCs)** that use OCR to convert text images obtained with an **optical scanner** into **word processing** files. Their claimed accuracy of 99.5% or better still leaves an annoying number of misinterpreted characters, but the time taken to correct them is far less than the time it would have taken to retype the document.

Additional reading: *Encyclopedia of Computer Science*, pp. 1326–1333.

Optical computing

COMPUTER TYPES

Optical computing is the use of optical components in a **digital computer**. John von Neumann considered this possibility as early as the 1940s, and he might well have done so had **lasers** been available at the time. By the 1960s, optical technology was used for computing the **Fast Fourier Transform (FFT)** of military images in matched filtering operations. *Synthetic aperture radar* (SAR) **data** fed to a **digital signal processor (DSP)** is an optical **pattern recognition** application that matches stored images with others read as input. Spectrum analysis is another application that is performed with acoustoöptic signal processing. These applications are performed optically when the need for high **bandwidth** exceeds what can be delivered electronically. The fastest computers have cycle times of the order of 1 nanosecond (ns), but optical elements switch on the order of 5 picoseconds (ps), 200 times faster. The development of suitable optical **logic circuits** has historically been the most critical obstacle to achieving an all-optical digital computer, although the **SEED** (Self-Electroöptic Effect Device) has shown promise. There are a number of ways that optics can supplement or replace electronics in computing. **Fiber optics** is a preferred medium for long-distance transmission because of low losses and high information-carrying capacity. Fibers can also be used for distances on the order of a few centimeters for connecting **integrated circuit** boards. Although optical digital computing has the potential to achieve a greater level of performance than electronic digital computing, the impact of optics is far greater for communications than for computation.

Additional reading: *Understanding Computrs: Alternative Computers*, pp. 69–97.

Optical scanner HARDWARE

An *optical scanner* is a **digital computer** input device that captures a bit-mapped image of a document (*see* **Bit map**). When the image scanned is a photograph, it may be stored for use as is or be subjected to **image processing**. When it is a drawing, that drawing might be imported into a paint **program** for editing. When it is a document, it might be analyzed with **optical character recognition (OCR)** software that formats the recognized text in accord with a **word processing** program. The

most widely used scanners are called *flat-bed scanners* because they are used like photocopy machines with regard to placement of pages to be scanned. But there are also handheld scanners that are better adapted to scanning the columns of a newspaper. In 1951, Benedict Cassen invented a rectilinear scanner, but it was prone to jamming. The following year, George E. Mueller developed a more reliable system by arcing each amplified pulse through layered paper to burn black spots that formed an image. A much better optical scanner was invented at EMI, Ltd. in London in 1955 but not perfected until 1972 (*see* **Computerized tomography [CT]**). An optical scanner consists of optical mirrors, light-sensitive diodes, and *analog to digital converters* that produce the final bit-mapped image.

Additional reading: White, *How Computers Work*, pp. 182–187.

Optical storage MEMORY

In contrast to various kinds of magnetic **memory**, an *optical storage* device uses a **laser** to etch and later detect microscopic pits in the surface of its recording medium. There are two principal kinds of optical storage, both of which had their origin as audio or audiovisual entertainment media. CD-ROM (*compact disc read-only memory*) is an optical storage medium used primarily with **personal computers (PCs)**. Information is recorded on a 12-cm disc by using a laser to etch billions of tiny pits into a thin metallic layer that is very much closer to the top surface of the disc than the bottom. In contrast to the concentric tracks used with **hard disks** and **floppy disks**, the pits are arranged in a three-mile-long spiral that is read from the center of the disc to the edge. The audio compact disc (CD) was invented by James T. Russell of the Batelle Memorial Institute in 1970. CD-ROM was adapted from the form of audio CD developed in the mid-1970s by **Sony** and Philips, and CD-ROM drives have been available for PCs since the mid-1980s. The capacity of a CD-ROM is 650MB, enough to hold a thousand copies of this book. There are now CD-R (Compact Disc Recordable) drives available that record information on a conventional blank CD, but only once, and hence are a type of WORM device (Write Once, Read Many). But there are also CD-RW (Compact Disc ReWritable) drives available at higher cost that provide true **auxiliary storage** that can be recorded (up to about a million times) and read back and are thus a type of WMRM (Write Many, Read Many) drive.

A DVD (*Digital Versatile Disc*, formerly *Digital Video Disc*) has the same size as a CD. The first "videodisk," as he called it, was conceived by David Gregg in 1958 and patented in 1961 and 1969. DVDs use a higher-frequency (shorter wavelength) laser to etch pits than is used to make a CD, so the pits are much smaller. Furthermore, the DVD's spiral tracks are more dense than those of a CD, both circumferentially—0.4 μm rather than 0.83 μm—and radially—0.74 μm rather than 1.6 μm. These attributes and **data compression** techniques allow a DVD to contain up to 8.5GB, 13 times as much as a 650MB CD (and twice again as much—17GB—when recorded on both sides). In keeping with the claim of versatility, a DVD drive can read CDs. Like CD-ROM, an ordinary DVD is DVD-ROM; that is, its prerecorded pit-encoded information can be read but not rewritten. But DVD-RAM drives are now available which, through use of technology similar to that described earlier for use with CD-RW, are WMRM devices.

Additional reading: *Encyclopedia of Computer Science*, pp. 1336–1339.

ORACLE COMPUTERS

ORACLE (Oak Ridge Automatic Computer and Logical Engine), a copy of the **IAS computer**, was installed at the U.S. Oak Ridge National Laboratory in 1953. The machine was named by J. Wallace Givens, who worked with Alston Householder on the **numerical analysis** of problems for the machine that were encoded by programmer Virginia Carlock Klema. The ORACLE engineers were Earl Burdette, Rudolph Klein, William Gerhardt, and James Woody. ORACLE remained in operation for about seven years; a piece of its **arithmetic-logic unit** (**ALU**) is in the Deutsches Museum in Munich, Germany.

Additional reading: http://www.csm.ornl.gov/ORLongWay.html

Oracle Corporation INDUSTRY

The *Oracle Corporation* was founded as *Software Development Laboratories* (SDL) in June 1977 by Bob Miner, Ed Oates, and Larry Ellison, the latter as president. The corporation soon changed its name to *Oracle* to conform to that of its major product, a **relational database** software system whose queries are expressed in **SQL**. *Oracle*, a name boldly borrowed from a Central Intelligence Agency (CIA) project of that name, was originally a **mainframe** product but now runs on **workstations** and **personal computers** as well. The company's rapid ascent to over 10 billion in sales was due partially to how quickly it converted Oracle for use on the **IBM PC** and to the fact that **IBM** also chose SQL as the language to support its own System R **database** software. That decision boosted Oracle over its major competitor, *Ingres*, which used a proprietary query language.

Additional reading: Freiberger and Swaine, *Fire in the Valley*, pp. 197–199.

ORDVAC COMPUTERS

ORDVAC (ORDnance Variable Automatic Computer), a copy of the **IAS computer**, was built at the University of Illinois along with its twin, the **ILLIAC**, and installed at the U.S. Army Ballistic Research Laboratory (BRL) at Aberdeen Proving Ground in Maryland. The machine's architecture was that of the IAS, but it used original circuitry designed by Abraham Taub, James Robertson, Edwin Hughes, and chief engineer Ralph Meager. David Wheeler and Donald Gillies wrote and debugged the **programs** that were used to test the machine. ORDVAC was provisionally accepted by the Laboratory in November 1951 at the university and then dismantled and shipped to the BRL. On 6 March 1952, ORDVAC successfully performed its three final acceptance tests. The machine continued in operation for about ten years.

Additional reading: James E. Robertson, "The ORDVAC and the ILLIAC," in Metropolis, Howlett, and Rota, *A History of Computing in the Twentieth Century*, pp. 347–364.

Osborne portable COMPUTERS

The *Osborne portable* of 1981, designed by Lee Felsenstein at the request of Adam Osborne, was billed as the first **portable computer**. But since it weighed 28 pounds when packed into its large suitcase, it is more accurately described as a *luggable computer*. The $1,795 Osborne ran the **CP/M** operating system, had two 91K 5.25-inch built-in disk drives, could run on an optional battery pack, and used the **Zilog Z-80** processor with a 64K **random access memory (RAM)**. The screen of its 5-inch **monitor** was too small to show a full 80 columns of text at a time, so it displayed selected 52-column portions of a screen of standard size in a movable "window." Despite some initial sales success that reached a level of 10,000 machines per month, the Osborne company remained in business for only about two years.

Additional reading: Kidwell and Ceruzzi, *Landmarks in Digital Computing*, pp. 99–100.

P

Packard Bell   INDUSTRY

Packard Bell (PB), which has no connection with either **Hewlett-Packard (*hp*)** or Bell Telephone, was founded in 1986 in Sacramento, California, by a group of Israeli immigrants including Beny Alagem as chief executive officer. The new company adopted the name of a defunct 1950s **television** manufacturer. PB quick-started its **personal computer (PC)** business plan through purchase of Zenith Data Systems. In 1995, Packard Bell was the leading retail PC company in the United States with a 15% market share. But by a year later, **Compaq** had passed it, and Packard Bell became a subsidiary of the **Nippon Electric Corporation (NEC)** in order to market NEC PCs outside Japan. As recently as 1999, Packard Bell sold over a million PCs in Europe. But its share of U.S. PC sales continued to decline in the late 1990s, to the point where it withdrew from that market. According to its quiescent Website, PB has offices in 24 countries but none in the United States.

    Additional reading: http://news.com.com/2100-1001-232363.html?tag=bplst

Packet switching   NETWORKS

Packet switching, a term coined by Donald Davies, describes the internal operations of a particular type of data communications **network**. A *packet* is a unit of information—usually 128 **bytes** or less—that is routed from one *packet switching exchange* (PSE) to another until it reaches its destination. The destination address is contained in the packet header. Ingeniously but counterintuitively, each packet of a group that comprises a message may go by a different route, but all packets are collected at their common destination in order to reassemble a coherent message. A variant of packet switching called *cell switching*, used in *Asynchronous Transfer Mode* (ATM) and *broadband ISDN* services, is based on fixed-size 48-byte packets. When a packet arrives at a PSE, the exchange determines whether it is a transit node, in which case it passes it on, or a destination node. This is called *store-and-forward* transmission. In packet switching systems, the store-and-forward operations generally occur in tens of milliseconds. Source to destination transmission delay is typically about 100 milliseconds for transcontinental packets, though it can be a second or more on slower portions of the **Internet**.

    Two alternative strategies have evolved in the implementation of packet switching systems—*datagrams* and *virtual circuits*. In the datagram strategy, packets are independently routed (at some slight risk of loss or duplication of packets), and transmission order between packets is neither preserved nor necessary. The Internet uses datagrams and the *Internet protocol* (IP) for controlling routing and transmission (*see* **TCP/IP**). In the virtual circuit (VC) approach, a logical path is created between source and destination. The VC allows the network to maintain order, discard duplicates, and detect missing packets. Public data transmission networks are usually based on virtual circuits.

For quite some time, credit for the invention of packet switching was shared jointly and amicably by Donald Davies of England and Paul Baran of the United States, who independently proposed the method in the mid-1960s. In 2001, Leonard Kleinrock of the United States, who worked on the **ARPAnet** and is noted for his work in **queueing theory**, claimed that he had priority of conception based on his 1962 Ph.D. dissertation.

Additional reading: *Encyclopedia of Computer Science*, pp. 1345–1348.

Page Description Language (PDL)

LANGUAGE TYPES

A *Page Description Language (PDL)* is a **problem-oriented language** that defines the layout of a document to be printed or displayed by a computer **program**. The root of PDL development is considered to be a 1967 remark of William Tunnicliffe, chair of the Graphic Communication Association (CGA), who said that there should be "separation of the information content of documents from their formatting." This inspired the 1969 creation of GML (Generalized Markup Language) by Charles F. Goldfarb, Ed Mosher, and Ray Lorie of **IBM**. GML, not coincidentally the successive initials of its creators, was not so named until 1973. In 1978, the IBM group began to work with a committee of ISO (International Standards Organization) led by Tunnicliffe to develop a standard for GML, renamed SGML (Standard Generalized Markup Language), that was issued in 1986. **HTML** (HyperText Markup Language) is a derivative of SGML invented by Tim Berners-Lee for the layout of **World Wide Web** pages, and **XML** (eXtensible Markup Language) is an extension of both SGML and HTML. **PostScript** is a specialized PDL used for image composition

Additional reading: http://www.sgmlsource.com /history/roots.htm

Palm computer. *See* **Handheld computer**; **Personal Digital Assistant (PDA)**.

Parallel computer

COMPUTER TYPES

A *parallel computer* is a **digital computer** that has multiple processing units that allow it to work on different portions of the same problem simultaneously. The **Flynn taxonomy** classifies the several different **computer architectures** that can be used to effect parallelism. The most significant early parallel computer was the **ILLIAC IV** *array processor* of 1975, which used 64 identical processing elements (PEs) arranged in an 8×8 array. A number of **supercomputers** have been *vector computers*, computers capable of adding the elements of a one-dimensional vector in parallel and using such operations to facilitate various **matrix** operations. A parallel computer necessarily costs more than a conventional serial *uniprocessor*, and it is a challenging problem to devise **algorithms** that divide a **program** into pieces that can be distributed among multiple processors in a way that achieves maximum theoretical speedup (*see* **Amdahl's law**; **Concurrent programming**). Development of specialized languages to automate the process have had only moderate success. In addition to the array processor and vector computer, parallel computers or computer types discussed herein are the **Beowulf** computer **cluster**, the **cellular automaton**, the **Cosmic Cube**, the **Connection Machine**, the **dataflow machine**, the **perceptron**, the **pipelined computer**, the **systolic array**, and the **Very Long Instruction Word (VLIW)** computer.

Additional reading: *Encyclopedia of Computer Science*, pp. 1349–1365.

Parameter

DEFINITIONS

As used in computing, a *parameter* is a numerical value entered as input to a computer **program** and passed by the program to a **function**, **procedure**, or **subroutine** that uses it to compute something that depends on that value. In keeping with the usual meaning of *parameter* as a *limit*, each parameter fed to a program must lie within lower and upper bounds that make sense for the computation in progress.

When a subprogram is written, an **identifier** used to represent a parameter is called a *dummy* or *formal parameter*. The corresponding value passed to the subprogram at the time it is invoked is called an *actual parameter*.

Parity
CODING THEORY

A **bit** sequence has even *parity* if it contains an even number of 1 bits; otherwise it has odd parity. The parity of a **set** of bits, such as **data** recorded on a **hard disk** or transferred over a **network**, may be used for a simple form of **error-correcting code**. A *parity bit* is a check bit whose value (0 or 1) depends upon whether the sum of bits with value 1 in the unit of data being checked is to be odd or even. An error caused by the detection of incorrect parity is called a *parity error*. The unit of data to which a parity check is applied may be a **character**, a **byte**, a **word**, etc. Since two or any even number of errors in the unit of data cannot be detected by a single parity bit, the smaller the unit of data to which the check is applied, the higher the probability that multiple errors will not occur.

Pascal
LANGUAGES

The **procedure-oriented language** *Pascal* was devised in the late 1960s by Niklaus Wirth and named after the famous French philosopher and **calculator** inventor Blaise Pascal. The project followed some earlier work by Wirth and Anthony Hoare to improve **Algol 60** with regard to its treatment of **data types**. Pascal was the first widely adopted **programming language** to embrace the principles of **structured programming** and was designed to be a good vehicle for teaching programming. When typeset in accord with the standards of Wirth's classic book *Algorithms + Data Structures = Programs*, Pascal is a physically beautiful language whose statements look much like straightforward English; no other language developed before or since allows the creation of more readable programs. Pascal was designed to be compiled efficiently in a single scan, or *pass*, through a program's **source code**; its **compilers** are relatively small and fast. Wirth's research group wrote and distributed a portable "Pascal-P"

compiler that produced an **intermediate language** called P-code for a hypothetical **zero**-address (stack-oriented) computer (*see* **Pushdown stack**). In 1977, Kenneth Bowles, Stephen Franklin, and Alfred Bork at the University of California at San Diego produced what became a very popular version of Pascal called *UCSD Pascal* for the *Terak* and other **microcomputers**. Their compiler was a descendant of one called *P2*, written by Urs Ammann, who had in turn adapted it from an early Wirth compiler. UCSD Pascal was highly valued for its portability. Then, in 1982, a flamboyant French mathematician and student of Wirth's, Phillipe Kahn, formed a company called *Borland International*. He and employee Anders Hejlsberg proceeded to write a version of Pascal for **IBM PC**–compatible **personal computers**, and later the **Apple Macintosh**, that emphasized speed of compilation. On 15 November 1983, Borland created a sensation by announcing that its new product, *Turbo Pascal*, would be sold for $49.95, a then-unheard-of price for a full-featured language compiler comparable to what **Microsoft** was selling for ten times as much. Over 125,000 copies were sold in the first year. Some version of Pascal held sway as the language of choice for introductory **computer science** courses for almost two decades until the ascendancy of the **object-oriented programming** paradigm caused pedagogy to give way to fashion, and the much more complex **C++** began to percolate down from more advanced courses. Turbo Pascal 6.0 survives as the language of Borland's *Delphi* development system. The origin of several other languages can be traced to Pascal, including Concurrent Pascal, Object-Pascal, **Ada**, **Modula-2**, and Oberon, the last two of which were also created by Wirth.

Additional reading: Reilly and Federighi, *Pascalgorithms*.

Pascal calculator
CALCULATORS

In 1642, the French mathematician and philosopher Blaise Pascal invented and marketed an 8-digit mechanical **calculator** that he called *Pascaline*. Within his short life of 39 years, he also invented the hydraulic press, the barometer, and the wheelbarrow, and in mathematics he cooperated with Fermat in developing the foundation of probability theory. Pascaline was literally just an adding machine; subtraction could be done only by adding the 9's **complement** of the minuend. To avoid straining gears, carries were effectuated through the action of falling weights. Numbers were entered through use of a stylus that moved eight dials, one by one, until each was in proper position to represent a digit. Results were displayed in a sequence of eight little windows at the top of the machine. Samuel Morland tried to make a better Pascaline in 1666, as did René Grillet in 1678, but neither effort was particularly successful. Jean Lepine, watchmaker to King Louis XV, also tried to improve Pascaline in 1725 by implementing the carry mechanism through use of a flexed spring rather than falling weights. Although Lepine's machine was not commercially successful either, one copy was repaired almost a hundred years later by Charles Xavier Thomas, who invented his very successful **Arithmometer** in 1820. Thus the calculational legacy of Pascal was passed on through the better part of three centuries.

Additional reading: Kidwell and Ceruzzi, *Landmarks in Digital Computing*, pp. 28–29.

Figure 30. The photo shows the interior of Pascal's calculator, which he marketed as Pascaline. The machine could add (and subtract with slightly more effort), but not multiply or divide. Surprisingly, it could operate in any of bases 10, 12, or 20 (*see* **Positional number system**). Of the 50 machines that Pascal made, there are eight known survivors, six in France and one each in Germany and the United States. (Bettmann/Corbis)

Pattern matching   ALGORITHMS

As used in **computer science**, *pattern matching* refers to **string processing** operations that answer the question as to whether one **string** is present in (is a *substring* of) another. Such a question has become of vital importance when the strings involved are very long DNA sequences, but even before that application came to the fore, pattern matching had been of great interest in **computational complexity** and the analysis of **algorithms**. When a substring is a sequence of specific **characters**, searching for it among a longer substring, possibly an entire document, is straightforward, but more complicated algorithms are needed when the substring is expressed as a pattern, such as a *palindrome* (a string that reads the same forward or backward) or an even more complex relationship such as "words that start and end with the same vowel and contain a third instance of that vowel followed by two consonants" (e.g., "elective"). Now-classic algorithms for pattern matching were invented by Robert Boyer and J Strother Moore in 1977; by Donald Knuth, James Morris, and Vaughan Pratt, also in 1977; and by Michael Rabin and Richard Karp in 1981. The Rabin-Karp algorithm has good average running time complexity and is extensible to two-dimensional character **arrays**. The Knuth-Morris-Pratt algorithm has excellent worst-case running time of O($m + n$), where m and n are, respectively, the lengths of the pattern and the text being searched. (For an explanation of "Big O" notation, see **Analysis of algorithms**.) The Boyer-Moore algorithm stands out when the pattern is long and the character alphabet in use is extensive.

Additional reading: Cormen et al., *Algorithms*, pp. 853–885.

Pattern recognition   APPLICATIONS

Computerized *pattern recognition* is the use of a **digital computer** to identify and classify patterns in a possibly noisy environment. Examples of patterns are the objects of interest in a photographic image, the tune being played in the background of a movie, the scents of flowers, or the tastes of particular foods. In the first two examples, technology exists to gather samples for examination; in the second two, appropriate input devices are yet to be developed (and little need is perceived for them). In the dawn of **artificial intelligence (AI)** in the mid-1950s, its proponents defined three major goals: that a computer would become the chess champion of the world (*see* **Computer chess**), that computers would be able to translate one **natural language** into another (*see* **Machine translation**), and that facial images would be as familiar to a computer as they are to a human (*see* **Computer vision**). The first goal has been achieved, although the event occurred far later than had been predicted, and reasonable progress has been made toward the second, but the third is proving very elusive. One great impediment is that humans see in three dimensions (3D), whereas a computer can "see" in only two (2D). While development of 3D viewers is awaited, those working in pattern recognition have set the more modest goal of computer identification of simple shapes in a 2D image. As it turns out, even "teaching" a computer to tell the difference between arbitrarily oriented triangles and rectangles is a very difficult problem, either with a conventional computer or with a **perceptron**. But **optical character recognition (OCR)**, **pattern matching** of character **strings**, and **speech recognition** are all kinds of pattern recognition, and good progress has been made in those areas.

Additional reading: *Encyclopedia of Computer Science*, pp. 1375–1382.

PDA. *See* **Personal Digital Assistant (PDA)**.

Pen computer. *See* **Handheld computer**; **Personal Digital Assistant (PDA)**.

Pentium. *See* **Intel 80×86 series**.

Perceptron COMPUTER TYPES

In 1958, psychologist Frank Rosenblatt of the Cornell Cognitive Systems Laboratory designed his first **neural net**work and called it a *perceptron* because he considered it a "perceiving and recognizing **automaton**" capable of **pattern recognition**. Before the name was coined, Gordon Pask and Stafford Beer had made what were essentially electrochemical analog percep-

trons in the early to mid-1950s. Rosenblatt's first perceptron, the electromechanical but digital Mark I, was demonstrated on 23 June 1960. Rosenblatt's inspiration was a thesis propounded in 1949 in the book *The Organization of Behavior* by Donald Hebb, namely, that a brain could learn a new task only by a physiological change occurring within it. A perceptron is a signal transmission **network** that can be "trained" to recognize and distinguish patterns such as circles, squares, and triangles. The receptor of the perceptron is analogous to the retina of the eye and is made of an array of photocells. In 1962, Rosenblatt proposed a number of variations of a procedure for training perceptrons. The set of given patterns of known classification are presented sequentially to the retina, with the complete set being repeated as often as needed. The output of the perceptron is monitored to determine whether a pattern is correctly classified. If not, certain statistical weights are adjusted according to an error correction procedure. The device followed Hebb's rule that "links between neurons grow stronger with use." If there exists a set of weights such that all patterns can be correctly classified, the pattern classes are said to be *linearly separable*. It was conjectured by Rosenblatt that when pattern classes are linearly separable, the error correction "learning" procedure will converge to a set of weights that correctly classifies all the patterns. Many proofs of this perceptron convergence theorem were subsequently derived, the shortest by Albert Novikoff in 1962.

The perceptron approach to **Artificial Intelligence (AI)** was severely criticized by AI pioneers Seymour Papert and Rosenblatt's friend Marvin Minsky in their 1969 book *Perceptrons*. They claimed that 2- and 3-level perceptrons such as Rosenblatt used could have only very limited success at pattern recognition. Rosenblatt had only two years to defend his approach because he and two of his students drowned when his sloop capsized in 1971. In the mid-1980s, interest in 3-level perceptrons was revived by Caltech physicist John Hopfield and his student Terrence Sejnowski. Hopfield was convinced that application of a certain backward error propagation **algorithm** invented by Paul Werbos could do wonders to improve the performance of a perceptron. The algorithm was improved by Geoffrey Hinton, David Rumelhart, James McClelland, and Ronald Williams, and this led to some breakthrough experiments at the University of California at San Diego in 1985. Experimentation with neural nets continues both with special-purpose perceptrons and with simulations on the massively **parallel computer** called the **Connection Machine**, which contains 65,536 separate processors. In 1988, Papert and Minsky revised their book with generous praise for those continuing to work with perceptrons. The revision was dedicated to the memory of Frank Rosenblatt.

Additional reading: *Understanding Computers: Alternative Computers*, pp. 48–60.

Peripheral   DEFINITIONS

A *peripheral device*, or just *peripheral*, is a component that facilitates the operation of a **digital computer**. A peripheral may be either *online* (directly attached) to the computer or *offline*, in which case it must either accept output **data** recorded by the computer on removable units of **auxiliary storage** or record data on such units for use as input to the computer (*see* **Input / output [I/O]**). The earliest computer peripherals were offline key-to-**punched card** devices (*keypunches—see* **Hollerith machine**), key-to-**punched paper tape** machines, and online card readers, card punches, printers, and **magnetic tape** units. After **magnetic drums** were no longer commonly used for main **memory**, they lingered on as peripheral devices used for auxiliary storage. **Cathode ray tube (CRT)** monitors were not common until the second generation of **transistor**-based computers (*see* **CDC 6600**). The typical peripherals used with **personal computers** are **keyboards**, joysticks to control **computer graphics** games, speakers for sound and **speech synthesis**, microphones for **speech recognition**, a **mouse** or **trackball** to control **cursor** movement on a **monitor** or flat-panel display, an **optical scanner**, an inkjet or **laser printer**, magnetic auxiliary storage devices such as **floppy disks** and **hard disks**, and **optical storage** devices for reading and sometimes "burning" (recording) compact discs (CDs) and DVDs (*see* **Optical storage**).

Perl   LANGUAGES

Perl (Practical Extraction and Report Language) is a **high-level language** first posted for use on 18 October 1987. Perl is **open source software** created by Larry Wall, and although it is a general-purpose language, it is particularly good for **string processing**. Partially for this reason, Perl became the language of choice for writing **Common Gateway Interface (CGI)** scripts, applications that run on the **server**-side of a **client-server computing network** like the **Internet**.

Additional reading: http://www.cise.ufl.edu/perl/

Persistent computing. *See* Ubiquitous computing.

Personal Computer (PC)

COMPUTER TYPES

A *personal computer (PC)* is a **digital computer** "no bigger than a bread box" (or more seriously, a desktop) that is intended for one user at a time and whose cost is within the reach of a typical family. The term "personal computer" is believed to have first been used in a **Hewlett-Packard** advertisement for a desktop **calculator** in 1968, but the concept has antecedents that go back to the **memex** of Vannevar Bush, to plans for Edmund Berkeley's **Simon**, of 1949, and to the concept of a *Dynabook* elaborated by Alan Kay at **Xerox PARC** and exemplified by that laboratory's development of the **LINC** of 1962 and the **Alto** of the mid-1970s. The first PC marketed commercially was the **MITS Altair 8800** of 1975, but since it had to be assembled from a kit, it appealed mainly to hobbyists. The first PCs that were close to ready-to-use appliances were the Processor Technology **SOL**, the **Commodore PET**, the **Radio Shack TRS-80**, and the **Apple II**, all of which made their debut between 1976 and 1977, and the Sinclair ZX Spectrum of 1980. The **IBM PC** of 1981 was a relative latecomer, but since it was marketed by the best-known computer company and featured **open architecture** that allowed it to be cloned by other vendors, it set the stage for the phenomenal growth of the PC industry over the next two decades. In 1982, "only" 4.8 million PCs were sold in the United States. In 2001, 124 million were sold. **Dell** held 25% of that market, followed by **Compaq** at 12%, Hewlett-Packard at 9%, **Gateway** at 8%, and **IBM** at 6%, leaving 40% for all others.

Additional reading: Freiberger and Swaine, *Fire in the Valley.*

Personal Digital Assistant (PDA)

COMPUTER TYPES

A *personal digital assistant (PDA)*, a term introduced in conjunction with the marketing of the **Apple** *Newton* in 1990, is a small **handheld computer** that allows one to store, access, and organize information. Most PDAs use either a version of *Microsoft PocketPC 2002* or *PalmOS* as an **operating system**, and most are screen-based; that is, they do not have a physical **keyboard**. Input is entered into a screen-based PDA by writing or tapping on its screen with a stylus, giving rise to the term *pen computer*. In 1992, Jeff Hawkins and Donna Dubinsky formed *Palm Computing*, and in 1994 the company introduced the *Palm Pilot*, a handheld pen computer small enough to fit in a shirt pocket. The success of the Palm Pilot led to widespread use of yet another synonym for handheld computers, namely *palm computers*. The Palm Pilot has limited **memory**, but enough to hold a proprietary operating system called *PalmOS* with enough left over to write or install user **programs**. Some similar products use a concise version of **Microsoft Windows**. Built-in options include monthly and daily appointment calendars, memo forms, and a **calculator** screen. The Palm Pilot also has a "Hot Sync" option that allows synchronization of its **data** in either direction with a **personal computer (PC)**.

In 1998, **3Com** acquired Palm Computing, and because it initially refused to spin off its acquisition, Hawkins left to form a new company called *Handspring*. But after selling millions of Palm Pilots under its own name, 3Com did make Palm Computing an independent company in 2000. April 2002 saw the introduction of the Sharp *Zaurus SL-5500*, a handheld computer with a bright color screen, 64MB (megabytes) of **random access memory (RAM)**, and two expansion slots for a **modem** card, a **digital camera**, or other accessories.

Perhaps more interesting, the Zaurus uses the **Linux** operating system and can run **Java** programs. Handspring, **Compaq**, and **Hewlett-Packard** also sell PDAs with color screens.

Additional reading: http://www.pencomputing.com/palm/Pen33/hawkins1.html

Pervasive computing. *See* Ubiquitous computing.

Philco Transac S-2000

COMPUTERS

The *Philco Transac S-2000* was the first transistorized **digital computer** to be marketed commercially. Although the Philco Corporation of Philadelphia, Pennsylvania, was a leader in the development and production of the **transistors** that were used in the early models of the **DEC PDP series**, the company did not become a computer vendor until 1957. Over the ten-year period 1957–1966, Philco marketed a very well designed and powerful upward-compatible family of computers beginning with the first *Transac* of November 1958. This computer was given the added designation Model 210 when the still faster Model 211 debuted in March 1960. After Philco was sold to the Ford Motor Company in 1961, the most successful member of the family, the Model 212, began shipping in February 1963. The 212 was an excellent scientific computer with 65,536 48-bit **words** and a comprehensive **instruction set** optimized for **floating-point arithmetic**. Because its block-formatted **magnetic tapes** could be read backward as well as forward, it was also an excellent **data processing** machine. Its processing speed was twice that of the rival IBM 7094, and when its initial 6-microsecond access **memory** was replaced by 2-microsecond memory in 1964, its speed exceeded that of the **Stretch**. But despite the performance advantage of the 2000 series, the Philco/Ford sales force was no match for that of **IBM**. The Philco division of Ford oversaw the design of one final model, the 213 of October 1966, but by that time even the most loyal users of the 2000 family had either migrated to the new **IBM 360 series** or become more interested in the latest **supercomputers**, principally the **CDC 6600**, and Ford withdrew from the computer business.

Additional reading: Chandler, *Inventing the Electronic Century*, pp. 40–41.

Pi (π)   MATHEMATICS

The symbol π, the first letter of the Greek word for "periphery," was first used to represent the ratio of the circumference of a circle to its diameter by William Jones in 1706. To 15 places after the decimal point, $\pi = 3.141592653589793$. The Babylonians of 2400 B.C. estimated pi to be $3\ 1/8 = 3.125$, and in the Rhind Papyrus of 1650 B.C., the noted Egyptian Ahmes the Scribe used $\pi = 256/81 = 3.160493827$, after which the nine digits after the decimal point repeat indefinitely. But since Johann Lambert proved in 1761 that pi is an *irrational number* (not a ratio of integers), no sequence of its digits of any length can ever actually repeat indefinitely, and this is necessarily true in any number base. In the third century B.C., Archimedes of Syracuse stated that pi could be approximated ever more closely by inscribing polygons in a circle and computing their perimeters, but the best he could do was to use a 96-sided polygon to obtain $\pi = 3.14159$. Mid-1980s formulas derived by Jonathan and Peter Borwein that allow calculation of pi to billions of decimal places are modifications of ones found by the great Indian mathematician Srinivasa Ramanujan. Pi has long been of interest to computer scientists who wondered whether the digits of pi are "apparently" random (they cannot really be because we know how to compute them). That is, does pi contain relatively more 3s than, say, 7s, or does each digit appear about 10% of the time? Based on the 400-hour computation of pi to over 1.24 trillion decimal places by Yasumasa Kanada on a Hitachi **supercomputer** in 2002, the latter seems to be the case. When repeated, such calculations are of use in showing that a given computer yields reproducible results and, when compared against similar calculations done on another computer, can lend confidence that both computers are working reliably. But whatever sport was involved in extending the computations beyond a trillion digits was, or should have been, forever dampened when, in 1995, David Bailey, Peter Borwein, and Simon Plouffe derived and proved the validity of a for-mula that computes the *n*th **hexadecimal** (base 16) digit of pi without having to compute all or any of its prior hex digits!

Additional reading: Blatner, *The Joy of* π, p. 119.

Figure 31. By many accounts, Srinivasa Ramanujan (1887–1920) was the most enigmatic and fascinating person in the history of mathematics. Born in India and self-taught until he was brought to England by mathematicians intrigued with results that he sent to them, Ramanujan filled several notebooks with conjectures whose origin was mystifying but most of which have now been proved. He devised some of the most efficient formulas known for computing pi to arbitrarily high precision. The computer algebra program *Mathematica* contains virtually all there is to know about applied mathematics. Ramanujan seemed to have its equivalent content embedded in his memory 70 years before it was written. But the culture shock of his move to England proved severe, and he died at the tragically young age of 33. (From the collection of the author)

Pilot ACE. *See* **ACE / Pilot ACE.**

Pipelined computer   COMPUTER TYPES

A *pipelined computer* is a **parallel computer** in which multiple functional units operate on a single **data** stream on an assembly-line basis. Thus a pipelined computer is a multiple-**instruction** stream, single-data stream (MISD) machine in the **Flynn taxonomy**. Each functional unit is responsible for execution of a specialized part of the instruction flowing through it; a pipelined computer maintains several partially completed instructions in process at one time. For example, a pipelined computer with five functional units might assign one each to instruction fetch, effective-address calculation, **operand** fetch, execution of the basic operation specified by the instruction, and storage of results. Pipelining was originally associated with **supercomputers**, the first of which to use the principle being the IBM **Stretch** of 1962, but all current high-performance **microprocessors** use pipelines to some extent. These include the **Intel** *Pentium III and IV*, the **Sun Microsystems** *Sparc*, the **Compaq** *Alpha*, and the **Motorola / IBM** *Power PC*.

Additional reading: *Encyclopedia of Computer Science*, pp. 1405–1408.

Pixel

DEFINITIONS

A *pixel*, short for *picture element*, is the smallest component of an image that a digital device can set or modify. With a monochrome **monitor**, a pixel can ordinarily assume any one of 256 shades of grey through specification by an 8-**bit** (one **byte**) **code**. With high-resolution color monitors, a 24-bit code can cause mixture of primary colors associated with each pixel— red-green-blue (RGB), perhaps—to create 16.8 millions of shades of color. A display grid with $1,000 \times 1,000$ pixels contains a *megapixel* (1 million pixels), the unit of measurements of the resolution of a **digital camera**. Typically, a monitor pixel is about 0.25mm wide (about a hundredth of an inch) and slightly taller, although some devices use square pixels.

Planimeter

HISTORICAL DEVICES

A *planimeter* is a device that can compute the area inside a closed and possibly irregular planar curve by tracing the outline of the curve with a wheeled stylus of some kind. The area of a lake, for example, can be computed by tracing its boundary as shown on a map. Planimeters are based on *Green's theorem*, which states that the area of a closed planar figure is just the *line integral* over its boundary (and, similarly, that the volume of a three-dimensional body is a certain integral over its surface). Although George Green discovered and proved his theorem in 1828, it was not widely known until Lord Kelvin popularized it in 1846. The first planimeter may have been that of the German Johannes Oppikofer who

exhibited it in Paris in 1836. A German planimeter patent was granted in 1849 to Kaspar Wetli and Georg Christoph Starke, and a polar planimeter was invented by the Swiss mathematician Jacob Amsler in 1854, but we do not know whether any of these inventors had knowledge of Green's theorem. J. H. Hermann of Munich is reputed to have built a planimeter as early as 1814, but there is no record as to how it functioned. The **slide rule** manufacturer Keuffel & Esser also made high-quality planimeters and sold them up through 1987.

Additional reading: Aspray, *Computing before Computers*, pp. 166–171.

Plankalkül

LANGUAGES

Plankalkül (Plan Calculus) was a **high-level language** designed by the German engineer Konrad Zuse beginning in 1938 but not published until 1946. The language is historically significant because Zuse perfected it during World War II while isolated from contact with those working in computing in England and the United States. Nevertheless, Plankalkül anticipated many of the features of modern **programming languages**. It supported **procedures** with a choice of **parameter** passing mechanisms; elaborate **data structures**; and the **control structures** such as repetition, **iteration**, and compound **statements** that are necessary for **structured programming**. The language did not have **pointers** and did not support **recursion**, but it did capture certain aspects of **list processing** and **object-oriented programming**. And since it included certain **operators** of formal logic and **set** theory, it would have been very effective for many problems for which **SETL** and **Prolog** are now used. Also, Plankalkül programs were laid out in a two-dimensional form suggestive of a **spreadsheet**. To test the design of his language, Zuse wrote out a complete program for playing **computer chess**. Plankalkül was a most remarkable intellectual achievement. Because of their limited memory and speed, Zuse did not even try to implement the language on his **Zuse computers**. But it is quite surprising that although he lived for 50 years beyond the end of the war, he never implemented Plankalkül for use on any of the many later computers whose capabilities would have been equal to the task.

Additional reading: F. L. Bauer, "The Plankalkül of Konrad Zuse," in Rojas and Hashagen, *The First Computers*, pp. 277–293.

Platform
DEFINITIONS

In computing, a *platform* is the combination of **digital computer** and **operating system** used to support particular **software** applications. Thus the combination of a computer using an **Intel 80×86** computer chip with **Microsoft Windows** is a different platform than one using the same chip with the **Linux** operating system, as are architecturally incompatible computers each of which uses **Unix**. The etymology of the term is uncertain, but it has proved useful since software written for a particular platform will not run on a different one unless written in a **high-level language** with great care given to interaction with the overlying operating system.

PL / I
LANGUAGES

PL / I (Programming Language I) was a **high-level language** developed in the mid-1960s by an **IBM** team led by George Radin. The intent was to support the new **IBM 360 Series** with a new **programming language** that would be as good as **Fortran** for scientific work and the equal of **Cobol** for business **data processing**. The result was a language so rich in syntax that it was difficult to make efficient **compilers** to cope with its complexity. That syntax was carefully defined, however, through use of a notation called the *Vienna Definition Language* (VDL) devised by Peter Lucas and Kurt Walk of the IBM laboratory in Vienna, Austria. VDL is a more elaborate version of **Backus-Naur Form** that attempts to capture the semantics (meaning) of language constructs as well as their syntax (grammar). Because PL / I is structured (*see* **Structured programming**) and was the first to support the concept of a **pointer** as a primitive **data type** (*see* next article), PL / I textbooks were used for a while in support of **computer science** courses called *data structures*. Nevertheless, PL / I had a relatively short lifetime, as did the next attempt to design an all-purpose language, **Algol 68**. The later **Ada** has fared much better in this regard.

Additional reading: Wexelblat, *History of Programming Languages*, pp. 551–599.

Pointer
PROGRAMMING

In general terms, a *pointer* is just another name for a **machine language** *address*, which can be said to "point" to a quantity of interest. In this sense, a pointer can point to only a **word** or a **byte**, depending on the relevant **computer architecture**. But more specifically, now, a *pointer* is a **high-level language** construct that can point to more complex **data structures** such as a **record** or a node of a **graph** or **linked list**. The first *general-purpose* **procedure-oriented language (POL)** to support pointers as a primitive (intrinsic) **data type** was **PL / I** in 1964. The concept is credited to Harold W. ("Bud") Lawson, who was named an **IEEE Computer Society** "Computer Pioneer Award" winner for 2000 for his invention. More recent languages that support pointers include **Pascal**, **Ada**, and **C++**. For reasons cited in its article, **Java** does not.

Additional reading: Reilly and Federighi, *Pascalgorithms*, pp. 378–411.

Polish notation
MATHEMATICS

In 1929, the Polish mathematician Jan Lukasiewicz invented a way of writing a mathematical **expression** of any complexity in a parenthesis-free notation now called *Polish notation*. He found that by consistently placing **operators** before their **operands** (*prefix notation*) or after them (postfix notation) instead of between them (conventional *algebraic* or *infix notation*), no parentheses are needed to insure that the expression is unambiguous. For example, in usual notation $b/a \times c$ is ambiguous unless written as $(b/a) \times c$ or $b/(a \times c)$, depending on whether the variable c is intended to be in the numerator or the denominator. But in postfix form the expression would be $ba/c\times$ in the first instance and $bac\times/$ in the second. Postfix Polish notation, the form most often used in computing, is also called *reverse Polish notation* (RPN). For example, the **HP-35** calculator used RPN in contrast to its competitor, the **TI SR-50**, which used normal infix algebraic notation. **Compilers** typically convert infix expressions to RPN internally as a prelude to generating their corresponding **machine language** code. Such *code generation* was very simple on computers such as the **Burroughs B5000** that did its arithmetic in a **pushdown stack** rather than in **registers**. This is so because the language tokens that comprise an RPN expression form successive groups that can be placed into one-to-one correspondence with the machine language instructions of a zero-address **digital computer**.

Additional reading: *Encyclopedia of Computer Science*, p. 1409.

Portable computer   COMPUTER TYPES

A *portable computer*, as the name implies, can be easily carried. The earliest such machine was the **Osborne portable** of 1981, but since it weighed 28 pounds when packed into its suitcase, it was more "luggable" than portable, and the same was true of the machines made by **Compaq** two years later and by George Morrow in 1985. By now, even a desktop **personal computer** is reasonably portable if one is willing to disconnect and later reconnect several cables, but **laptop** PCs of briefcase size are genuinely portable without such fuss. The most easily portable computers are **handheld computers** that are sometimes called **personal digital assistants (PDAs)**.

Additional reading: *Encyclopedia of Computer Science*, pp. 1414–1417.

Positional number system

NUMBER SYSTEMS

The invention of the *positional number system* is a milestone on a par with the concept of an **algorithm** as a candidate for the single most important development in computing. In a positional number system, a number is expressed as a sequence of single-digit coefficients of a polynomial in an implied number base b. The most familiar positional number system, decimal, uses the base 10 in conjunction with the digit symbols 0, 1, 2, 3, 4, 5, 6, 7, 8, and 9. Thus 76032, for example, stands for $7 \times 10^4 + 6 \times 10^3 + 0 \times 10^2 + 3 \times 10^1 + 2 \times 10^0$. The use of superscripts to indicate powers of numbers was introduced by the French mathematician Nicolas Chuquet in the fifteenth century. By the rules of algebra, $10^1 = 10$ and $10^0 = 1$. Since the number base does not show in

the 76,032, the sequence has the familiar value 76,032 only if its writer and its reader have agreed to use base 10. If they had used the base 8 **octal number system**, then the expansion shown would have had an 8 wherever a 10 is listed, and the decimal value of octal 76032 would be 31,770. The **ternary number system** is of interest because 3 is the base closest to the theoretical "base" of minimal **hardware** cost, which is $b = 2.7181828459045 \ldots$ The minimum base with which we can convey information is base 2, the **binary number system**, whose symbols are usually taken to be 0 and 1. The first five integers expressed in binary notation are 0, 1, 10, 11, 100, and 101. When the base b exceeds 10, additional symbols beyond our familiar ten are needed to represent numbers. In base 16, the **hexadecimal number system**, the usual choice is to use the letters A through F to stand for the decimal values 10 through 15, respectively. Thus 3B72AC5 is a perfectly good number in that system. Algorithms exist for converting numbers expressed in one number base to that of another, and these are now embedded in even inexpensive handheld **calculators**.

Although invented in both the old and new worlds centuries earlier—the Hindu mathematician Aryabhata used a positional number system in India in the sixth century—they were not widely known in Europe until the end of the first millennium; up until then the awkward Roman numeral system was still in use. The Babylonians initially used a *sexagesimal* (base 60) positional system whose vestige survives today in our use of 60 seconds to the minute and 60 minutes to the hour of time or angle of degree. For example, to convert 3:07:21 hours to seconds, we must compute $3 \times 60^2 + 7 \times 60^1 + 21 \times 60^0$ to obtain 11,241 seconds. By 300 B.C., they had switched to a decimal system and had begun to use placeholders of no intrinsic value in the interior of a number so as to be able to distinguish, say, 53 from 503 by writing the latter as 5□3, but they did not recognize **zero** as a digit in its own right usable in the units place for another thousand years. But when Muslim conquests reached as far east as China in the ninth century, they picked up the true **zero** from India along the way, and word of the complete decimal system as we know it finally reached Western Europe a mere four hundred years later through the efforts of the world traveler Leonardo of Pisa, the famous Fibonacci. Only well after the expeditions of Columbus was it eventually learned that the Mayans of Central America had a fully realized *vigesimal* (base 20) positional system in about A.D. 300. And by numbering the 20 days of each of their 18 months 0 through 19, they did not shrink from zero. If only the monk Dionysius Exiguus of the sixth century had known, he would not have saddled us with a system that provoked the silly millennium controversy, the true **Y2K problem.** *Any* consecutive period of 1,000 years is a millennium, so most computer scientists consider that the "first" millennium began in the year 0, the one that Dionysius called 1 B.C., and ended in 999. (In essence, Dionysius shorted the so-called "first" millennium by a year, an appropriate mistake for a man whose full name means, literally, "Dennis the Short.") Accordingly, the only reasonable correction is to consider that the third millennium began in 2000 despite the historical purists who think we should continue to live with an awkward system at variance with all other work in computation.

Additional reading: Seife, *Zero: The Biography of a Dangerous Idea.*

PostScript LANGUAGES

PostScript is a device-independent **page description language (PDL)** based on work originally done by John Gaffney at *Evans and Sutherland* in 1976. Martin Newell and Gaffney, now at **Xerox PARC**, developed an improved version called "JaM" ("John and Martin"), and PostScript was then implemented in its current form in 1982 by John Warnock and Charles Geshchke of **Adobe Systems, Inc**. Page description languages are **programming languages** that are optimized to render document images on display devices such as **monitors**, **laser printers**, film recorders, **facsimile transmission (fax)** machines, or phototypesetters. PostScript programming is especially applicable to situations where a user has a mathematical description of an image and wants to see what it looks like. Adobe Systems

licensed the technology to printer manufacturers, most notably to **Apple** for its LaserWriter introduced in January 1985. PostScript and *Adobe PageMaker* are essentially responsible for the popularity of do-it-yourself **desktop publishing** on **personal computers** and **workstations**. An excellent **freeware** version of PostScript called *Ghostscript*, written by Peter Deutsch, is available on the **World Wide Web**; see www.ghostscript.com. PostScript **source code** is expressed in **ASCII** representation and a postfix **Polish notation** syntax that makes it easy to generate with a simple **text editor**. Like **Forth**, its stack-oriented operations are interpreted rather than compiled (*see* **Pushdown stack**). PostScript descriptions are device independent. Closed-path objects can be filled and paths can be stroked. *Stroking* a path produces an image on the page consisting of a line of specified thickness. Filling an object paints an area inside a closed path with a specified gray level. Paths and outlines can be specified as sequences of lines, circular arcs, and curves called *Bezier cubic splines*. PostScript source code is presented to **software** that processes it as a specially formatted *encapsulated PostScript file* with the unique **file** extension .eps. The invention of PostScript is an important milestone because it has become the language of choice for encoding artistic illustrations that do not originate as photographs.

Additional reading: *Encyclopedia of Computer Science*, pp. 1417–1419.

Precision DEFINITIONS

Precision relates to the number of **bits** or digits in the representation of a number in a **positional number system**, in contrast to **accuracy**, which is freedom from error. For example, 3.142857143 (approximately 22/7) is precise to ten digits but is not a very accurate approximation of **pi** since it errs beginning in the fourth digit. On the newest **digital computers**, a 32-bit integer is said to be a *single-precision* integer in contrast to a 64-bit *double-precision* integer. Double-precision integers have the equivalent of 19-decimal-digit precision, but the accuracy of the result of an extended sequence of double-precision arithmetic operations can be ascertained, if at all, only through careful **numerical analysis**.

Pretty Good Privacy (PGP)

APPLICATIONS

Pretty Good Privacy (PGP) is a **data encryption** program written by Phil Zimmerman in 1991 for encrypting **electronic mail (e-mail)** PGP is also the name of a company founded by Zimmerman and a trademarked brand of encryption products offered by the U.S. firm *Network Associates*. The PGP program is partially based on the **RSA algorithm**. But because pure RSA ran too slowly on **personal computers** for use with long messages, Zimmermann wrote his own e-mail encryption routine and called it *Pretty Good Privacy*, now considered an understatement. To achieve the speed he desired, Zimmerman decided to use RSA, a **public-key cryptosystem (PKS)**, only to exchange the relatively short key needed for the heart of his system, an **algorithm** called IDEA. At about the same time, the U.S. Senate was considering an omnibus anticrime bill that included language that would outlaw the use of "strong" encryption systems that did not permit the government to obtain the plaintext contents of voice, **data**, and other communications upon request. Fearing that use of his program would soon be illegal, Zimmermann gave copies of PGP 1.0 to some friends, and one of them uploaded the program to a few **bulletin board systems**. In late 1992, a much strengthened version, PGP 2.0, was released on the **Internet**. In February of 1998, PGP, the company, was sold to Network Associates, a company formed just a year earlier through the merger of Network General and the McAfee company known for its **computer virus** protection **software**. PGP, the program, incorporated into several popular software products, continues to be widely used.

Additional reading: Levy, *Crypto*, pp. 186–225.

Prime number

MATHEMATICS

A *prime number* is an integer that has only two distinct factors, itself and 1. Thus, by the definition, 1 is not a prime number. The only even prime is 2. Numbers greater than 1 that are not prime are called *composite*. Euclid proved that there are an infinite number of prime numbers.

In 1752, the 15-year-old Carl Friedrich Gauss conjectured that the primes were distributed among the integers such that for very large n, the number of primes less than n is approximately $n / \ln n$, which can be inferred by examining the amount of arithmetic needed to compute primes by the **Sieve of Eratosthenes**. In 1896, Jacques Hadamard and Charles de la Vallée Poussin, each of whom lived to age 97 $\pm$ 1, independently gave the first formal proof of this conjecture, now called the *prime number theorem*.

Primes are important in computing in several ways. They play a role in optimum **algorithms** for searching by the method called *hashing*. By encoding the successive letters of the alphabet with the first 26 primes, two **strings** of length n can be tested to see if they are *anagrams* (like "solemn" and "lemons") in O(n) (linear) time— just multiply the constituents of the encoded strings to see if their product is the same. The search for large primes of a special kind called *Mersenne primes* is the object of a popular project in **cooperative computing**. And perhaps most important, composite numbers that are the product of two large primes are used to construct secure keys in the **public-key cryptosystem (PKC)**.

Additional reading: Ribenboim, *The Little Book of Big Primes*.

Printed circuit

HARDWARE

In 1903, the German Albert Hanson, living in London, filed a "printed" wire patent aimed at solving the need for compact **telephone** exchange circuitry. Although not a true printed circuit, Hanson's method produced conductive metal patterns on a dielectric by cutting or stamping copper or brass foil patterns and bonding them to paraffin paper with adhesive. A true *printed circuit*, invented by the Austrian Paul Eisler in 1936 but not widely used until 1948, is created by lithographically printing "tracks" of copper or other conductor on one or both sides of an insulating board. Components such as diodes, resistors, and capacitors can be soldered to the surface of the board or, more commonly, attached by inserting their connecting pins or wires into holes drilled in the board. Thus printed circuit technology was intermediate between that used with discrete-component **digital computers** of the early 1950s and the **integrated circuit** technology used since the 1970s.

Additional reading: http://www.pcbuk.com/pcb-history.asp

Printing. *See* Typography.

Problem-Oriented Language (POL)

LANGUAGE TYPES

A *problem-oriented language (POL)*, also called a *domain-specific language*, is a **programming language** that supports special **data structures**, **control structures**, and other syntactical constructs that make solution of problems in a specific domain easier to solve. By "solve" is meant the ability to encode appropriate **algorithms** in a more straightforward way, using, perhaps, the technical language characteristic of the domain. The first widely used POL was **APT**, developed in 1957 to facilitate control of machine tools. Others are **Dynamo** for simulating an **analog computer**, **Simula** for general simulation, T$_E$X for designing typefonts, and **PostScript** for drawing diagrams.

Additional reading: *Encyclopedia of Computer Science*, pp. 1433–1440.

Procedure

PROGRAMMING

In programming, a *procedure* is a **subroutine** that does something, such as **sorting** an **array** of numbers, in contrast to a **function**, which returns a single value that can be used in an **expression**. **Pascal**, in fact, uses the keywords *procedure* and *function* in exactly this way. So does **Fortran**, although it uses the keyword *subroutine* rather than *procedure*. Predefined procedures (and functions) may be inserted into a **program** from a public *program library*, and new ones may be defined by the programmer for use only with the program that contains their definitions. In **Algol**, Pascal, **Ada**, and C++, but not the earliest versions of Fortran, procedures and functions may be recursive (*see* **Recursion**).

Additional reading: Reilly and Federighi, *Pascalgorithms*, pp. 68–88, 215–260.

Procedure-Oriented Language (POL)

LANGUAGE TYPES

A *procedure-oriented language (POL)* is a general-purpose language with a sufficiently rich grammatical structure to encode any conceivable **algorithm**. Examples are **Algol**, **Pascal**, and C++. The earliest extensive POL was Konrad Zuse's **Plankalkül** of 1946, but it was never implemented. A **programming language** becomes implemented when a **compiler** is created for it that translates its **statements** into **machine language** code. Early compilers such as Grace Hopper's A-0 and A-2 and the Laning-Zierler compiler of the early 1950s processed languages that had no independent name. The first POL that was widely used was the **Fortran** of 1957, whose descendants such as Fortran 95 are still used extensively. **Cobol**, for *Common Business Oriented Language*, is only three years younger but borders on being a **problem-oriented language**, a related kind of POL since most problem-oriented languages are also procedural.

Additional reading: *Encyclopedia of Computer Science*, pp. 1441–1451.

Program

SOFTWARE

A *program* is a sequence of **instructions** that implements an **algorithm**. The first program of any kind must have been one prepared for the **Jacquard loom**. Computer programs may be written in low-level **machine language** or in a high-level **programming language** that is either a **procedure-oriented language** or a **problem-oriented language** (POL in either case). A portion of a program is called **code**. A person who writes programs is called a *coder*, or more prestigiously, a *programmer*. The first computer programmer was undoubtedly Charles Babbage [1791–1871] who wrote programs for his **Analytical Engine**. The second (and first female) programmer was Lord Byron's daughter and Babbage assistant Augusta Ada Byron King, Countess of Lovelace [1815–1852], for whom the language **Ada** is named. A program for computing Bernoulli numbers is included in her expanded 1843 translation (French to English) of a paper by Luigi F. Menabrea that describes the Analytical Engine (*see* Bowden's *Faster Than Thought*). The program contained a **bug** (mistake) that she likely would have found, but the Analytical Engine was never completed. The first **digital computer** program that ran to completion and produced correct results was written in 1948 by Tom Kilburn for the **Manchester Mark I**. The program, which an untrue legend claimed was the only one he ever wrote, used repetitive subtraction to simulate the division needed to identify the largest factor of a given number (see Figure 32). The first book on programming was the 1951 classic *The Preparation of Programs for an Electronic Digital Computer* by Wilkes, Wheeler, and Gill.

Additional reading: Ceruzzi, *Reckoners*, pp. 73–103.

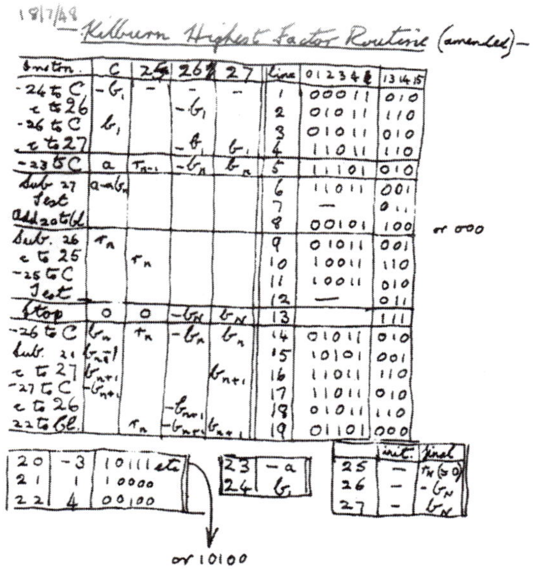

Figure 32. The source code for the first program to run successfully on a stored-program digital computer, a program to find the highest proper factor of any number a by trying every integer b from $a - 1$ downward until one is found that divides exactly into a. The program, written by computer pioneer Tom Kilburn, was run on the prototype of the Manchester Mark I on 21 June 1948. This document of a month later is from the notebook of Kilburn's assistant, Geoff Tootill; Kilburn's original has been lost. (Reprinted with permission of the Department of Computer Science, The University of Manchester)

Program verification

THEORY

As initially written, a computer **program** is essentially a *conjecture* that its logic, when executed, will produce correct results. As was first said by Edsger Dijkstra in 1972, debugging and vigorous testing can reveal the presence of **bugs** but never their absence. Only through *program verification*, rigorous proof that the program must necessarily meet its specification, can it be raised to the level of a mathematical *theorem*. The demonstration that such proofs are possible was first enunciated in the 1960s by Robert Floyd, a Turing Award winner in 1978, and additional early work was done by the English computer scientists C. A. R. Hoare and Robin Milner. The hardest parts of a program to prove correct are its **loops**. A fundamental principle of program verification is that all loops have a key *invariant relation*, an assertion that is true before the loop starts and remains true throughout its **iteration** and after it terminates. Of possibly many candidate invariant relations that pertain to a loop, one must find the particular one that, in conjunction with the status of the control **variable** that determines loop termination, proves that it fulfills its purpose.

Additional reading: *Encyclopedia of Computer Science*, pp. 1458–1461.

Programmable calculator

HARDWARE

For consistency with the articles **computer** and **calculator**, a *programmable calculator* must be considered an oxymoron. The devices that have been marketed as such for the last several years can be used as mere calculators, but they become, when programmed, **handheld computers**. The first programmable calculator, the HP-65 of 1974, appeared just 18 months after **Hewlett-Packard** introduced the **HP-35**, the first pocket calculator with exponential, logarithmic, and trigonometric **functions**. The HP-65 memory was volatile, so **programs** were recorded on small magnetic cards, but later programmable computers used nonvolatile **flash memory**. The HP-65 was housed in a case that looked much like that of the HP-35, but it was slightly larger to accommodate its card reader. The machine had 100 lines of addressable space. Valid label names were the **hexadecimal** digits 0–9 and A–F. Six-bit commands allowed for 64 distinct operations. The HP-65 was the first HP pocket calculator to support octal / decimal base conversion and arithmetic in units of degrees, minutes, and seconds. The first **Texas Instruments** programmable calculator was the SR-52 of 1975, but the company made a major advance when it introduced the TI-59 on 24 May 1977. The TI-59 allowed up to 960 **instructions** packed into 100 **registers**, almost a kilobyte of **random access memory (RAM)**. This was augmented by solid-state **read-only memory (ROM)** modules and a magnetic card reader. At $300, many were sold in preference to early **personal computers** such as the **Radio Shack TRS-80** and **Commodore PET** that cost a hundred dollars more.

Additional reading: *Encyclopedia of Computer Science*, pp. 192–193.

Programming language   SOFTWARE

As contrasted to a **natural language** like English or French, a *programming language* is a highly constrained human-made language in which **programs** for a **digital computer** can be expressed unambiguously. Every digital computer contains a fixed **machine language**, but although one can use it to encode programs numerically, this is not what is usually meant by a "programming language." **Assembly language** in which machine instructions can be expressed symbolically is the lowest-level programming language, but the term usually connotes a **high-level language** that is either a **procedure-oriented language** or a **problem-oriented language**. The difference between these two is subtle and is developed in the highlighted articles. Assembly language programs are translated into executable **object code** (numeric machine language) by an **assembler**, and high-level languages by a **compiler**.

Project MAC   NETWORKS

Project MAC (Machine-Aided Cognition, and Multiple-Access Computer) was an MIT research project of the mid-1960s funded by the U.S. Defense Advanced Research Projects Agency (DARPA). The project was suggested by the visionary J. C. R. Licklider, an early advocate of **time sharing** and an "Intergalactic Network" that led first to the **ARPAnet** and then to the **Internet**. Project MAC was a large and well-funded effort. Its initial DARPA grant of about $2 million per year was raised to $4.3

million in 1969, dropped to under $3 million in 1973, and rose again in the late 1970s. Project MAC's research staff peaked in 1967 at 400. The Project, whose current name is the MIT Laboratory for **Computer Science**, developed the **CTSS** time sharing system for the IBM 7094 (*see* **IBM 700 Series**) and the **Multics** time-shared **operating system** for the **GE 600 series**.

Additional reading: Waldrop, *The Dream Machine*, pp. 217–236.

Prolog   LANGUAGES

Prolog, developed in 1972 by French scientists Alain Colmerauer and Philippe Roussel of the University of Aix-Marseille for research in **natural language** processing, has become the language of choice for **logic programming** in general and **automatic theorem proving** in particular. The first attempts to incorporate logical methods into programming were based on J. Alan Robinson's *resolution principle* of 1965. The resolution principle, an integral part of Prolog, makes the derivation of logical conclusions from a **set** of axioms computationally possible. Some version of Prolog is now available for virtually all **digital computers** in current use, most of which are descended from a 1977 implementation by David H. Warren.

Additional reading: *Encyclopedia of Computer Science*, pp. 1026–1030.

Protocol   COMMUNICATIONS

A *protocol* is an agreement between two or more communicating entities as to how to conduct a conversation. Whereas most **algorithms** are executed by a single person or computer and hence require no cooperation, a protocol is an algorithm whose steps must be carefully orchestrated among its alternating executors. Protocols are usually structured in layers and implemented in a combination of **hardware** and **software**. The lowest layer is typically concerned with electronics. The middle layers deal with grouping **bits** into well-defined units (frames, packets, messages, etc.) and transmitting them hop by hop over several intermediate machines to their destination. The upper layers have to do with the meaning of the information sent, such as the **file transfer protocol (FTP)** and protocols used to exchange **electronic mail**. The **Open Systems Interconnection (OSI)** is a set of protocols that cover the entire spectrum of networking applications from the lowest layer to the highest. Standardizing a protocol makes it possible for multiple vendors to produce products that can interact successfully. The **Ethernet** protocol makes it possible for **personal computers** and **workstations** from differing vendors to communicate, and the **TCP / IP** is at the very foundation of the **Internet**. *See also* **Packet switching**; **Router**.

Additional reading: *Encyclopedia of Computer Science*, pp. 1486–1487.

Public-Key Cryptosystem (PKC)

COMMUNICATIONS

Conventional cryptographic systems, including the **Data Encryption Standard (DES)**, are called *symmetric systems* because they use the same key for both encipherment and decipherment. In 1976, the Americans Whitfield Diffie, Martin Hellman, and Ralph Merkle proposed an asymmetric *public-key cryptosystem (PKC)* in which the enciphering and deciphering keys are different and not computable one from the other. Although not known until certain documents were declassified in the late 1990s, three British cryptologists—James Ellis, Clifford Cocks, and Malcolm Williamson—had also discovered what Ellis called "Non-secret Encryption" six years earlier. To use a PKC, one creates a matched pair of such keys and distributes copies of just the encipherment key to those with whom he or she wishes to communicate, keeping the decipherment key secret. The recipients can then send enciphered mail to the creator of the encipherment key that only that person can decipher. Even if an eavesdropping cryptanalyst obtains a copy of the enciphering key, it does no good. This demonstrates the flexibility of a public-key cryptosystem for *key distribution*, an area where conventional cryptosystems are awkward because all keys must be kept secret. The PKC was not practical until the invention of a suitable *one-way trapdoor function* such that deduction of the private decipherment key from the public encipherment key would be computationally intractable. This was done in 1978 by Ron Rivest, Adi Shamir,

and Leonard Adleman, who developed the **RSA algorithm** based on the **computational complexity** of factoring the product of two large **prime numbers**. Public-key cryptosystems can also be used to provide *digital signatures*: a user can create a signature for a message by enciphering it with a private key. Someone else can check the validity of the signature by checking that it deciphers to the message using the signer's public key. This capability of public-key cryptosystems has important applications in **electronic funds transfer (EFT)** and to the secure transmission of credit card information over the **Internet**.

Additional reading: Levy, *Crypto*, pp. 66–89, 313–330.

Punched card

I/O DEVICES AND MEDIA

A *punched card* is a piece of cardboard stock into which encoded **data** can be punched in a prescribed format. The earliest known use of punched cards to control a programmed device was to direct the weaving of patterns on the **Jacquard loom**. Babbage planned to use punched cards as the input medium for his **Analytical Engine**, but that computer was never finished. The **Hollerith machine** used in conjunction with the U.S. Census of 1890 processed punched cards that had 34 and 37 columns, but eventually Hollerith settled on a 45-column card with round holes. A brief description of the use of punched cards in the 1900 census appeared in an article by F. H. Wines in the January 1900 issue of *National Geographic*. **IBM**'s earliest rival was the Powers Accounting Machine Company founded by James Powers in 1911 and acquired by **Remington Rand** in 1927. The Powers punched cards, like the earlier Hollerith cards, were 45-column cards with round holes that were sensed mechanically. In 1928, IBM introduced the higher-capacity 80-column "IBM card," which used rectangular holes that were sensed electromechanically. The ability to store 80 **characters** on one card made it tempting to cram all the vital statistics of a person or item into one card-based "unit record," leading to the term *unit record machine* for any of the various devices invented to process the cards. But since each of the 80 columns was precious, the temptation may have been great to effect the 2.5% saving afforded by storing years in only a 2-digit rather than a 4-digit format, thus sowing the seeds of the **Y2K problem** of 72 years later.

In conjunction with the **UNIVAC** in 1952, Remington Rand introduced a 90-column card that used round holes. Such cards were always much less popular than the 80-column format even though all 90 columns of the longer card could be punched from a local **memory** at once, after errors had been corrected. These cards were used on the New Jersey Turnpike in the l960s. In 1969, IBM introduced a small punched card with its System/3 computer. The new card held 96 characters arranged in three tiers of 32 characters. Each character was represented by a 6-bit **code**. By this time, however, the use of punched cards was already in decline, and the 96-column card never achieved anything like the penetration of the 80-column card. From the beginning of the 1970s, the advent of key-edit equipment and **interactive computer** terminals caused a rapid falloff in punched card usage to the point where few computers are equipped to process them. But there are still three U.S. companies that make punched cards, primarily for use in lotteries and as timecards for payroll systems. The U.S. Card Corporation of Tiffin, Ohio, for example, still produces 30 million punched cards per month.

Additional reading: Bashe et al., *IBM's Early Computers*, pp. 1–24.

Punched paper tape

I/O DEVICES AND MEDIA

While the **ENIAC** used **punched cards** for in-**put / output (I/O)**, the **EDSAC**, the **stored program computer** built by Maurice Wilkes at Cambridge University, used *punched paper tape*. This form of input and storage of **pro-grams** and **data**, adapted from **telegraph** equipment, was quite common on early university computers. Up through the beginning of the **microcomputer** era, devices called Flexowri-

ters (and later, Model 33 Teletypes) were used to punch programs and data encoded in **Baudot code** into 5-channel paper tape. Paper tape continued as a form of input up through the beginning of the microcomputer era. In 1975, for example, Bill Gates and Paul Allen delivered their first **Basic** interpreter for the **MITS Altair** on punched paper tape. The medium is seldom used now in conjunction with general-purpose computers, but it continues to have application to specialized laboratory equipment.

Additional reading: http://www.srcf.ucam.org/~jsm28/ECMA-10/

Pushdown stack

DATA STRUCTURES

A *pushdown stack* is a linear **data structure** to which items may be added or removed only at one end, traditionally called "the top of the stack." New items are *pushed* into the stack and the current top of the stack may be *popped* into a **register**, forming a *Last-In, First-Out* (LIFO) storage structure akin to cafeteria plates in a self-leveling spring-loaded dispenser. LIFO is an accounting term that antedates **digital computers**, and while earlier programmers used the term *pushdown list*, Allen Newell, Herbert Simon, and Cliff Shaw used *stack* in conjunction with their IPL language of 1956 (*see* **IPL-V**), as did Edsger Dijkstra in Europe, and *stack* stuck. Pushdown stacks have several applications in **computer science**. They play a role in the *parsing* (grammatical analysis) of **expressions** by a **compiler**, the evaluation of reverse **Polish notation** (RPN) expressions as is done in the **HP-35** calculator and the **Burroughs B5000** computer, the formation of the *activation records* needed to implement **recursion**, and in the implementation of the Federighi *gerund function* (*see* **Structured programming**). A computer whose only **memory** is a push-down stack, a *pushdown stack* **automaton**, is one of the four classes of computer that comprise the **Chomsky hierarchy**.

Additional reading: Reilly and Federighi, *Pascalgorithms*, pp. 431–442.

Q

COMPUTER TYPES

Quantum computing is based on the use of **logic circuits** that obey the laws of quantum mechanics rather than those of classical physics. That this would eventually be possible was envisioned in the early 1980s by Paul Benioff and later Richard Feynman. A two-state quantum system that is used to store a single unit of information is called a *qubit*. But in contrast to an ordinary **bit** that can be either a 1 or a 0, a qubit can represent both a 1 and a 0 simultaneously, like the now-famous Schroedinger cat that is both dead and alive until an observer opens the box that houses it. Similarly, once measured, a qubit in a superposition state will always be found to have a definite 0 or 1 value, with only the *probability* of each result being predictable by quantum mechanics. In 1986 David Deutsch realized that a *quantum computer* could use quantum superposition in conjunction with a continuously valued **fuzzy logic** to provide massive computational parallelism. In a quantum computer, a sequence of n qubits can contain the superposition of all possible 2^n values at once, providing a highly parallel **memory**. In 1994 Peter Shor showed how qubit operations could be used to perform a quantum Fourier transform (QFT) operation with exponentially fewer operations than the classical **Fast Fourier Transform (FFT)** and also showed that factoring composite integers could be solved efficiently on a quantum computer.

In 1994 a 129-digit number known as RSA-129 was factored in eight months, requiring 10^{17} instructions performed on over 1,000 networked computers (*see* **Cooperative computing; RSA algorithm**). But a hypothetical quantum computer with a clock speed of a mere 100 MHz could have factored this number in a few seconds, requiring a memory of about 2,000 qubits and 10^9 quantum logic circuits. As of now, it is only the concept of a quantum computer and a few suitable **algorithms** suitable for use on one that are milestones, but the actual construction of even a 10-qubit quantum computer would become a candidate for the most significant milestone in the history of computing.

Additional reading: Johnson, *A Shortcut Through Time: The Path to the Quantum Computer.*

Queue DATA STRUCTURES

The first recorded use of the French word *queue* in English to mean "waiting line" dates to 1837. In **computer science**, a queue is a **data structure** in which entries may be made only at one end of the list and removed at the other. Queues are used by **operating systems** to hold pending jobs whose only priority is First In, First Out (FIFO). There are two kinds of related structure, the *deque* ("double-ended queue"), which allows addition or removal at either end, and the *priority queue*, a queue in which an occasional entry is removed from or entered into the interior of the waiting list. A queue and a deque are most easily implemented with a linear **array** and two **pointers**, but the priority queue's need to accommodate interior changes suggests implementation as a **linked list**.

Additional reading: Reilly and Federighi, *Pascalgorithms*, pp. 442–446.

Queueing theory   THEORY

Queueing theory is the mathematical theory of waiting lines (**queues**); it is concerned with the analysis of systems that provide service to random demands. Queueing theory originated in the early 1900s with the work of Danish engineer Agner Erlang of the Copenhagen Telephone Company. Several important formulas for telecommunications engineering (and a **programming language**) bear his name. Queueing theory is applicable to both the communications and computational aspects of **computer science**. A computer's **input / output (I/O)** devices compete for service from the **channels** that service them; queueing theory can be used to configure the arrangement of devices and channels that will optimize system performance through **multiplexing**. Similarly, in **operating systems** that support **multiprogramming**, jobs awaiting processing can be ranked in accord with their expected running time or storage requirements and distributed among queues with different priorities of access determined by queueing theory. The maturation of queueing theory as applied to **digital computers** is due principally to the work of Leonard Kleinrock, who published *Queueing Systems*, a very influential two-volume work on the subject, in 1976.

Additional reading: *Encyclopedia of Computer Science*, pp. 1496–1498.

Quicksort   ALGORITHMS

In 1962, C. A. R. "Tony" Hoare invented an easily programmed **algorithm** for **sorting** that he called *Quicksort*. Hoare reasoned that the principal defect of an exchange sort (*bubblesort*) was that the **records** that were exchanged never moved very far; each exchange moved them only one position closer to their ultimate location. What if, he pondered, records were partitioned with respect to a chosen *pivot*, a particular key taken from one of the records in the data set being partitioned? All records having a

key less than or equal to the pivot would be placed in a left partition; all other records would be placed in a right partition, and the pivot record would lie between them. Neither the left nor right partition would then necessarily be sorted, but the pivot record would be in its final location, never having to be moved again. Next, the left and right partitions are partitioned in the same way, using **recursion**. When all partitions have reached size 1, the sort is complete. Since the running time for Quicksort is proportional to n log n, where n is the number of elements to be sorted, the algorithm has become the method of choice for sorting by comparison of elements.

After graduation from Oxford, Hoare invented the Quicksort algorithm during a postgraduate year at Moscow State University and perfected it after joining Elliott Brothers Limited in London in 1960 as a programmer. Initially, he had difficulty convincing his boss that it was superior to the Shell sort he had been asked to write for the **Elliott 803**. But he made his point by showing how succinctly the algorithm could be described when expressed recursively and that it really would "sort quickly" when compiled and executed using the one-pass **compiler** for **Algol 60** he had written for the 803. Speaking of Algol and recursion in his brilliant Turing Award address of 1980, "The Emperor's Old Clothes," Hoare said, "I have regarded it as the highest goal of **programming language** design to enable good ideas to be elegantly expressed." That address, cited in the Elliott 803 article above, should be required reading for every student of **computer science**.

Additional reading: *Encyclopedia of Computer Science*, pp. 1653–1654.

Quipi   HISTORICAL DEVICES

From ancient times, knotted strings have been used as memory devices. Some of the most striking records of this sort were kept on *quipi* by the Inca, who governed what is today Peru and part of Ecuador, Chile, and Argentina. The Incas, who were conquered by the Spanish in the sixteenth century, had no written language, but they used quipi to keep accounts of taxes paid and to maintain a census of their population. The quipi had a main cord, held horizontally, to which several hanging cords were attached. Knots were tied in the hanging cords at regular intervals. The shape of the knot indicated the digit a knot represented (1, 2, 3, etc.), while its distance from the main cord indicated its place value (ones, tens, hundreds, etc.). The color of the cord might indicate the kind of person or object represented. Thus the quipi was an early form of technological **memory** capable of storing digital information.

Additional reading: Kidwell and Ceruzzi, *Landmarks in Digital Computing*, p. 14

R

Radio

Radio, a shortened form of "radiotelegraphy" called *wireless* in many countries, is a communications milestone that became a computing milestone with the development of the **Internet**. Many of the "hops" that an **electronic mail** message takes to reach its destination proceed via **wireless connectivity**, and **World Wide Web** users can listen to many radio stations across the country and the world over the Net. Completely wireless Internet connections are becoming increasingly available and affordable.

The story of radio begins with the electromagnetic theory of James Clerk Maxwell of Scotland whose equations of 1864 predicted that it should be possible to send wireless signals through the atmosphere. The theory was confirmed experimentally by Heinrich Hertz of Germany in 1888, but the first person who succeeded in transmitting wireless messages over a reasonably long distance was Mahlon Loomis, a dentist. In 1865, he flew two kites carrying wires from mountain tops 14 miles apart. The wire from one kite was grounded through a **telegraph** key; the other kite-wire was grounded through a sensitive galvanometer. When he operated the key at one site, detectable changes of current occurred in the wire at the other. He obtained a patent on his system in 1872. In 1894, Sir Oliver Lodge of England sent **Morse code** signals a half mile away. In 1895, Guglielmo Marconi of Italy transmitted a signal beyond the horizon, the same year in which Aleksandr Popov of Russia built a radio that was demonstrated publicly the next year. In 1901 Marconi transmitted across the Atlantic. Marconi's achievements would not have been possible without the *coherer* (signal detection device) that he used, one that he used without acknowledgment to its inventor, Acharya Jagadish Chandra Bose of India. And on 21 June 1943, the U.S. Supreme Court overturned Marconi's radio patents in favor of the competing claims of Nikola Tesla, who had died five months earlier. The Canadian Reginald Fessenden, who worked as a chemist for Thomas Edison during the 1880s and later for Westinghouse, invented a "heterodyne principle" to modulate radio waves so as to allow voice transmission rather than telegraphic code. But since an element crucial to the success of his system was still lacking as of 1904, Fessenden asked Ernst F. W. Alexanderson of **General Electric (GE)** to design a high-frequency generator of 100 kHz and output measured in kilowatts. He was successful, and on 24 December 1906 Fessenden was able to broadcast the world's first radio program with song and music. While doing research for the U.S. Navy during World War I, the American electrical engineer Louis Hazeltine invented the neutrodyne circuit that neutralized the noise that had plagued all radio receivers up to that time. In

1924, Hazeiltine formed the corporation that bears his name. The first practical car radio was developed for Motorola in 1929 by William Lear, who, years later, marketed the Learjet airplane.

Radio was initially based only on *amplitude modulation* (AM); the much more clear *frequency modulation* (FM) was invented by Edwin H. Armstrong in 1933. Earlier, Armstrong had invented the *regenerative circuit* in 1912, the *superheterodyne circuit* in 1918, and the *super-regenerative circuit* in 1922. His inventions and developments form the backbone of radio broadcasting as we know it.

Additional reading: Coe, *Wireless Radio.*

Radio Corporation of America (RCA)

INDUSTRY

The *Radio Corporation of America (RCA)* was established as a subsidiary of **General Electric (GE)** in 1919 when GE vice-president Owen D. Young purchased the assets of the Marconi Wireless Company of America. RCA inherited **radio** patents from GE, capital from co-investor Westinghouse, and the 28-year-old David Sarnoff, who had joined American Marconi in 1911 and become its commercial manager. Sarnoff became general manager in 1921, and under his leadership RCA became a highly successful communications company. He persuaded GE and Westinghouse to let RCA buy the Victor Talking Machine Company, which occurred on 4 April 1930. The GE/Westinghouse/RCA alliance then became so dominant that in November 1932 GE and Westinghouse were forced by the U.S. government to accept a consent decree that freed RCA to become an independent company with Sarnoff as president and CEO. RCA's first **digital computer** venture was the *Bizmac* of 1956. The U.S. Army Ordnance Tank-Automotive Command in Detroit, Michigan, purchased the $4 million **vacuum tube** machine for inventory management. The **transistor** circuitry RCA 501 was introduced in 1958. In December 1964, after only modest success in computing, the company decided to challenge the supremacy of **IBM**. The resulting **RCA Spectra series**, upward compatible from the **IBM 360 series**, did capture some of IBM's share of the **mainframe** market,

but in the long run RCA was no more successful than was the **Amdahl Corporation** with its similar business plan. In September 1971, Robert Sarnoff, who had succeeded his father as CEO on 1 January 1968, sold RCA's computing venture to the Univac division of **Sperry Rand**. In 1987, GE's president Jack Welch reacquired RCA, and its subsidiary NBC with it, for $6.28 billion. Because of overlap with GE's own endeavors, Welch then proceeded to divest most of RCA's assets—all but NBC.

Additional reading: Chandler, *Inventing the Electronic Century*, pp. 15–49.

Figure 33. The first model of the Radio Shack TRS-80 of 1977. Despite being called the "Trash 80" by fans of competitive machines, it was beloved by many users and its architecture was the forerunner of what became the IBM PC and its successors. (Library of Congress)

Radio Shack TRS-80

COMPUTERS

The *Radio Shack TRS-80*, introduced by the Tandy Corporation in 1977 at a base price of $400, was one of the first three fully assembled **personal computers (PCs)** to attain widespread use. Its competitors were the **Commodore PET** and the **Apple II**. The TRS-80 used the **Zilog Z-80** chip, which was slightly more powerful than, but compatible with, the **Intel 8080** of that era. Main **memory** options were 4, 8, or 16KB (1 kilobyte = 1,024 **bytes**), and an audio cassette was used for **auxiliary stor-**

age. The computer's **monitor**, a black and white TV set without a tuner, sat upon an expansion interface that allowed installation of additional memory or circuit cards that served external devices such as an 8-inch **floppy disk** drive or a **hard disk**. **MS-DOS** was used as an **operating system**, and a **Basic** interpreter stored in 4 kilobytes (4,096 bytes) of **read-only memory (ROM)** was immediately available upon startup. The TRS-80 went through models 1 to 4 and then was succeeded by the powerful Tandy 2000, which used an Intel 80286 chip (*see* **Intel 80×86**) before the IBM AT did, but the Tandy marketing effort was no match for that of **IBM**. Radio Shack stores began to carry **IBM PCs** in the early 1990s but switched to those of **Compaq** in 1998.

Additional reading: Kidwell and Ceruzzi, *Landmarks in Digital Computing*, pp. 96–99.

RAID   HARDWARE

RAID (Redundant Array of Inexpensive Disks) is an architectural concept developed to transform an array of relatively slow but inexpensive **hard disks** into a faster, more reliable **auxiliary storage** system. RAID was devised in 1987 by David Patterson and Randy Katz of the University of California at Berkeley and Garth Gibson of Carnegie Mellon University in Pittsburgh. RAID systems derive their speed from "striping" data across multiple disks, placing successive pieces of a **file**, *stripes*, on different disks to allow parallel data access. Reliability is generally achieved through replication ("mirroring") or by using **error-correcting code** across the disk array.

Additional reading: *Encyclopedia of Computer Science*, pp. 1512–1514.

RAM. *See* **Random Access Memory (RAM)**.

RAMAC   COMPUTERS

The IBM 305 *RAMAC* of 1955, initially called DRAM, for *Direct Reference Accounting Machine*, was renamed at the suggestion of Jerrier Haddad, manager of Engineering, Programming, and Technology at **IBM** corporate headquarters. RAMAC originally meant *Random Access Memory Accounting Machine*, but copy-

right problems led to a change to *Random Access Method of Accounting and Control*. The RAMAC was developed in response to the announcement of **Remington Rand**'s *Univac File Machine*. Though RAMAC was still a **vacuum tube** computer, the product is historically significant for its use of the model 350 Disk Storage Unit, the first **hard disk** ever delivered. Its capacity was only five megabytes, but this was a breakthrough innovation at the time. Over 1,000 RAMACs were sold over the period 1956 to 1961.

Additional reading: Bashe et al., *IBM's Early Computers*, pp. 297–300.

RAND Corporation   INDUSTRY

The company that has been the not-for-profit *RAND Corporation* since 1 November 1948 began as Project RAND on 1 October 1945, a contract awarded to the Douglas Aircraft Company in Santa Monica, California, at the suggestion of Henry H. "Hap" Arnold, then Commanding General of the U.S. Army Air Force. General Arnold was concerned that the end of World War II would leave too few scientists and engineers available for military research. Frank Collbohm was the first director of RAND, an acronym formed from "research and development"; the name bears no relation to a person or to **Remington Rand**. By the time of its change in status in 1948, RAND had grown to 200 staff members, including many applied mathematicians and engineers whose work gradually transformed them into *computer scientists*, a term not yet born. The establishment of RAND is an important milestone because for over 50 years its scientists have made, and continue to make, significant contributions to **computer science** and **information technology (IT)**. In furtherance of its defense contracts, RAND mathematicians have made important contributions to **game theory**. Researcher Paul Baran's work on **packet switching** provided the basic building blocks for the **Internet**. RAND staff designed and built one of the earliest **digital computers**, the **JOHNNIAC**, developed an early online interactive terminal-based computer system called **JOSS**, and invented the **TCP/IP** telecommunications

protocol that has become the basis for modern computer **networks**.

Additional reading: http://www.rand.org/history/

Random Access Memory (RAM)

MEMORY

A *random access memory (RAM)* is an address-able **memory** in which the time to fetch the contents of a unit of information—a **byte** or a **word**, say—is the same for any address. Few applications actually fetch words at random; the name merely implies that location does not matter when words are accessed in any sequence whatsoever. The term arose to distinguish such memories from early main memories such as **magnetic drum** and **ultrasonic memory** whose access times depended on the position of the device in space or time when a fetch was initiated. The first random access memory was the *Williams tube*, an electrostatic storage device of 1947. This was followed by **magnetic core** memory, which held sway for almost three decades, and then by the various kinds of **solid state memory** still in use today.

The acronym *RAM* can also mean *random access machine*, an entirely different concept. A random access machine is a hypothetical computing device equivalent in computing power to a **Turing machine** but which uses an addressable memory and **instruction set** similar to that of a simple **stored program computer**. This RAM is hypothetical for the same reason as is a Turing machine; namely, both have an indefinitely expandable memory.

Additional reading: Dewdney, *The Turing Omnibus*, pp. 96–102.

Random number generation

ALGORITHMS

Random number generation is needed to produce the prodigious supply of "random" numbers needed for **Monte Carlo** calculations. Such numbers might be truly random if they are produced by monitoring a physical source that fluctuates rapidly and erratically, **radio** static perhaps. But since such methods have proved problematic, scientists must settle for the availability of number sequences that are generated algorithmically. Since such sequences are deterministic—anyone who knows or can deduce the **algorithm** can compute the next number from the current one—the numbers are called *pseudo-random*. At a symposium in 1951, John von Neumann said that "anyone who considers arithmetical methods of producing random digits is in a state of sin" (Goldstine, *The Computer from Pascal to von Neumann*, p. 297), but even he accepted the need to do so. Pseudo-random number generators produce number sequences that pass most, but never all, of the various statistical tests of apparent randomness. The successive digits of **pi** do, but they cannot be generated fast enough to keep random number generation from being the most time-consuming part of Monte Carlo calculations. And even being able to recognize a "random" number on sight is quite difficult. The most useful definition of a random sequence to date was enunciated in 1965, independently by Gregory J. Chaitin of **IBM** and Russian mathematician Andrei N. Kolmogorov. Their definition is based on a comparison of the length of a random sequence to the length of the computer **program** needed to generate it, both measured in **bits**. A sequence may reasonably be considered random if the program producing it must be at least as long as the sequence itself, a view quite consistent with the concept of *entropy* in Shannon's **information theory**.

What is usually desired is a source of random fractions f such that $0 < f < 1$. Then, if random integers over some range are preferred—1 to 6 perhaps to simulate the roll of a die—one can compute the integer part of $1 + 6*f$, which can never be less than 1 nor greater than 6. An early but now obsolete method of generating the fractions is to start with an arbitrary nonzero n-digit f called a *seed*, square it to produce a number of $2n$ digits, and take its middle n digits as the next f of the sequence. For example, using a six-digit decimal example with a seed of 0.781093, the next two iterates would be 0.106274 and 0.294164. It is easy to see that this sequence must eventually cycle back and repeat itself—there are only a million different six-digit numbers in the range .000000 to .999999 inclusive. Even if this random number generator were to produce all of them in succession, its cycle would be just one million, and if .000000 (or a number that has occurred previously) comes up, its cycle will be much shorter. All pseudo-random number generators are necessarily cyclic. The search for ever better random number generators is an active area of research in **numerical analysis**. The foremost authority on this subject is George Marsaglia of Florida State University.

Additional reading: Dewdney, *The Turing Omnibus*, pp. 49–55.

RCA. *See* **Radio Corporation of America (RCA)**.

RCA Spectra series   COMPUTERS

The *RCA Spectra series* was the first series of commercial **digital computers** to use monolithic **integrated circuits (ICs)**. The initial family consisted of models 70/15, 70/25, 70/35, 70/45 and 70/55, all announced between October 1965 and July 1966. The **Radio Corporation of America (RCA)** designed and marketed the series to compete with the **IBM 360 series** by deliberately numbering the computers in between those of **IBM** and offering better price / performance ratios. The Spectra computers used the same internal **computer architecture** as the 360 series but not the same **input / output** facilities or **operating system**, both of which inhibited compatibility and thus sales. The Spectra proved less able to compete with the IBM 360 Series than were comparable products made by the **Amdahl Corporation**, and RCA's faith in the product line began to wane. To service the initial Spectra users through the mid-1970s, RCA brought out models 70/46, 70/60, and 70/61 between November 1968 and February 1971, seven months before the company sold its computer division to **Sperry Rand**.

Additional reading: Chandler, *Inventing the Electronic Century*, pp. 38–39.

Read-Only Memory (ROM)   MEMORY

A *read-only memory (ROM)* is computer **memory** whose content can be read and reread any number of times but cannot be altered. ROM was used as early as the late 1940s to store the **instruction set** of microprogrammable **digital** computers (*see* **Microprogramming**). On the earliest **personal computers (PCs)** of the late 1970s, it was customary for vendors to store a **Basic** language **interpreter** in ROM to make it instantly available when the machines were *booted* (started up). The current most prevalent form of ROM on PCs is CD-ROM (Compact Disc ROM), which is the medium of choice for the distribution of **software**. *See* **Optical storage**.

Real-time system   COMPUTER TYPES

A *real-time system* is a computing system that can respond to input from a real-world event fast enough to give human monitors the opportunity to affect that event. The term real-time *computer* is avoided here because any general-purpose **digital computer** can respond to some events in real time but not to others. Any current computer can monitor the progress of, say, the flight of a radio-controlled model airplane, but even the fastest **supercomputers** cannot compute next month's weather in less than a month, or compute the outcome of most chemical reactions faster than they actually occur. A real-time system, then, is one that runs on a computer that is primarily dedicated to the control of a well-defined real-world task that can be monitored in real time. The first real-time system that meets this definition was the **Whirlwind** of 1948, which was programmed as a flight simulator to train bomber crews.

Additional reading: Foster, *Real Time Programming*.

Record   DATA STRUCTURES

Since the early thirteenth century, the English verb *record*, literally *to cord again*, has meant "to commit to memory." In computing, the noun *record* is the **data structure** that forms the constituent elements of a **file**, the place where records are recorded. A record generally holds data that pertains to a particular item or person, such as name, address, height, and weight. Since the first two of these attributes are normally recorded as **strings** of **characters** and the last two as numbers, a record is typically an *inhomogeneous* data structure. The languages **Cobol**, **Pascal**, and **C++** all support the record as a primitive (inherent) **data type**, though the latter calls it a *struct*. Thus a record may be prepared or modified in main **memory** and then used to update (replace) a corresponding record in an external file. The other operations that qualify *record* as a data structure are the addition or deletion of a record to or from a file. Disk file *tracks* are of uniform length, and **magnetic tape** is often organized into fixed-length *blocks*. The integral number of records that can be stored in one block or track is called the *blocking factor*.

Additional reading: Reilly and Federighi, *Pascalgorithms*, pp. 336–348.

Recurrence relation   MATHEMATICS

A *recurrence relation* relates a mathematical entity to be computed to one or more similar entities of lower order. Perhaps the best-known recurrence relation is the *Fibonacci sequence* 0 1 1 2 3 5 8 13 21 . . . for which $F_n = F_{n-1} + F_{n-2}$ for $n > 1$ (provided we are told that $F_0 = 0$ and $F_1 = 1$). A recurrence relation may be, but need not be, computed by **recursion**. **Newton's method**, for example, is stated as a recurrence relation that is better suited to calculation of roots of equations by **iteration**. And for the Fibonacci sequence, successive values of F_n for $n > 1$ can be computed and stored in the **array** Fib$[0 . . . n]$, with n as large as desired, provided that we pre-store the values Fib$[0] = 0$ and Fib$[1] = 1$. In the "Big O" (order of) notation of the **analysis of algorithms**, this algorithm is O(n); that is, solution time grows linearly with the problem size n. The heart of a recursive solution would be

    if ($n < 2$) **then** return(n)
    else return(F($n-1$)+F($n-2$)),

but a recursive solution of a recurrence relation that is based on two or more entities of lower order, as this one is, is very inefficient because so many intermediate values are computed more than once; the recursive computation of the Fibonacci sequence has the exponential running time O(ϕ^n), where ϕ is the *Golden ratio* $(1 + \sqrt{5})/2 = 1.618033989 . . . (see$ **Discrete mathematics**).

Recurrence relations exist for (certain) continuous **functions** as well as for integer se-

quences. An example is the trigonometric half-angle identity $\cos(A) = 2\cos^2(A/2) - 1$. Thus the cosine of 60°, say, can be computed if one knows the cosine of 30°. By extension, then, the formula can be used as the basis for a recursive function for computing the cosine of an angle of arbitrarily large size, provided it can be repetitively halved by recursion until it reaches a small enough size, say x, that its cosine can be computed by the approximation $\cos(x) = 1 - x^2/2$, the first two terms of the **Taylor series** for the cosine function.

Additional reading: Abramowitz and Stegun, *Handbook of Mathematical Functions*, pp. xiii.

Recursion   PROGRAMMING

Recursion is to writing **programs** as *induction* is to proving theorems. In 1838, Augustus DeMorgan rigorously defined and introduced the term *mathematical induction*, although the process had been used intuitively for centuries. Induction, in turn, led to the formulation of *recursive function theory*, a field of mathematics formalized from about 1931 to 1936 by Alonzo Church, Kurt Gödel, Stephen Kleene, Emil Post, and Alan Turing. Once **stored program computers** became prevalent, it became obvious to those who created **Algol 60**, **IPL-V**, and **Lisp**, which is based on Church's **lambda calculus**, that recursive functions could be expressed and calculated through use of a **high-level language**. **Fortran** did not provide for recursion, but most languages invented since 1960 do. Perhaps the simplest example of a function that may be (but does not have to be) computed recursively is the *factorial function*, which, for a given integer n greater than or equal to 2, is the product of n and all successively smaller integers down to 2. (Down to 1, by definition, but the final multiplication by 1 is wasteful. Also, factorial 0 is defined to be 1.) The heart of the recursive definition of n factorial is

    if ($n < 2$) then return(1)
    else return ($n*$fact($n - 1$))

To the student who first encounters such a function, the process seems like magic. How, for example, does the logic, once set into play, stop? And along the way, where are the intermediate products stored? When n is either 0 or 1, we can see that the correct answer, 1, will be returned. And if we assume for the moment that there is a good answer to the second question, then we can see that for a high value of n, the process will stop when the initial n is counted down to 2. The process will be correct if we can have faith that, before a final answer is returned, the intermediate values of n have been saved and will participate in the multiplicative chain. Finally, the answer to our second question is that these intermediate values of n are indeed saved and, more specifically, are saved in exactly the right **data structure** that best implements recursion—a **pushdown stack**. The *Ackermann function*, a superexponential function devised by the German mathematician Wilhelm Ackermann, is often used to test the efficiency of the recursive mechanism of a **procedure-oriented language** or as a benchmark of the speed of a particular computer. Recursion is one of the most powerful techniques available to those who develop **algorithms** for solving computer problems, and its appearance on the programming scene in the late 1950s was a very significant milestone. *See also* **Recurrence relation**.

Additional reading: Roberts, *Thinking Recursively*.

Reduced Instruction Set Computer (RISC)

COMPUTER TYPES

Until the mid-1970s, the prevailing mode of **computer architecture** was exemplified by the **complex instruction set computer (CISC)**, that is, a **digital computer** with an extensive **instruction set**, some of which were so specialized that they would seldom be generated by a **compiler** for a **high-level language**. The only notable exception was Seymour Cray's **CDC 6600**, which had a very small instruction set. In 1975, **IBM** Fellow John Cocke advocated making all **instructions** so simple that their computer could execute them at the rate of one every machine cycle, where the cycle would be very fast. The idea was subsequently and enthusiastically promoted by John Hennessy at Stanford University and by David Patterson at the University of California at

Berkeley, the person who first used the term *Reduced Instruction Set Computer (RISC)* for the new philosophy. RISC embraces a small instruction set consisting of simple, fixed-length, fixed-format instructions that execute in a single machine cycle; a large number of fast **registers** to minimize the need to access slower main **memory**; and use of an optimizing **compiler** to manage certain resources that had previously been managed by **hardware**. A RISC program will necessarily be much longer than a corresponding CISC program, but memories are now so large and inexpensive that this is not a factor and, even though longer (that is, consist of more instructions), may very well execute faster. For some time in the late 1980s through the late 1990s, RISC **workstations** became the norm. The PowerPC chip developed jointly by IBM and **Motorola** and used in the latest **Apple** products is a RISC. But gradually, RISC designs became more complicated, and CISC designs began to incorporate some of the more important RISC concepts, so that contention between the two schools of design has cooled considerably. *See also* **Very Long Instruction Word (VLIW)**.

Additional reading: *Encyclopedia of Computer Science*, pp. 1510–1511.

Register

COMPUTER ARCHITECTURE

A *register* is a unit of **memory** that holds a sequence of digits—**bits** in a binary **digital computer**—whose length is sufficient to hold an integer or a single-precision **floating-point** number; four **bytes** is common, and certain *double-length registers* have eight. With most **computer architectures**, registers supplement main memory, but in a few, registers have addresses that are part of the main memory addressing structure. Registers are usually constructed with higher-performance and more costly switching elements than main memory, and hence register-to-register data transfers are faster than memory-to-register transfers (or the reverse). The concept of *register* is not in itself a milestone because even the earliest **stored program computers** had to have at least two special-purpose registers called, respectively, the *accumulator* (the place where sums are accumulated) and the ***program*** counter (or *pro-*

gram address register) that keeps track of the address of the next **instruction** to be executed. Storing a number (address) in the program counter is equivalent to telling the computer to "jump" (transfer control) to a new memory address, presumably the beginning of an alternative branch of program logic. A double-length register that holds the product of a multiplication or the dividend and later the quotient and remainder of integer division is called the *Multiplier-Quotient (MQ) register*. This article is included as an umbrella term to point to the articles on special-purpose registers such as the **index register** and the **shift register** and to the concept of a **Reduced Instruction Set Computer (RISC)** having many registers and a small **instruction set**.

Additional reading: *Encyclopedia of Computer Science*, pp. 1515–1518.

Relational database

INFORMATION PROCESSING

A *relational database* is a **database** in which **data** and the relations among them are represented in forms of two-dimensional *flat tables*, each of which consists of a **set** of rows called *tuples*. As with mathematical sets, no two rows may be alike. The columns of a table are called *attributes* and may be labeled with names such as ITEM, STYLE, COLOR, MODEL NUMBER, PRICE, and so on. The entries in at least one column must be unique so that the corresponding attribute can be used as a *primary key* that can be used to fetch a particular tuple. The relational database model was formulated by Edgar F. Codd of **IBM** in 1970 and clarified in 1985 through his publication of 12 rules that a well-formed relational database must follow. For example, one rule says that table entries in general may be null but that no element of a primary key may be null. A more difficult rule to enforce is that a user should not have to be aware as to whether all data is stored at a single site or whether the data is scattered throughout a **distributed system**. In principle, queries may be presented to a relational database in the form of either the *relational algebra* or the *relational calculus*. Most implementations of the model, *dBase*, and **Oracle Corporation**'s *Oracle*, for example, use relational algebra, or more partic-

ularly, relational algebra as embodied in the language **SQL**, although Michael Stonebraker's *Ingres* uses a proprietary query language. Queries that can be answered by reference to a single table are rather easily posed and answered, but a relational algebraic language also supports several **matrix**like operations that allow queries to be answered by combining multiple tables. The relational database model is a very significant milestone because its structure and flexibility have proved so much more robust than that of earlier hierarchical and **network** database models that, for all practical purposes, only the relational model has survived.

Additional reading: *Encyclopedia of Computer Science*, pp. 1519–1524.

Remington Rand

INDUSTRY

Remington Rand, named for Eliphalet Remington and James Rand, Sr., was formed in 1927 through the merger of the Remington Standard **Typewriter** Company; the Rand Kardex company owned jointly by the Rands, senior and junior; the (James) Powers Accounting Machine Company (*see* **Punched card**); the Safe Cabinet Company; and the **Dalton Adding Machine** Company. In 1950, Remington Rand entered the **digital computer** business by buying the **Eckert-Mauchly Computer Corporation (EMCC)**. Two years later, Remington Rand bought **Engineering Research Associates (ERA)**, gaining with it its principal assets: J. Presper Eckert and William Norris. In 1955, the company merged with the (Elmer) Sperry Corporation to form **Sperry Rand** with Harry Vickers as chief executive officer (CEO) and former General Douglas MacArthur as chairman of the board. Prior to 1954, Sperry had 30 large computers installed, while **IBM** had only four. In 1954, the latter started taking orders for the **IBM 700 series** computers, which were more powerful than anything Sperry could offer. By 1956, IBM had installed about 75 computers to Sperry Rand's 50 or fewer. Once IBM took the lead, Sperry faded as a strong competitor. The company was sold to the **Burroughs Corporation** in 1986; the combined companies continue to operate as **Unisys**.

Additional reading: Chandler, *Inventing the Electronic Century*, pp. 96–99.

Remote Job Entry (RJE)

INFORMATION PROCESSING

Remote Job Entry (RJE) is a mode of **digital computer** operation that allows execution of jobs received from a remote site and return of their output to the same or a different remote site via a communications link. Remote job submission or monitoring was demonstrated remarkably early; George Stibitz demonstrated his Complex **Calculator** over a **telephone** line in 1940 (*see* **Bell Labs relay computers**). In 1955, a National Security Agency (NSA) **Alwac III-E** under supervision of the author supported four remote terminals that were allowed to control the central computer in round-robin fashion for up to 15 minutes at a time. RJE was initially devised to relieve the bottlenecks of **batch processing**, but it proved much less effective in doing this than the development of **time sharing**, and the term is gradually fading into disuse.

Report generator. *See* **RPG**.

RISC. *See* **Reduced Instruction Set Computer (RISC)**.

Robotics

APPLICATIONS

Robotics, a term coined in 1940 by Isaac Asimov, is the study of robots and the machines used in factory **automation**. *Robot*, of course, is an older term that dates to Karel Čapek's 1921 play *R.U.R.*, or *Rossum's Universal Robots*. *Robot* is derived from the Czech word *robota*, meaning "slave," and indeed the play features machines created to simulate human beings and which rebel against their creators. Such action is contrary to Asimov's *Three Laws of Robotics*:

1. A robot may not injure a human being, or, through inaction, allow a human being to come to harm.

2. A robot must obey the orders given it by human beings except where such orders would conflict with the First Law.

3. A robot must protect its own existence as long as

such protection does not conflict with the First or Second Law.

Asimov claimed that the laws were originated by John W. Campbell in a conversation they had on 23 December 1940. Campbell in turn maintained that he derived them from Asimov's science fiction and that all he did was to number them. The first Asimov story to state the Three Laws explicitly was "Runaround," which appeared in the March 1942 issue of *Astounding Science Fiction*. *Robot* implies a machine that has the general form of a human. Thus all robots are **automata**, but not all automata are robots. Robots abound in fiction and movies. Some of the more famous movies that featured robots are Fritz Lang's *Metropolis* of 1926; *The Day the Earth Stood Still*, a 1951 film by Robert Wise adapted from a story by Harry Bates; *Forbidden Planet* of 1956 by Irving Block and Allen Adler, a film that featured *Robby the Robot*; and of course, George Lucas's *Star Wars* of 1977 that featured *C3PO* and *R2D2*. The Westinghouse robot *Electro* was a big hit at the New York World's Fair of 1939.

Robots are still being built but principally for amusement; no company yet markets an affordable "personal robot" that can do anything useful. The serious side of robotics relates to the construction and application of robotic devices that do not necessarily resemble humans but which can perform specific tasks that were either traditionally performed only by humans or could have been, had the tasks not been too dangerous. An example of the latter is the use of mechanical hands for the remote handling of radioactive substances. On the home front, it is now possible to buy robotic swimming pool cleaners and lawn mowers. But by far the most widespread use of robotic devices is in factory **automation**. The mass production of automobiles is highly automated, and there are many smaller products that are produced entirely by robotic machines. Considerable effort is being expended at industrial laboratories and universities to devise ever better robotic devices. The work of a vibrant center of such research, the MIT *Artificial Intelligence Laboratory* directed by Rodney Brooks, is described in the reference.

Additional reading: "Robots: Intelligence, Versatility, Adaptivity." Six-paper theme of the *Communications of the ACM*, **45**, *3* (March 2002), pp. 30–63.

Figure 34. Elektro and a friend who strings along make beautiful music together at the New York World's Fair of 1939. Sixty-four years later there are still no affordable personal robots (PRs), although there are many PR men (and women). (Hulton Archive)

ROM. *See* **Read-Only Memory (ROM)**.

Router   COMMUNICATIONS

A *router* is a communications interface that interconnects otherwise incompatible **hardware** devices. A router can interconnect two otherwise incompatible **digital computers**, or two **packet switching** networks. This is possible because although the underlying technology and formats of packets may be different, their essential service **protocols** are the same. Alternative terms for routers or devices similar to routers are **gateway**, *switch*, and *network level relay*. A gateway that verifies user identity and refuses network access to unauthorized users is called a **firewall**. A router is distinct from a *repeater* or a *bridge*, but the logical functionality of a bridge and a router can be combined

into a single physical unit, commonly known as a *bridging router*, bridge-router, or *brouter*. The principal router is use today was invented by Leonard Bosack and Sandra Lerner in 1984 (*see* **Cisco**).

Additional reading: *Computing Encyclopedia*, vol. 3, p. 223.

RPG   LANGUAGES

RPG (Report Program Generator) is a nonprocedural **problem-oriented language (POL)** developed by **IBM** in 1959 for use with its **IBM 1400 series** and later used with the **IBM 360 series** and the **IBM AS/400**. At a time when access to professional programmers was still scarce, the syntax of RPG enabled users to generate specially formatted reports by specifying *what* they wanted rather than *how* to produce them. The concept has endured to the point where there are now many kinds of report generators. MathWorks now includes a report generator with its popular MATLAB **program**. There is also an **open source software** program called *GNU Report Generator* (GRG) that reads **record** and field information from any one of a *dBase3+* **relational database** file, a comma-delimited **ASCII** text **file**, or the output of a **SQL** query and produces a report listing. The program was loosely designed to produce T$_E$X formatted output, but plain ASCII text, **PostScript**, **HTML**, or any other kind of ASCII-based output format can be produced just as easily.

Additional reading: *Computing Encyclopedia*, vol. 3, pp. 209–210.

RSA algorithm   ALGORITHMS

Even after the general principle of the **public-key cryptosystem (PKC)** was invented, it could not be implemented until someone devised a suitable "one-way trapdoor" **algorithm**, a **function** that is very easy to compute in one direction but whose inverse is computationally intractable. Such an algorithm was discovered in 1977 by MIT students Ronald Rivest, Adi Shamir, and Leonard Adleman. The algorithm, called RSA for the initials of that trio, depends for its security on factoring a very large number, perhaps 100 digits or more, which is the product of two **prime numbers** known only to the sender of a message, not even to any recipient. Rivest, Shamir, and Adleman received the 2003 Turing Award for their achievement. RSA calculations involve high-**precision modular arithmetic** and the **Euclidean algorithm**. When RSA was described by Martin Gardner in a *Scientific American* column of August 1977, he displayed a 129-digit number called RSA129 that was guaranteed to be the product of two primes and offered a modest prize for the first person or team that could factor it. It was 17 years before a group of 600 **personal computer (PC)** users scattered over two dozen countries used **cooperative computing** to factor the number over an eight-month period ending in April 1994. Those whose encrypted messages need to remain secret for longer than 17 years—hundreds of thousands, perhaps—

need only base their **data encryption** on, say, a 200-digit number.

In 1999, the then-16-year-old author of the reference, Sarah Flannery of Ireland, wrote and delivered a prizewinning paper on the *Cayley-Purser algorithm*, one that she discovered and named for mathematician Michael Purser, a colleague of her father, and Arthur Cayley, the nineteenth-century inventor of **matrix** algebra. The algorithm is demonstrably faster than RSA but, though originally thought to be as secure as RSA, proved not to be. Since the flaw made Cayley-Purser useless as a **public-key cryptosystem** but not for one where a prior exchange of keys can be arranged, the Flannery result was nonetheless a remarkable achievement for such a young scientist.

Additional reading: Flannery, *In Code*, pp. 63–64, 168–175.

S

S-100 bus

HARDWARE

When the **MITS Altair 8800** was introduced in 1974, it came with an expansion **bus** with a published specification that used commonly available parts and connectors. Through the efforts of George Morrow, this **open architecture** soon became a standard. One hundred pins were provided for various signals on this *S-100 bus*, a term coined by Harry Garland and Roger Melen. This meant that new video cards, more **memory**, and serial and parallel **input / output (I/O)** ports could be added to the computer as needed. At the height of its popularity, more than 100 companies manufactured products compatible with the S-100 bus.

SABRE

NETWORKS

In 1953, due to a chance encounter between two Smiths on an airplane, **IBM** seized on an opportunity to capitalize on the experience it had gained in real-time computation with **SAGE**. When Blair Smith, a top IBM salesman, found that his seatmate on a flight from Los Angeles to New York was the (unrelated) president of American Airlines, C. R. Smith, the talk quickly turned to AA's need for a wholesale replacement of its antiquated Reservisor airline reservation system. The application turned out to be just what Thomas Watson, Jr., was looking for, and he placed SAGE (and SAGE) engineer Perry Crawford, newly hired away from the Office of Naval Research, in charge of developing a **distributed system** to meet the needs of the airline. IBM tided AA over with improved **punched card** equipment until the project could get rolling in 1957. In 1960, system design was far enough along to warrant the acronym SABRE, and although this endured, its awkward name, Semi-Automatic Business Research Environment, was soon dropped. The system was fully implemented by 1963, the culmination of the largest commercial **data processing** operation undertaken as of that time. Housed in Briarcliff Manor, New York, SABRE consisted of 10,000 miles of leased **telephone** lines that allowed 1,100 agents in 50 cities to make 10 million reservations per year in an average time of less than three seconds each. The heart of the system was a pair of IBM 7090 computers, duplexed for reliability. Over the period 1965 to 1968, Delta, Pan Am, and Eastern Airlines purchased similar systems, and in 1987, United Airlines installed a successor system based on eight S/370 model 3090 machines.

Additional reading: Bashe et al., *IBM's Early Computers*, pp. 516–522, 557–558.

SAGE

NETWORKS

In 1949, the U.S. government declared an urgent need to expand air traffic control technology into a full-fledged air defense monitoring system. The government solicited bids from a number of private contractors to expand the

Whirlwind project at MIT to form *SAGE* (Semi-Automatic Ground Environment), whose development work was to be done at the newly formed MIT Lincoln Laboratory. Bidding were Raytheon, **IBM**, and **Remington Rand**, which had just acquired **Engineering Research Associates (ERA)**. The contract was awarded to IBM, who placed Morton Astrahan in charge of systems planning and part of the development. In implementing SAGE, IBM expanded and improved the Whirlwind. Each new machine, which carried the military prefix AN/FSQ (Army-Navy Fixed Special eQuipment), was actually two identical computers that operated in duplex mode for greater reliability. The AN/FSQ-7 was a 75KHz (kilohertz) single-address machine with 8,192 32-bit **words**, the first IBM computer to use **magnetic core** memory. **Auxiliary storage** included 12 **magnetic drums** of 12,288 words each. Each AN/FSQ-7 was capable of supporting 100 display consoles while sending and receiving **data** from 12 remote sites. Beyond its contribution to national defense, SAGE was notable for major technological advances in computer technology, among them a very large **real-time system**, the first **distributed system**, a highly disciplined **program** structure, **multiplexing** via I/O **channels**, digital **data transmission**, **cathode ray tube (CRT)** monitors, **light pen** interaction, and duplexed operation for reliability. The benefits of the experience gained with the project showed up rapidly in the **IBM 700 series** computers that followed shortly after conclusion of IBM's intensive participation in SAGE.

Additional reading: Redmond and Smith, *From Whirlwind to Mitre: The R & D Story of the SAGE Air Defense Computer.*

Scanner. *See* **Optical scanner**.

Schickard calculator CALCULATORS

As far as is now known, the first mechanical **calculator** was that of mathematician and linguist Wilhelm Schickard, a professor and Protestant minister in the German town of Tübingen. His 6-digit machine could add and subtract, and since he and his friend Johann Kepler were quite familiar with both **logarithms** and **Napier's bones**, he built the latter

into his machine to facilitate the accumulation of the partial products produced during multiplication. On 20 September 1623, Schickard wrote to Kepler: "You would burst out laughing if you were present to see how it carries by itself from one column of tens to the next or borrows from them during subtraction." Only a rudimentary sketch of Schickard's machine, which he called a "calculating clock," remains, and that sketch became the basis for a postage stamp issued by what was then West Germany in 1973 (*see* Figure 35).

Additional reading: Aspray, *Computing Before Computers*, pp. 35–39.

Figure 35. Schickard's "calculating clock" antedates Pascal's calculator by 20 years and that of Leibniz by almost 50, but in some ways it may have been superior to both. But since the original and all copies have been lost, this 1973 postage stamp of West Germany depicts a plausible reconstruction made from the only sketches known to historians. (From the collection of the author)

SEAC COMPUTERS

SEAC (Standards Eastern Automatic Computer), three-address **digital computer** built in 1950 for the National Bureau of Standards (NBS) in Washington by a team led by Samuel Alexander, was the first **stored program computer** to use solid state diode logic rather than **vacuum tubes**. The SEAC used **ultrasonic memory**, mercury delay lines purchased from the supplier to the **EDVAC** project. **Magnetic tape** was used for external storage. The first **program** for the SEAC was written by Ethel

Marden. The SEAC remained in service for 14 years. NBS also commissioned the **SWAC** (Standards Western Automatic Computer) that was built at the *Institute for Numerical Analysis* in Los Angeles. In 1953 a SEAC clone called *MIDAC* (MIchigan Digital Automatic Computer) was built at the University of Michigan under the direction of John Weber Carr III, later president of the **Association for Computing Machinery (ACM)**.

Additional reading: Ralph J. Slutz, "Memories of the Bureau of Standards' SEAC," in Metropolis, Howlett, and Rota, *A History of Computing in the Twentieth Century*, pp. 471–477.

Search engine　　　　　　　SOFTWARE

A *search engine* is an **information retrieval** program at a specific Website that is invoked through a **browser** in order to satisfy queries posed by a user at a **personal computer**. It is the combination of browser and search engine that gives the **Internet**, or more particularly the **World Wide Web**, such power. A true search engine continually searches ("crawls through") the full text of Websites, new and old, in order to build an ever-expanding directory of key words and phrases that someone may later use in a query. The first of these was *WebCrawler*, the 1994 creation of Brian Pinkerton at the University of Washington. Other services such as *Yahoo.com*, built by Jerry Yang and Dave Filo in that same year, do not search but rather look up keywords in a directory compiled by humans. Search engines of either kind are reached through specification of a *Uniform Resource Locator* (URL) such as www.google.com, www.northernlight.com, or www.altavista.com, which pertain to three of the most popular engines. *Google*, founded in 1998 by former Stanford University **computer science** doctoral students Sergey Brin and Larry Page, had indexed the most Web pages as of 2002 and was the sixth most frequently referenced Website of any kind based on its speed of response, the large number of pages indexed, and the relevancy of its responses based on a *Page rank* **algorithm** (playfully named for Page, not "page"). A typical query to any search engine might ask for references to Websites that contain information about *Eskimos* AND *penguins*, or about *lions* OR *tigers* (or both). More complicated queries can be built using multiple connectives from **Boolean algebra**. The response is in the form of a list of links (URLs) to the Websites that meet the request, each containing a few words from the site that help the user decide whether to visit it by clicking its link. Search engines attempt to rank responses in decreasing order of likely relevance.

Additional reading: Glossbrenner and Glossbrenner, *Search Engines for the World Wide Web*.

Sector　　　　　　　HISTORICAL DEVICES

The *sector* is a simple but effective mechanical **analog computer** dating back to the sixteenth century. The device consists of two legs of equal length and identical linear calibration connected to form the shape of the letter A without its crossbar. The legs are fastened at the apex in such a way that the legs can be moved to form any desired angle and maintain that angle once set. Multiplication and division are done by measuring ratios of the sides of similar triangles, one inside the other. Measurement of lengths is done with auxiliary calipers. For example, suppose we want to compute the product, *p*, of 2.7 and a given multiplicand *m*. To do so, the sector is configured as in the following diagram.

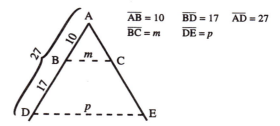

Choose a point B on one leg that is 10 units from the apex, and open the legs so that at that point their separation is *m*, the base of the smaller triangle. Now find point D, 27 units from the apex along that same leg, because 27 / 10 is the desired multiplier, 2.7. Then, by the geometric theorem that relates corresponding sides of similar triangles, the length of the longer base at that point, *p*, as measured with the calipers, must be 2.7*m*, the desired product. As with any analog device, the result will have reasonable **accuracy**, but low **precision**.

In 1606, Galileo described the operation of the sector in his very first publication, and then built and sold several of them. But the English mathematician Thomas Hood had also published a description of a sector, a term he may have originated, in 1598. Still earlier, the Frenchman Jean Errard depicted what appears to be a sector in his book of 1594, and at almost that same time, the Swiss mathematician Joost Bürgi, the later coinventor of **logarithms**, invented the *proportional compass*, a device somewhat similar to that of a sector. The authors of the reference estimate that the true sector was invented by an unknown military officer about 20 years earlier.

Additional reading: Michael R. Williams and Erwin Tomash, "The Sector: Its History, Scales, and Uses," *IEE Annals of the History of Computing*, **25**, *1* (January–March 2003), pp. 34–47.

SEED   HARDWARE

SEED is an acronym for Self-Electro-optic Effect Device, a light-amplifying switch invented by David Miller in 1986 that established the feasibility of **optical computing** for the first time. At low voltages, a SEED is transparent, but at high voltages, it is opaque. SEEDs and similar devices called *optical transistors* are used in large numbers to construct experimental **parallel computers**.

Additional reading: *Understanding Computers: Alternative Computers*, pp. 94–97.

Semiconductor   HARDWARE

A *semiconductor* is a solid chemical element or compound that can conduct electricity under some conditions but not others, making it a good medium for the control of electrical current. Certain nineteenth- and early-twentieth-century milestones in semiconductor physics laid the groundwork for the invention of the **transistor** in 1948, a very important milestone in both physics and **computer science**. In 1874, Carl Ferdinand Braun of Germany described semiconductor action in the form of *rectification*. In 1900, Paul Drude formulated a useful model of electrical conductivity that explained why certain materials were good conductors, others were good insulators, and still others had intermediate properties. But it was left to J.

Figure 36. A mathematician calculating with a sector. (Reprinted from D. Henrion, *L'usage du compas de proportion*, Paris, 1681. Illustration provided through the courtesy of the editors of the *IEEE Annals of the History of Computing*)

Königsberger and J. Weiss of Germany to coin the word *Halbleiter* (semiconductor) in 1911.

The conductance of a semiconductor varies depending on the current or voltage applied to a control electrode. The specific properties of a semiconductor depend on the impurities, or *dopants*, added to it. An N-type semiconductor carries current in the form of negatively charged electrons like the conduction of current in a wire. A P-type semiconductor carries current as electron deficiencies called *holes*. "Holes" have a positive electric charge, equal and opposite to the charge on an electron, and flow in a direction opposite to the flow of electrons. Silicon is the best-known semiconductor, forming the basis of most **integrated circuits (ICs)**. Common semiconductor compounds include gallium arsenide, indium antimonide, and the oxides of most metals. Of these, gallium arsenide (GaAs) is widely used in low-noise, high-gain, weak-signal amplifying devices. A semiconductor device such as a transistor can perform the function of a **vacuum tube** having hundreds of times its volume.

Additional reading: *Encyclopedia of Computer Science*, pp. 892–902.

Server COMPUTER TYPES

A **server** is a *host computer* in a **client-server computing** environment. The host may actually be a **cluster** of servers consisting of a principal **digital computer** devoted to computation and communications assisted by one or more auxiliary servers dedicated to a special purpose. The most common special-purpose servers are the *file server* and the *mail server*. A file server handles all matters pertaining to the **files** its **memory** contains—open and close a file, transmit a copy to a client, update the file, and so on. A mail server handles all aspects of the **electronic mail** system supported by the overall system.

Additional reading: *Encyclopedia of Computer Science*, pp. 710–711.

Set DATA STRUCTURES

In **computer science**, a *set* is a **data structure** consisting of a collection of related but unordered entities no two of which are alike. Thus we may speak of a set of coins or a set of playing cards (a Bridge hand, perhaps). A mathematical set may be infinite, as is the set of all **prime numbers**, but the sets used in computing are necessarily finite. A set that has no members is called the *null set*. The set is the simplest of all data structures because the stated interconnection of its constituents is none. Typical operations performed on sets are the testing to see if a certain entity is or is not a member of a given set; creation of new sets through the *union*, *difference*, or *intersection* of other sets; and formation of subsets of a given set. The first and apparently the only **problem-oriented language** that was invented to facilitate set operations was **SETL**, but certain **procedure-oriented languages** such as **Pascal** support

primitive set operations as an integral part of their grammar.

Additional reading: *Encyclopedia of Computer Science*, pp. 1569–1570.

SETL LANGUAGES

SETL (SET Language) is a **problem-oriented language** developed by Jacob Schwartz of New York University (NYU) in 1973. SETL is available for many systems, including the **IBM 360 series** and many **Unix** systems. ISETL (Interactive SETL), a variant developed by Gary Levin while at Clarkson University, is intended for teaching and learning mathematical concepts. SETL2, also from NYU, is intended for prototyping or building large applications. It extends SETL by including a package and library system similar to that of **Ada** and supports **object-oriented programming**. Both ISETL and SETL2 are available for most common **personal computers**.

Additional reading: *Encyclopedia of Computer Science*, pp. 1569–1570.

SGI. *See* **Silicon Graphics, Inc. (SGI)**.

SHARE. *See* **User group**.

Shareware SOFTWARE

Shareware is **freeware** for which an honor-system charge is considered due from those who choose to continue using the **software** beyond the end of a stated trial period. The concept was originated by Jim Knopf, a.k.a. Jim Button, who wrote the rudimentary **database** program *PC-File* for the **IBM-PC** in 1982 and who co-founded the *Association for Shareware Professionals* in 1987. The term *shareware* was coined in 1983, apparently independently, by Jay Lucas, writing in *InfoWorld*, and by Bob Wallace. Wallace, former **Microsoft** Employee #9, marketed his **word processing** program *PC-Write* on that basis, the first reasonably successful shareware product.

Additional reading: *Encyclopedia of Computer Science*, pp. 733–735.

Sharp Corporation INDUSTRY

The *Sharp Corporation* is the current name of a metal works company founded in Tokyo in

1912 by Tokuji Hayakawa. In 1915, Hayakawa invented a mechanical pencil that he called the *Ever-Sharp*, a name that quickly elided to Eversharp and shortened to Sharp as the company name. In 1964, Sharp produced the first all-**transistor** desktop **calculator** and in 1973 the first such calculator to use a **liquid crystal display (LCD)**. In keeping with its motto "From Sharp minds come sharp products," Sharp introduced an **Internet** ViewCam based on MPEG4 technology in 1999. Sharp is a worldwide developer of digital technology such as optoelectronics, infrared sensors, **semiconductors**, LCD displays, and **flash memory**. The Sharp Corporation employs 66,700 people, nearly half outside Japan, and operates from 64 bases in 30 countries. Annual sales exceed $12 billion.

Additional reading: http://www.sharp.co.uk/companyinformation/history

Shift register   COMPUTER ARCHITECTURE

A *shift register* is a **register** whose content can be moved laterally (shifted) left or right by n places. There are variations of the shift operation each of which replaces **bit** positions vacated during shifting in accord with a different **algorithm**. With a *circular shift*, bits shifted out of one end of the register reappear at the opposite end. With a *logical shift*, **zeroes** replace vacated positions. With an *arithmetic shift* to the right on a **digital computer** that represents negative numbers in complement form (*see* **Binary number system**), the leftmost bit (the sign bit) is propagated to the right. On early computers, the *accumulator* doubled as a shift register in the sense that it was the only register whose contents could be shifted, and shifting was serial—the time to shift by n places was n times longer than a shift by one place. As faster performance digital computers evolved, their shift register, seemingly a singular entity, became an **array** of m **words** (bit sequences of register length m), a two-dimensional $m \times m$ array of bits with conduction paths between rows such that any word brought to the 0-th row would instantly create $m - 1$ images below it, each successively left shifted one bit position. Then, when a running **program** called for a circular shift of n places,

the computer could just use **table lookup** to access the nth row of the shift register for a left shift or the $(m - n)^{th}$ row for a right shift, thus making shifting time independent of the number of places shifted.

Additional reading: *Encyclopedia of Computer Science*, pp. 1572–1573.

Short Order Code   LANGUAGES

Short Order Code was a small interpretive **programming language** devised by John W. Mauchly for use on the **BINAC** of 1949. The language did little more than let the user use symbolic rather than numeric **operation codes (opcodes)** for the command portions of **instructions**, but the Short Order Code is nonetheless considered the antecedent of subsequent **assembly language** and **high-level language**.

Additional reading: http://zeus.fh-brandenburg.de/history/19.html

SIAM. *See* Society for Industrial and Applied Mathematics (SIAM).

Siemens-Nixdorf   INDUSTRY

The Nixdorf company was founded in 1952 in Paderborn, West Germany, by Heinz Nixdorf to produce **vacuum tube** electronic **calculators** and grew to become the fourth largest computer company in Europe. Nixdorf captured a good share of the European **minicomputer** market before the **DEC PDP series** began to impact sales. In 1990, four years after Nixdorf died, Nixdorf Computer AG was acquired by Siemens AG for $350 million and renamed *Siemens-Nixdorf Information Systems* AG. The company enjoyed rapid growth through the 1990s, and in 1998 Siemens-Nixdorf Retail and Banking Systems was founded as an independent entity. On 1 October 1999, Siemens-Nixdorf split into two **information technology** companies, **Fujitsu**-Siemens with headquarters in Amsterdam and Wincor Nixdorf, an independent company with headquarters in Paderborn and offices in 28 other countries including the United States. Fujitsu-Siemens is principally a **hardware** company with a strong line of **servers**, **personal computers**, and **handheld computers**. Wincor Nixdorf specializes in **software** support for **electronic commerce**.

Additional reading: Chandler, *Inventing the Electronic Century*, pp. 185–189.

Sieve of Eratosthenes   ALGORITHMS

The *Sieve of Eratosthenes* is an **algorithm** invented by Eratosthenes of Cyrene [276–196 B.C.], the third librarian of the **library** at Alexandria, to compute the first *n* **prime numbers** for any value of *n*. For example, if $n = 100$, write down (or place in an **array** in **memory**) the integers from the first known prime, 2, to 100. Then cross out (or zero out, in memory) all multiples of 2: 4, 6, 8, . . . 100. Next, note the first nonzero number after the last prime isolated, the 3 following the 2 at this step, and cross out all multiples of 3. Continue in this way until the first nonzero number in the list is equal to or exceeds the square root of *n*. The only numbers left must be prime. Writing a program to implement the algorithm is a common assignment in many elementary programming courses. And, of course, the primes generated are useful in real-world applications of mathematics and computing. Sieves are actually a general class of algorithms that may be adapted to the solution of other problems in **discrete mathematics**. In his essay "A History of the Sieve Process" in Metropolis, Howell, and Rota (*see* Listing of Cited References), D. H. Lehmer describes several mechanical devices of the period 1926 to 1946, some built by himself and his students, for solving sieve problems from the realm of so-called "pure" mathematics. Beginning in about 1946, such efforts were directed to writing sieve programs for general purpose **digital computers**, which is very much an *applied mathematics* endeavor. Thus, after a century or more of divergence, the two aspects of mathematics began to coalesce. And we owe it all to the venerable Eratosthenes.

Additional reading: Petzold, *CODE*, pp. 359–360.

Silicon Graphics, Inc. (SGI)   INDUSTRY

Silicon Graphics, Inc. (SGI) of Mountain View, California, was founded in 1981 by James Clark after he left Stanford University. Cofounder Marc Regis Hannah was named vice president and chief scientist. The company specializes in high-performance **workstations** for use in **computer graphics** in general and **computer animation** in particular. SGI products were used to create the dinosaurs in the movie *Jurassic Park*. Clark stayed with SGI until 1994 when he left to co-found **Netscape Communications** with Marc Andreesen. SGI bought **Cray Research** in 1996 to further augment its strength in scientific **data visualization** and computer animation equipment that facilitates production of quasi-three-dimensional motion pictures.

Additional reading: http://www.sgi.com

SILLIAC   COMPUTERS

The *SILLIAC* (Sydney ILLIAC) was a copy of the **ILLIAC**, and hence of the **IAS computer**, installed at the University of Sydney, Australia. Work on the machine began early in 1954 with Brian Swire as chief engineer and ran its first program on 24 June 1956. John Bennett joined the SILLIAC project early in 1956 as chief numerical analyst to the Adolph Basser Computing Laboratory, named for the benefactor who funded the project. Bennett later became one of the pioneers of the **EDSAC** as Maurice Wilkes's first research assistant at the Mathematical Laboratory in Cambridge, England. The SILLIAC remained in active use for 12 years and was decommissioned in May 1968 when it was replaced by an **English Electric KDF9**.

Additional reading: http://www.austehc.unimelb .edu.au/tia/589.html

Simon   COMPUTERS

In his classic 1949 book *Giant Brains, or Machines That Think*, American computer pioneer Edmund Berkeley outlined **logic circuits** of his design that, when implemented, would form a small programmable **digital computer** called *Simon*, for "simple Simon." He went on to describe the circuitry in more detail in issues of *Radio Electronics* in 1951 and 1952, and eventually sold 400 copies of the complete plans. By the criteria of the reference, Simon was the first **personal computer (PC)**, but to make one, hobbyists had to gather their own parts, in contrast to the **MITS Altair 8800** of 1975 that was sold in ready-to-assemble kit form and is more often cited as the first true PC. Nonetheless, Berkeley can be credited with a significant milestone: At a time when machines like the **ENIAC** and the **EDVAC** filled whole rooms, he created the expectation that the average person would eventually be able to afford and house a computer of his or her own.

Additional reading: http://www.blinkenlights. com/pc.shtml

Simplex method   ALGORITHMS

The *simplex method* is an **algorithm** devised in 1947 by George Dantzig for solving *linear programming* problems. Linear programming (LP) is not a kind of computer programming but rather a method of solving practical real-world problems in which a particular quantity is to be minimized or maximized. Given a suitable algorithm, a computer **program** can be written to do this, but LP antedates the development of **digital computers**. The quantities to be optimized are usually subject to the constraint that they must be nonnegative integers. Consider this sample problem due to Temple University mathematics professor David Zitarelli:

A clothing manufacturer makes two types of suits, wool and a blend of wool and polyester. The wool suit sells to the retail stores for $160 each and the blend for $90 each. The cost of each blend suit is $40. If the manufacturer makes no more than 120 suits per week and budgets no more than $6000 per week, how many of each type should be made to maximize profit?

Since this problem involves only two variables, it has a simple geometric (graphical) solution (20 wool suits and 100 blends), but in his work as a mathematical adviser to the U.S. Air Force, Dantzig was confronted with problems involving hundreds of variables. The simplex method that he invented to solve them has become the standard algorithm in the field and hence is a major milestone of **computer science** in general and **numerical analysis** in particular. In the reference, Dantzig himself modestly points out that although not known to him at the time, algorithms that were essentially identical to his simplex method had been published and forgotten at least four times, by Joseph Fourier in 1824, Charles de la Vallée Poussin in 1911, Leonid Kantorovich in 1939, and Frank Hitchcock in 1941. But their methods applied only to spe-

cial cases of LP, so Dantzig is appropriately to be credited for the breakthrough that is so widely used in business, industry, and government.

Additional reading: George Dantzig, *"Origins of the Simplex Method,"* in Nash, *A History of Scientific Computing*, pp. 141–151.

Simpson's rule   ALGORITHMS

Simpson's rule is an **algorithm** for numerical integration, also called *numerical quadrature*, published in 1750 by Thomas Simpson [1710–1761] and based, with acknowledgment, on earlier work of Newton and Roger Cotes [1682–1716], who chose not to publish it within their lifetimes. It is not the most accurate method of quadrature, but it is easily understood and programmable by beginning students of **numerical analysis**. Basically, the region of integration is divided into a *mesh* consisting of an even number of uniform size intervals of width h, and the area of successive pairs of intervals is approximated by fitting a parabola through the three points that define the top of the pair. The error of approximation can be shown to be proportional to h^5, so it can be made as small as desired (at the expense of greater computational time, of course) by choosing a very small value for h (a large number of mesh points). Simpson's rule is just one of a sequence of Newton-Cotes formulas of increasing accuracy (errors proportional to higher powers of h). These formulas in general and Simpson's rule in particular are a milestone because they gave confidence that complicated **functions** that could not be integrated in so-called *closed form* (expressed in terms of known functions) could nevertheless be approximated with a reasonable amount of arithmetic.

Additional reading: Ralston and Wilf, *Mathematical Methods for Digital Computers*, vol. 1, pp. 242–248.

Simscript   LANGUAGES

Simscript is one of the three most widely used discrete-event **simulation** languages, the others being **GPSS** and **Simula 67**. Simscript has gone through several revisions over several decades, but the original version was written as a **Fortran** preprocessor in 1962 by **RAND**

Corporation scientists Harry Markowitz (a Nobel Prize winner in Economics), Bernard Hausner, and Herbert Karr. Simscript II.5 is still in use.

Additional reading: http://cgibin.erols.com/ziring/cgi-bin/cep/cep.pl?_alpha=s

Simula 67 LANGUAGES

Simula 67, the successor to *Simula 1*, was the first **object-oriented programming (OOP)** language. Simula 67 introduced most of the key concepts of OOP such as *objects*, *classes*, and *virtual procedures*. The Simula languages were developed at the Norwegian Computing Center, Oslo, Norway, by Ole-Johan Dahl and Kristen Nygaard. Nygaard's work in operations research in the 1950s and early 1960s showed the need for precise tools for both the description and **simulation** of complex human–machine systems. In 1961 work began on the design of a language usable for both purposes. Simula 1 was initially a **problem-oriented language** influenced by the earlier discrete-event simulation language **Simscript**. But when the concept of inheritance was invented in 1967 and added to it, the resulting Simula 67 became a general-purpose language usable for many tasks, including, but not limited to, system simulation. *Inheritance*, the most important innovation in Simula 67, is a mechanism for declaring subclasses that have the ability to inherit and use selected portions of a previously created parent class without having to incorporate the entire and possibly voluminous parent into a new **program**. Simula 67 **compilers** were soon written for their computers by **Sperry Rand**, **Control Data**, **IBM**, **Burroughs**, **Digital Equipment Corporation (DEC)**, and other vendors in the early 1970s. Simula 67 is still being used, but its main impact has been its foundation for object-oriented programming. Among the many object-oriented languages influenced by Simula 67 are *Beta*, *Eiffel*, *Clos*, *Self*, *C++*, and *Java*.

Additional reading: Wexelblat, History of Programming Languages, pp. 439–493.

Simulation APPLICATIONS

Simulation is a process by which the essential elements of a real-world system are modeled on a computing device. All **analog computers** are simulators by their very nature. A **digital computer** becomes a simulator when it is programmed to emulate the behavior of any one of a number of dynamic systems—economic behavior, weather, nuclear reactors, air- or spacecraft flight, stock market fluctuations, and so on. Special-purpose simulation languages are either continuous-event modeling languages such as **Dynamo** or discrete-event modeling languages such as **GPSS**. Output of a simulation is seldom just a list of numbers; more usually elaborate **computer graphics** displays are used to monitor conditions that vary as input **parameters** are changed at an operator's console. Perhaps the earliest digital simulation to come to widespread public attention was the CBS electoral model of 1952. Running on a **UNIVAC** and with only 7% of the vote total in, the model correctly predicted that Eisenhower would be elected in a landslide, projecting that he would receive 438 electoral votes. The final total was 442 for Eisenhower and 89 for Adlai E. Stevenson.

Simulation can also be a tool of **computer science** rather than merely one of its applications. It was once fashionable to simulate one digital computer on another of quite different **computer architecture** by writing an appropriate *simulator* for it. The simulator, running on a *host computer*, would read instructions encoded for a *target machine* one by one and, for each, execute the one or (usually) more of its own **instructions** needed to produce the intended **bit**-faithful result. In this sense, the simulator, though it may itself have been written in a **high-level language**, was an **interpreter** for the **machine language** of a computer that might not yet have been operational. The simulator would necessarily run much more slowly than the anticipated machine, but it would at least let programmers debug **programs** in advance of its delivery (*see* **Bug**). In the other direction, a simulator could be used to run older *legacy programs* on a new computer of different architecture. To increase efficiency, **firmware** was often employed to raise simulation to the level of **emulation**. A third rationale for software simulation is educational. An especially entertaining simulator for the historic **EDSAC** computer written by Martin Campbell-

Kelly of Warwick University is available for most **personal computers** (*see* http://www. dcs.warwick.ac.uk/~edsac/ and the reference). Sometimes simulators are written for computers that never existed in order to teach the rudiments of **machine language** programming. The best-known example is the simulator for the MIX (1009) computer used in the **Knuth textbooks**.

Additional reading: *Encyclopedia of Computer Science*, pp. 1578–1587.

Sketchpad COMPUTER GRAPHICS

Sketchpad, designed by Ivan Sutherland in 1963 for the TX-2 **digital computer** at MIT, is considered the first substantive **computer graphics** application (*see* **TX-0 / TX-2**). The system allowed users to work with a **light pen** to create two-dimensional (2D) graphics by first making simple primitives, like lines and circles, and then applying operations that constrain the geometry of the shapes created from them. Its **graphical user interface (GUI)**, whose displays could be modified from within Sketchpad, constitute its claim to having been the first **visual language**. By defining appropriate constraints, users could develop structures such as complicated mechanical linkages and move them about in real time.

Additional reading: Hiltzik, *Dealers of Lightning*, pp. 90–91.

Slide rule CALCULATORS

The *slide rule* is an analog **calculator** invented by mathematician William Oughtred in about 1628 based on some earlier work of Edmund Gunter. (It was Oughtred who, in his 1631 classic *Clavis Mathematica*, proposed the use of the symbol $\times$ for multiplication and coined the terms *sine*, *cosine*, and *tangent* for the three principal trigonometric ratios.) Richard Delamain, a student of Oughtred's, made a circular slide rule in 1629 that he described in his *Grammelogia*, but very little use was made of either his or Oughtred's invention for almost 200 years. James Watt, inventor of the steam engine, made the first well-made slide rule in the very late 1700s, and then 19-year-old French officer Amédée Mannheim added a movable cursor to the Watt design to form the

slide rule as we know it today. The operator of a slide rule effects multiplication by positioning a sliding scale against a fixed scale, each of which is ruled with gradations laid out in logarithmic proportion. Adding **logarithms** (portions of two properly aligned slide rule scales) and then looking up an *antilog* (read off one of the scales) produces the desired product. Division is done in the inverse way. The operator must, of course, keep track of decimal points. The accuracy of a slide rule is limited only by its length, but the length most conducive for use as a portable **calculator** yields answers that are accurate to only three or four decimal places. Beginning in the late nineteenth century, excellent slide rules were manufactured by the Eugene Dietzgen company of Chicago, a firm founded by a German immigrant of that name in 1885. The most popular slide rule used by engineering students of the twentieth century was Keuffel & Esser's *Log Log Duplex Decitrig*, the last of which was made in 1976. For another ten years or so, it was fashionable to mount them in wall cases in computing centers labeled "In case of emergency, break glass."

Additional reading: Aspray, *Computing before Computers*, pp. 27–33.

Figure 37. A typical Keuffel & Esser log-log-duplex-decitrig slide rule of the 1950s. The title is justified only by the scales on the reverse side. (Library of Congress)

Smalltalk LANGUAGES

Smalltalk, the first true **object-oriented programming** language, was conceived at **Xerox PARC** by Alan Kay in 1972 and implemented by Dan Ingalls and Adele Goldberg. The first version released was Smalltalk-80, but not until 1983. Kay credits several prior antecedents— the Burroughs B20 **file** system, the **Burroughs B5000** systems architecture, **Lisp**, **Sketchpad**, and the **class** concept of **Simula**. Smalltalk **procedures** are called *methods*, and a method may call another method by sending a message to that method's *object*, the abstract **data struc-**

ture that is packaged with its associated methods (*see* **Abstract data type**). After more than 20 years, the language still has a group of devoted users.

Additional reading: Kaehler and Patterson, *A Taste of Smalltalk.*

Smart card   APPLICATIONS

A *smart card* is a credit or debit card that contains **memory** in the form of a magnetic stripe and logic in the form of a tiny **integrated circuit (IC)** chip. The idea of incorporating an IC into an identification card was patented by German inventors Jurgen Dethloff and Helmut Grotrupp in 1968 and Kunitaka Arimura of Japan in 1970. In 1974 French inventor Roland Moreno mounted a chip on a card and devised a system to use the card for payment transactions. In 1979, Michael Ugon from **Machines Bull** in collaboration with **Motorola** created the first card containing a complete **microprocessor** and was the first to call it a "smart card." In 1984, the French Postal and Telecommunications Services carried out a successful field trial with **telephone** cards, and within a few years 60 million such cards were in circulation. Bank cards are more complex because of their need for **data encryption**; cryptography is essential to debit cards. By 1994 all French bank cards included IC chips. France Telecom and Carte Bancaire were early adopters of smart cards. VISA, MasterCard, and American Express now all issue smart cards.

Additional reading: http://www.getoranged.co.za/nemo_history.php

Snobol   LANGUAGES

Snobol was a **string processing** language designed and implemented in1963 at the Bell Telephone Laboratories by Ralph Griswold, Ivan Polansky, and David Farber with the assistance of programmer Laura White Noll. Snobol was followed by Snobol 2 in 1964, Snobol 3 in 1965, and Snobol 4 in 1967. Snobol 4—the work of Griswold, Polansky, and Jim Poage— has especially powerful **pattern matching** features that attracted a large body of users and that qualify this version as the actual milestone. Griswold went on to create **Icon**, a more structured string processing language.

Additional reading: Wexelblat, *History of Programming Languages*, pp. 601–660.

Society for Industrial and Applied Mathematics (SIAM)   GENERAL

The *Society for Industrial and Applied Mathematics (SIAM)* was established as a professional society in 1952 "to further the application of mathematics to industry and science." Its emphasis on computer applications increased over the years as is reflected in the list of distinguished computer scientists and numerical analysts who have served as SIAM president, a list that includes John W. Mauchly, Alston Householder, J. Barkley Rosser, J. Wallace Givens, Burton H. Colvin, and C. William Gear. SIAM is one of the three professional societies that **computer science** and engineering professionals tend to join, the others being the **Association for Computing Machinery** and the **IEEE Computer Society**. The SIAM publications of most interest to these professionals are the peer-reviewed *SIAM Journal on Computing*, . . . *on **Numerical Analysis***, . . . *on **Discrete Mathematics***, . . . on ***Matrix** Analysis and Applications*, and . . . *on Scientific and Statistical Computing*.

Additional reading: I. Edward Block, "Shaping the Evolution of Numerical Analysis in the Computer Age: The SIAM Thrust," in Nash, *A History of Scientific Computing*, pp. 199–205.

Soft computing   GENERAL

Soft computing is a newly coined term that encompasses applications of **chaos theory**, **fuzzy logic**, **genetic algorithms**, and the *evolutionary algorithms* discussed in the article on **artificial life (AL)**. Springer-Verlag now publishes an international journal called *Soft Computing*, and there exists a *World Federation on Soft Computing* which held its first conference in 1999.

Software   GENERAL

As used in computing, any applications or systems **program** is considered to be *software*, so its origin can only be correlated to that of the first program. But the word *software* is relatively new, having been coined by John W. Tukey in 1958 in response to the felt need for a broad generic term to refer to intangible programmatic information that is every bit as valuable as the very tangible **hardware** on which it resides. In the January issue of *American Mathematical Monthly* of that year, Tukey wrote, "Today the 'software' comprising the carefully planned interpretive routines, **compilers**, and other aspects of automative [*sic*] programming are at least as important to the modern electronic **calculator** as its 'hardware' of tubes, **transistors**, wires, tapes and the like." Software can be instantiated in hardware, as when recorded on a CD-ROM for distribution, but the CD is not the software; its stored information is. Software is now a major industry. Since software errors in critical systems are potentially (and have been actually) as catastrophic as hardware failures, the need to give careful attention to the construction of software has given rise to the new engineering discipline of **software engineering**.

Additional reading: *Encyclopedia of Computer Science*, pp. 1613–1620.

Software engineering   SOFTWARE

Software engineering is a branch of **computer engineering** that was founded on the premise that **software** intended for public use should be engineered to the same standards as prevail in traditional engineering disciplines. Poorly designed civil structures such as bridges and highways can kill, and so can bad software. The term *software engineering* was deliberately and provocatively chosen by the organizers of a NATO (North Atlantic Treaty Organization) conference on the subject held in Brussels, Belgium, in October 1968. More particularly, participant and historian Brian Randell believes that the term was coined by Friedrich L. "Fritz" Bauer. Two of the early stimulants to the development of software engineering were Harlan Mills's concept of the *Chief-Programmer Team* in 1971 and the publication of *The Mythical Man-Month* by Fred Brooks in 1975. The field has now matured to the point where there is a *Software Engineering Institute* (SEI) at Carnegie Mellon University in Pittsburgh, Pensylvania, sponsored by the U.S. Department of Defense (DoD); a *Software Engineering Laboratory* (SEL) created in 1976 at the NASA

Goddard Space Flight Center (NASA/GSFC) in partnership with the University of Maryland and the Computer Sciences Corporation (CSC); and many university degree–granting programs at both the undergraduate and graduate level. CASE (computer-aided software engineering), pioneered by James Martin, is the use of computer-assisted methods to organize and control the development of software for large complex projects.

Additional reading: *Encyclopedia of Computer Science*, pp. 1606–1611.

SOL   COMPUTERS

The *Processor Technology* company designed and sold a full line of boards for **S-100 bus** computers. In 1976, Lee Felsenstein and Bob Marsh of that company designed the **Intel 8080**–based *SOL* **personal computer (PC)** using company circuit boards. Because they wanted to advertise their computer as having the "wisdom of Solomon," they named their computer SOL in honor of Les Solomon, technical editor of *Popular Electronics* magazine. The SOL's **floating-point arithmetic** was handled by a separate circuit board designed by George Millard, and Charles Grant and Mark Greenberg were hired to write an **interpreter** for **Basic** for the machine. The SOL was packaged in a very attractive case with walnut wood sides. It sold well for a while, but Processor Technology did not survive in the highly competitive world of Silicon Valley startups and folded on 14 May 1979.

Additional reading: Freiberger and Swaine, *Fire in the Valley*, pp. 130–135.

Solid state memory   MEMORY

Solid state memory is **memory** whose constituent elements operate in accord with the behavior of electrons in solids. For quite some time after **transistors** and **integrated circuits** had replaced **vacuum tubes** in the **logic circuits** of **digital computers, magnetic core** continued to be used for main memory. In 1971, the Intel 3104 became the first solid state memory chip, and it and its successors quickly became the dominant form of **random access memory (RAM)**. The principal forms of solid state memory chips are *dynamic RAM* (**DRAM**), invented by Robert H. Dennard in 1968, and low-power *static RAM* (SRAM), invented at Integrated Device Technology, Inc. (IDT) in 1981. DRAMs are used in integrated circuits that contain capacitors, which must be refreshed. SRAMs are faster but more expensive; their use is primarily confined to **cache** memory. Use of RDRAM, a type of DRAM made by Rambus, Inc. and licensed to **Intel**, is now being used in the highest-performance **personal computers (PCs)** because it is even faster than SDRAM. **Bubble memory** and **flash memory** are types of solid state memory used for **auxiliary storage** and as the only memory of appliances such as **Automatic Teller Machines (ATMs)**, videogames, vending machines, and **digital cameras**.

Additional reading: *Encyclopedia of Computer Science*, pp. 1132–1136.

Sony

INDUSTRY

Sony, which is not an acronym, began business in 1946 as the *Tokyo Telecommunications Engineering Company, Ltd.* Its founders were Akio Morita and Masaru Ibuka. In 1953, their company became one of the first firms outside the United States to acquire a license to produce **transistors**, and they used them to make a small transistorized **radio** that they marketed under the brand name Sony. The success of that product in the United States prompted a change in the company's name to Sony in 1957. By 1975, Sony held a 5.8% share of the U.S. color **television** market, and its share increased steadily thereafter based on the quality of its Trinitron tube. In 1982, the alliance of Sony and Philips introduced the compact disc (CD), and in 1985, the CD-ROM (*see* **Optical storage**). Also in 1985, Sony acquired **Apple**'s **magnetic** disk division. In 1991, Sony withdrew from the **digital computer** business but reentered in 1998 with the introduction of a line of **IBM**-compatible **personal computers**, both desktops and **laptop computers**. Sony had entered the videogame market in 1993 and by the end of 1998 had sold over 50 million Sony PlayStations.

Additional reading: Chandler, *Inventing the Electronic Century*, pp. 54–62, 65.

Sorting

ALGORITHMS

Sorting is the process of rearranging an initially unordered sequence of **records** until they are ordered with respect to all of or that part of each record designated as its *key*. For use with one of his **Hollerith machines** of the late 1800s, Herman Hollerith developed an **algorithm** called *Radix sort* because it depends on multiple sort passes, one for each digit (radix) position of the maximum value number to be sorted. On a card-sorting machine, the radix is invariably 10—the digits are decimal digits—but any radix may be used from two upward. But the algorithm is not well suited for use on a **stored program computer**. The earliest published description of Radix sort is by Leslie J. Comrie in 1929.

Sorting algorithms may be classified as being either *comparative* or *distributive*. A comparative algorithm rearranges record order by comparing record keys. A distributive algorithm moves records to or close to their final correct position based on intrinsic key characteristics. No sorting algorithm *based on key comparisons* can have running time superior to $O(n \log n)$ (*see* **Analysis of algorithms**), but the four primitive algorithms that are logically most straightforward fail to achieve that level of performance. *Selection sort, bubblesort, insertion sort*, and *enumeration sort* are all $O(n^2)$; that is, when the number of items to be sorted is doubled, running time is four times longer.

As John von Neumann pointed out in 1945, repetitive *merging* can be made the basis of a very efficient $O(n \log n)$ sort strategy. Once we are able to form some initially sorted sublists, no matter how short—and that is not difficult because the sublists need only be one or two numbers long—we can merge, say, two lists of length two to make one of length four, two fours to make eight, and so on, until we have formed one ordered list. **Recursion** is used in order to make the machine's **memory** remember all currently unprocessed lists.

When conditions permit, an extremely fast linear time sorting algorithm can be based on *frequency counting*. The algorithm, first described by Harold Seward in 1954, has no standard name. It was called *Mathsort* by Wallace Feurzeig in 1960 and *Ultrasort* by Reilly and Federighi in 1989. A linear time distributive algorithm applicable to **data** of restricted range was described by Earl Isaac and Richard Singleton in 1956. The method was called sorting by *address calculation* in the days of **machine language** programming—each data value being directed to an address equal to itself—but could now be called sorting by *index calculation* when implemented in a **high-level language**.

A **data structure** usable with an improved insertion sort is a *binary search tree* (*see* **Tree**). The first unsorted item becomes the *root* of the tree. Each successive item is then inserted recursively into the left subtree if it is strictly less than the root and into the right subtree if it is equal to or greater than the root. If insertions are made according to this rule, then an *inorder traversal* of the search tree will produce the desired sorted sequence. This algorithm, known as *Treesort*, was first described by David J. Wheeler in 1957 and Conway M. Berners-Lee (Tim's father) in 1958. Another algorithm that improves the performance of sorting by insertion, now called *Shellsort*, was published by Donald L. Shell in 1959.

The *Radix exchange sort*, an improved version of exchange sort, was invented in 1959 by Paul Hildebrandt, Harold Isbitz, Hawley Rising, and Jules Schwartz. The basic idea is to look at the binary representation of the keys to be sorted **bit** by bit, a column (radix position) at a time, starting at the leftmost column that has at least one 1-bit. Radix exchange runs in $O(pn)$, where p (for **precision**) is the number of bit positions to be examined. In 1962, C. A. R. Hoare also invented an improved version of exchange sort called **Quicksort**.

In 1964, John Williams invented an improved selection sort called *Heapsort* because it is based on a data structure called a *heap*, a complete binary tree in which the value of every node is less than or equal to the value of either child node. By "complete" is meant that there are no missing nodes except, perhaps, for one or more leaves at the right of the bottom level. Heapsort running time is $O(n \log n)$, even in the worst case.

All of the foregoing algorithms apply only to the *internal sorting* of data all of which fits in **random access memory**. The earliest widely used external storage medium was **magnetic tape**, so it is not surprising that so many *external sorting* algorithms were devised for sorting **files** stored on tape. Three important tape sorts are called *multiway merge*, *cascade merge*, and *polyphase merge*. Tape sorting algorithms are now mainly of historic interest. Disk sorts use an efficient internal sort coupled with a balanced multiway merge.

Additional reading: *Encyclopedia of Computer Science*, pp. 1649–1664.

Source code DEFINITIONS

A **program** expressed in **assembly language** or a **high-level language** intended for compilation is called *source code*, in contrast to the executable **object code** into which it is translated by an **assembler** or a **compiler**. A program intended for execution by an **interpreter** is both source code and its own object code, so the terms are not normally used in such a context.

Speech recognition APPLICATIONS

Speech recognition by **digital computer** is a subfield of **artificial intelligence (AI)**. But since speech recognition is a form of **pattern recognition**, it is a very difficult problem, especially if the intent is to be able to recognize the same sentence when spoken by any of a large number of speakers, even if they all speak the same **natural language**. But the problem is simplified if the goal is recognition of only one voice, and the computer can be trained by some initial trial and error. It should also be understood that "recognition" means merely that spoken words are translated into their corresponding typed form. No attempt to teach the computer to "understand" what is being spoken is implied, for that would increase the complexity of the problem by an order of magnitude. Yet it is very useful to be able to dictate to a computer rather than having to type documents letter by letter. Of course, results must then be edited because only a very (artificially) intelligent computer **program** would be able to guess which form of "there" and "their" or "for" and "four" to use. The first device invented to analyze speech was the *Vocoder* invented in 1928 by Homer Dudley of Bell Labs, who later invented a **speech synthesis** machine called the *Voder*. A significant advance in speech recognition was made in 1988 by Teuvo Kohonen of Helsinki University in Finland. Kohonen used a **neural net**work approach with a computer that could be trained to recognize over 90% of the words spoken by half a dozen different speakers. A new speaker had only to dictate about a hundred words from a given list before the computer could identify most of what was said thereafter, whether or not the words spoken were in the training list. For **personal computers (PCs)** and **handhold computers** equipped with an online microphone, **IBM** now offers its *ViaVoice®* **software** that frees the user from dependence on a **mouse**, **keyboard**, or stylus for many applications.

Additional reading: *Encyclopedia of Computer Science*, pp. 1664–1667.

Speech synthesis
APPLICATIONS

Speech synthesis is the use of a machine, most usually a **digital computer** equipped with audio speakers, to generate artificial speech. The problem is not trivial but is easier than **speech recognition** because it is only necessary that the computer speak with a single voice in a single language, whereas recognition must cope with the attempt to understand a multiplicity of speakers who, in the most challenging applications, may not all speak the same **natural language**. One way to generate speech is to store a recorded version of, say, 100,000 English words and then program a computer to proceed through text and use **table lookup** to determine the word to utter next. The result is understandable to a human but not esthetically pleasing because it would sound **robotic**. A slight improvement can be made by looking ahead to find the punctuation marks in the stored text and then modify the pitch and volume of the words that precede them. Better methods try to emulate the naturally spoken transitions from one word to another. In 1936, Bell Labs scientist Homer Dudley invented the

first electronic speech synthesizer. Dudley called his "talking machine" a *voice coder*, which, shortened to just *Voder*, became a big hit at the New York World's Fair of 1939. Among the applications of text-to-speech synthesis are readers for the blind in which an **optical scanner** and a **personal computer (PC)** combine to provide voice output of books and documents. The first such device was invented in 1976 by Raymond Kurzweil and then improved through his collaboration with musician Stevie Wonder. Systems are also available that convert text displayed on a **monitor** to speech.

Additional reading: *Encyclopedia of Computer Science*, pp. 1664–1667.

Spelling checker
APPLICATIONS

A *spelling checker* is a **program** that checks each word in a document to see if it is contained in a stored list of reasonably common legal words of a **natural language** such as French, German, or American or British English. Allegedly incorrect spellings are reported to the user, who is given the opportunity to correct them. Possibly the first two spelling checkers were those of Les Earnest of MIT in 1966 and a program called SPELL written for the **Digital Equipment Corporation** DEC-10 in 1971 by Ralph Gorin of Stanford University. Spelling checkers are now expected to be an integral part of **software** such as **word processors**, **desktop publishing** systems, and Web page editors. A good spelling checker will not only flag a suspect word; it will try to suggest one or more similar words that may have been intended. Spelling checkers can only guarantee that a word is *some* correctly spelled word, not necessarily the one intended. If "from" is mistyped as "form" (a common transposition) or "affect" is misused for "effect," a spelling checker will not notice. Trying to detect these kinds of error requires a much more complicated and only marginally effective program called a *grammar checker*. As long as its word list has no incorrectly spelled words, a checker will never "miss" an incorrectly spelled word that is not, coincidentally, some other legal word. But spelling checkers often report correctly spelled words as possible errors. These may be proper names, technical terms, or uncommon words that are not in its main word list. Most systems allow a user to add words to their original list. Doctors, lawyers, and scientists, for example, generally use extensive auxiliary word lists designed for the specialized vocabularies of their fields.

Additional reading: *Encyclopedia of Computer Science*, pp. 1667–1668.

Sperry Rand INDUSTRY

Sperry Rand was formed in 1955 through the merger of the Sperry Corporation and **Remington Rand**. The company retained the latter's Univac division, which went on to produce the **Univac 1100 series** and other **digital computers**, and then changed its name again to **Unisys** when it merged with the **Burroughs Corporation** in 1986.

Spread-spectrum communications

COMMUNICATIONS

Although the term *spread-spectrum communications* was not coined for another 20 years, the concept of "frequency hopping" was patented in 1942 by American composer George Antheil and his Vienna-born wife, Eva Hedwig, the actress known in the United States as Hedy Lamarr. Their idea was that **radio**-controlled torpedoes could not be jammed by a targeted enemy if the frequency of transmission used was constantly varied in a way that was seemingly random to the eavesdropper but yet in accord with a pattern that was known to and could be synchronized with the receiving device. The principle could not be efficiently implemented at a size small enough for use on a torpedo until the development of the **transistor** but is now the basis for the **cellular telephone**, the **wireless connectivity** of computer **peripherals**, and **mobile computing**.

Additional reading: Hans-Joachim Braun, "Advanced Weaponry of the Stars," *American Heritage of Technology & Invention*, **12**, 4 (Spring 1997), pp. 10–16.

Spreadsheet SOFTWARE

A *spreadsheet* is an interactive two-dimensional **array** of **data** and formulas originally devised to facilitate financial and business modeling. The spreadsheet is the most significant **soft-**

Figure 38. Spread-spectrum communications. Austrian Eva Maria Hedwig, aka Hedy Lamarr, earned a significant patent for her co-invention, with her husband, George Antheil, of the concept now known as spread-spectrum communications. The article in the magazine depicted indicates that her contribution was just as significant as that of Antheil, an American composer. Thus she went from having nothing to hide in her first movie to showing how to hide information, visual or otherwise, from voyeurs who don't even know where to look. (Courtesy of American Heritage of Invention & Technology)

ware product developed for **personal computers** that had no prior incarnation on **mainframes**, although a form of noninteractive spreadsheet for scientific applications was incorporated in the language *Omnitab* of Joseph Wegstein and Joseph Hilsenrath in 1966. But the interactive cell-oriented spreadsheet as we now know it was invented by Robert Frankston, Daniel Bricklin, and Daniel Fylstra, who introduced their *VisiCalc* (*Visible Calc*ulator) **program** to the public on 11 May 1979 at the West Coast Computer Faire. Originally written in **assembly language** for a 32KB (kilobyte) **Apple II**, VisiCalc had a terse single-line menu. Its virtually instant popularity helped sell so many Apple IIs that other vendors had to react rapidly to license VisiCalc for use with their models as well. In 1981, *SuperCalc* was developed for the **Osborne portable** and became the primary spreadsheet for 8-bit **CP / M** computers. In 1983 Jonathan Sachs created *Lotus 1-2-3* for 16-bit **MS-DOS** computers. With online help, **macros**, sophisticated menus, and facilities for **computer graphics** and **database** management, it quickly became the leading spreadsheet product. Although spreadsheets originated as stand-alone programs, they are now typically packaged as components of integrated office programs that run under **Microsoft Windows**, **Linux**, and other **operating systems**. The leading spreadsheets are *Microsoft Excel*, Lotus 1-2-3, *Quattro Pro*, and a **Sun Microsystems** product distributed with Linux.

The spreadsheet format consists of a large rectangular array whose size is limited only by available **memory**. Only a portion of a large spreadsheet can be displayed, but the "window" through which it is viewed is movable. Spreadsheet columns are identified by letters, and rows by positive integers. For example, G7 refers to the cell in column G of row 7. After highlighting a cell by moving a **mouse** or arrow keys, the user may enter a number, a textual **string**, or a formula that refers to other cells. After completion of entry of a formula, the formula itself is hidden (and can be recalled for later inspection or editing), and its computed value is displayed in that cell. When the value or formula in any cell is changed, all spreadsheet formulas are recalculated and the display is updated, faster than the eye can follow. Spreadsheets are far more than mere accounting tools. They may be used as nonprocedural **problem-oriented languages**, and computer scientists have used them to model **cellular automata**, **perceptrons**, **systolic arrays**, and other **parallel computers**.

Additional reading: *Encyclopedia of Computer Science*, pp. 1670–1674.

SQL
LANGUAGES

SQL (Structured Query Language) was designed at **IBM** in 1974 for the express purpose of answering questions posed to a **relational database**. SQL is pronounced "sequel," its original name that had to be changed for rea-

sons pertaining to trademark and copyright. The language was adopted for use with **Oracle Corporation**'s *Oracle* in 1979. SQL **statements** are based on the *relational algebra* devised by E. F. Codd in 1972 and read very much like English. Despite a first attempt to standardize SQL in 1986 and an update in 1991, the various relational database **software** products use slightly different dialects, but a typical *dBase III +* SQL-like query is:

> list TITLE, FIRST, LAST for COMPANY ="GE" .and. CONTRIBUTION > 24.

The above pertains to a single **database** table; combining two or more tables requires more elaborate syntax.

Additional reading: *Computing Encyclopedia*, vol. 4, p. 72.

SSEC
COMPUTERS

In 1945, Frank Hamilton and Rex Seeber, disappointed that they could not convince Howard Aiken to make the Harvard Mark II a **stored program computer**, asked Thomas Watson of **IBM** for approval to begin design of a "super calculator" to be called the *Selective Sequence Electronic Calculator* (SSEC). Watson, still smarting from his perceived affront from Aiken at the dedication of the **Harvard Mark I**, gave quick and enthusiastic assent. A key member of the SSEC team was Byron Phelps, who had developed the electronic circuitry of the **Card Programmed Calculator (CPC)**. The machine was completed in Endicott, New York, in the summer of 1947 and then disassembled and shipped to the company's headquarters at 590 Madison Avenue in New York City. Reassembled and displayed in the window and dedicated in January 1948, the machine had 12,500 **vacuum tubes** and could do 14- × 14- digit decimal multiplication in 20 ms (milliseconds), division in 33 ms, and addition or subtraction of 19-digit numbers in just 300 μs (microseconds). The machine was 250 times faster than the Harvard Mark I, which took 6 seconds for multiplication and 11.4 seconds for division. Perhaps because it cost $1 million, a prodigious sum in those days, only one SSEC was ever built. Although its 13,000 vacuum tubes certainly must have done much to heat the building, evidence that it ever did much useful work is sketchy. But the machine was a public relations triumph for IBM. The SSEC became the source of many valuable IBM patents and was the training ground for two of Seeber's best programmers, John Backus, the driving force behind the later development of **Fortran**, and Edgar F. "Ted" Codd, who invented the **relational database**. The SSEC sat proudly in the window of IBM headquarters until, in an attempt to divert attention from the **UNIVAC**, it was replaced by an IBM 701 in 1952.

Additional reading: Bashe et al., *IBM's Early Computers*, pp. 47–59, 585–587.

Stack. *See* **Pushdown stack**.

Stanhope calculator CALCULATORS

Charles Mahon, 3rd Earl of Stanhope, who had earlier invented a cylindrical biconvex lens to eliminate spherical aberration, built a 13-digit mechanical **calculator** in 1775 that could multiply and divide by repeated addition or subtraction. The two **operands** of an arithmetic operation were dialed in on two sets of wheels, a fixed set at the front of the machine and a set at the rear that moved back and forth in proportion to the size of the multiplicand or divisor. The result was displayed on a third set of wheels at the top of the fixed front section. The machine was considered to be an experimental device and was never produced commercially.

Additional reading: Evans, *The Making of the Micro*, pp. 30–31.

Statement DEFINITIONS

The fundamental unit of a **program** written in **machine language** or **assembly language** is the **instruction**. The corresponding unit of a **high-level language** is the *statement*. Statements can be very simple, such as **declarations** or those that assign **expressions** to **variables**, or they can be rather involved **control structures**. The **compiler** used to process programs written in a particular high-level language must translate each statement encountered to the sequence of machine language instructions that will accomplish the intended objective. Statements that are grammatically incorrect are

flagged by the compiler, which will not run to completion until all such statements are corrected to conform to the syntax of the language at hand.

Steiger calculator CALCULATORS

In the 1890s Swiss engineer Otto Steiger invented a very reliable **calculator** called the *Millionaire*, the first commercially successful calculator capable of direct multiplication rather than multiplication through repeated addition. This was done by building mechanized versions of the multiplication tables into the upper lid of the machine and obtaining products through a form of **table lookup**. Somewhat earlier than Steiger, the Spaniard Ramon Verea had also devised a direct multiplication calculator, but it was never marketed. But more than 4,600 Millionaires were sold by the firm of Hans Egli and Son of Zurich between 1894 and 1935. The record survived until **Microsoft** generated even more millionaires in the 1990s.

Additional reading: Evans, *The Making of the Micro*, pp. 50–51.

Stirling's approximation MATHEMATICS

The *factorial* of a positive integer n is defined as being 1 for $n = 0$ or 1 and otherwise the product of n and all successively lower integers down to 2. (To 1, actually, but multiplication by 1 is obviously a waste of time.) Thus 3 factorial is 6 and 7 factorial is 5,040. Mathematicians use the notation $n!$ for n factorial, where the ! is not an expression of excitement but rather a *postfix* **operator** that is applied to the n that precedes it. By the courage of its inventor, Ken Iverson, the language **APL** uses ! as a *prefix operator* for consistency with the many other unary operators of mathematics. A *unary operator* is an operator that applies to one **operand**, as opposed to *binary operators* such as $+$, $-$, and $\times$ that apply to two operands, as in $a + b$ or $c \times d$. (Iverson calls these *monadic* and *dyadic* operators, respectively.) In 1730, James Stirling discovered the approximation

$n! \cong \sqrt{2\pi n}(n/e)^n$ + higher-order terms that are negligible for large n

where e is the base of the natural **logarithms**. It isn't that we need this formula to compute $n!$—using the formal definition would do—so why is the approximation a computer milestone rather than just a mathematical one? The answer is that in the **analysis of algorithms**, we often encounter $n!$ as part of an **expression** that tries to explain how the time for a certain **algorithm** increases with problem size n. For example, **sorting** n items by comparison of one to another is known to take time proportional to the logarithm of $n!$ So, just how fast does *that* grow? Ah, but if we compute the logarithm of Stirling's formula for $n!$, we find that running time to sort by comparison can be proportional to n times the logarithm of n, that is, faster than linear (proportional to n) but more slowly than proportional to n^2, the result of analyzing certain inefficient sorting algorithms. So, to theorists, at least, the Stirling formula is a dandy little gem that warrants inclusion in a book of computing milestones.

Additional reading: Weisstein, *Concise Encyclopedia of Mathematics*, p. 1739.

Stored program computer

COMPUTER TYPES

A *stored program computer* is a **digital computer** in which **programs** are stored in main **memory** in the same form as **data**. The term is usually associated with John von Neumann, the author of the highly influential 1945 "First Draft of a Report on the **EDVAC**," but the efficacy of storing programs in memory was described in a paper by Konrad Zuse in 1937, the year after Alan Turing's theoretical **Turing machine** stored **instructions** and data in the same form. J. Presper Eckert and John W. Mauchly were also fully aware of the benefits of a stored program computer but did not build the **ENIAC** as one, probably because its limited memory would not have been able to hold programs of sufficient length to be interesting. When programs are stored in main memory, the computer's **central processing unit (CPU)** can switch from one program to another as rapidly as a new one is loaded and can implement program **loops** controlled by conditional jump statements (*see* **Control structure**). Its self-referential nature is what makes a digital computer a *general-purpose* tool of such extraordinary power. There are four candidates for "first stored program computer." The ambiguity arises from whether one chooses the year of conception (EDVAC in 1945), the date the first stored program that ran successfully (21 June 1948 on the **Manchester Mark I**), the date the machine was declared "operational" (**EDSAC** on 6 May 1949), or the date the machine was delivered to and accepted by a waiting customer (**BINAC** on 22 August 1949).

Additional reading: *Encyclopedia of Computer Science*, pp. 1691–1693.

Stretch

COMPUTERS

The *Stretch* (IBM 7030) was a joint project of **IBM** and the Los Alamos Scientific Laboratory of the U.S. Atomic Energy Commission whose goal was a 100-fold advance in performance over the then-existing computer technology. The first machine, delivered to Los Alamos in 1961, did not "stretch" quite as far as the ambitious performance goal originally set, but at the time it was still the most powerful computer in existence. Stretch was the first major solid state computer developed by IBM, and its **transistor**, **magnetic core**, and **hard disk** storage technologies were applied extensively to other IBM computers. An **instruction lookahead** unit permitted up to six instructions at one time to be in various stages of execution, making Stretch the first **pipelined computer**. It was also the first computer to use an **interleaved memory**. The Stretch was very much a **Complex Instruction Set Computer (CISC)** with a 64-bit **word** and an **instruction set** that allowed extraction and processing of bit strings of length 1, 2, 4, 8, 16, 32, and of course, 64 bits. The reference, written by IBM 7030 project director Werner Buchholz, is a classic description of the practice of **computer architecture**. In all, seven Stretch machines were built. A specially enhanced version called *Harvest*, built for the National Security Agency (NSA) and delivered in 1958, was equipped with an advanced tape system called *Tractor* that could fetch any of 160 tape cartridges in a few seconds and then deliver its contents at a then-unbelievable transfer rate of a megabyte per second. The Harvest, whose **software** was written by Walter Jacobs, served NSA well for 14 years.

Additional reading: Buchholz, *Planning a Computer System: Project Stretch.*

String DATA STRUCTURES

A *string* is a sequence of **characters** stored in **memory**. The characters that comprise the string are usually represented in a one-**byte** character **code**, the most commonly used being **ASCII**. Strings of characters may be read as input and printed in recognizable form as output. Since a string is a collection of data values with a specified interconnection and may be modified in memory by the specific operations of **string processing**, a string is a **data structure**. In languages such as Turbo **Pascal** that allow their **declaration**, *string* is a **data type** as well as a data structure. When a specific string is expressed in a **procedure-oriented language**, it is typically quoted, as in "zephyr", "7 Up", or "supercalifragilisticexpialidocious." Memory permitting, a string may be arbitrarily long, or may be very short, a single character such as "Q" or even the *null string* " " consisting of no characters. Because long strings extend throughout several computer **words**, string processing operations are most easily programmed on *byte-addressable* computers, that is, ones whose **word length** is a byte. The concept of a string is too natural to be considered a milestone associated with a particular person, but there are milestones associated with string processing in the article that follows.

Additional reading: *Encyclopedia of Computer Science*, pp. 1694–1701.

String processing INFORMATION PROCESSING

String processing encompasses operations on **strings** of **characters** stored in **memory**. The most basic string processing operation is the *concatenation* (or just *catenation*—English being what it is, the words are exact synonyms) or joining of one string to another. An example would be concatenation of "data" and "base" to form "**database**" (the preferred rendition, incidentally). Other string processing operations are inserting, deleting, or extracting (copying) a substring with respect to a larger string; search-

ing to see if a given string is contained within another; and **pattern matching**, such as inspection of a string to see if it contains the sequence of digits and dashes characteristic of a social security number or a **telephone** number. String processing is heavily used in the humanities to do such things as compile concordances and compare literary texts in an attempt to ascertain whether they have the same author (*see* **Natural language**). String processing can be done in any computer language from **machine language** through **procedure-oriented languages** but is done most easily in certain **problem-oriented languages**. The earliest of these was **Comit**, which dates back to 1957. The later string processing languages that have been most widely used are **Snobol** and **Icon**. But since many general-purpose procedure-oriented languages such as **M** (**Mumps**), Turbo **Pascal**, and **C++** include good string processing capabilities, use of special-purpose languages for this purpose is waning.

Additional reading: *Encyclopedia of Computer Science*, pp. 1694–1701.

Structured programming

PROGRAMMING METHODOLOGY

Structured programming (SP) is a methodological style whereby a **program** is constructed by concatenating or coherently nesting logical blocks of **high-level language** code, blocks of **code** that are either themselves structured programs or else are of the form of one or another of a small number of well-understood **control structures**. Such a definition is inherently and deliberately recursive. Intense interest in the concept followed the publication of a letter to the editor in *Communications of the ACM* in March 1968 by Edsger Dijkstra. In this letter, entitled "GOTO Considered Harmful" (by the editors, not the submitter), Dijkstra reported his observation that the fewer the number of unconditional transfers of control (GOTOs) that a program contained, the easier it was to read. In other words, when someone reads a program trying to understand its logic, encountering a GOTO is just as annoying as encountering "continued on page x" in a newspaper.

In a 1964 paper, Corrado Böhm and Giuseppi Jacopini proved that every "**flowchart**" (program), however complicated, could be re-written in an equivalent way using only re-peated or nested subunits of no more than three different kinds—a *sequence* of executable **statements**, a *decision* clause of the form `if-then-else`, and an ***iteration*** construct, which repeats a sequence of statements `while` (or `until`) some condition is satisfied. Each control structure has a single entry point and a single exit, a key to their intelligible intercon-nectibility. So structured programming has a sound mathematical foundation, but, somewhat ironically, the original version of the scientific language **Fortran** lacked a sufficiently rich set of control structures to implement structured programming, but the business-oriented lan-guage **Cobol** does. **Algol 60** was the first rea-sonably widely used language that contained the necessary control structures, so **Pascal**, in some sense a derivative of Algol, also does. For example, every structured **data type** in Pascal has at least one corresponding control structure or language feature that supports that data type:

array: `for-do` and `repeat-until`
record: `if-then-else`
set: `case-of`
file: `while-do`
pointer: (recursive procedures)

The major contribution of structured program-ming has been twofold—the elevation of pro-gramming technique to something less of an art and more of a science and the demonstration that carefully structured programs can be crea-tive works of sufficient literary merit to deserve being read by humans and not just by comput-ers.

Additional reading: Wirth, *Algorithms + Data Structures = Programs*.

Subroutine

A *subroutine* is one of two types of *subpro-gram* usable within a stored **program**, the other being the **function**. Subroutines *do* something; functions return a single value for use in an **expression**. An *open subroutine*, more com-monly called a **macro instruction**, consists of the same sequence of **instructions** copied to two or more places within a program. A *closed*

subroutine consists of one and only one se-quence of instructions that is isolated ("closed off") from a main program (stored out of its flow of control) and invoked by jumping to it, with control returned to the calling program when the subroutine is finished. The role of a closed subroutine is twofold: it advances the progress of the routine to which it is subordi-nate, and its **code** segment is reusable else-where in the flow of control of that higher-level routine. Physically, a closed subroutine is in-voked through a *calling sequence*. That calling sequence places **parameters** that the subroutine will use in a place agreed upon, it will save a *return address* for use at the end of the subrou-tine so that the calling routine can be resumed, and then it will jump to the beginning of the subroutine.

Subroutines were envisioned for even the first **stored program computers** but were awk-ward to implement until the advent of the **UNI-VAC**, whose **instruction set** included a special subroutine jump, which, upon invocation of a subroutine, automatically saved the *program counter* for use in computing its return address.

Additional reading: *Encyclopedia of Computer Science*, pp. 193–194, 1365–1367, 1708–1710.

Sun Microsystems

Sun Microsystems was founded in 1982 by Stanford M.B.A.s Scott McNealy and Vinod Khosla, with McNealy as chief executive offi-cer (CEO). Their business plan was to market a powerful **workstation** designed by Stanford computer engineer Andreas Bechtolsheim. Their school's influence is reflected in the com-pany's name, since SUN originally meant *Stan-ford University Network*. McNealy quickly recruited Bill Joy, an **operating system** expert from the University of California at Berkeley, to design a version of **Unix** for the first Sun workstation. Production began In 1984. The machines used the MIPS **integrated circuit** chip, a RISC **microprocessor** made by MIPS Computer Systems, Inc. (*see* **Reduced Instruc-tion Set Computer [RISC]**). MIPS had been incorporated as the outgrowth of another Stan-ford project named for the customary unit of processing speed, millions of **instructions** per second (MIPS). Sun emphasized the ease with which multiple workstations could be networked through use of **Ethernet** interconnections. By 1987, Sun held 29% of U.S. workstation revenue, edging out the former leader Apollo at 21%. McNealy had hoped that his company's version of Unix would become an industry standard, but that plan was thwarted when his competitors formed the *Open Software Foundation*. Nonetheless, Sun has prospered and now derives revenue and widespread admiration for Bill Joy's invention of the popular **Internet** language **Java**.

Additional reading: Chandler, *Inventing the Electronic Century*, pp. 148–151.

Supercomputer COMPUTER TYPES

A *supercomputer* is a **digital computer** that is considerably faster, and hence much more expensive, than the **mainframe** computers of its era. Of course, even a **personal computer** of today is, in turn, much faster than the supercomputers of even a decade ago and sell for a small fraction of their cost. Articles devoted to computers that were considered supercomputers in their day are **CDC 6600, Connection Machine, Cosmic Cube, Cray-1, ILLIAC IV, LARC, NORC,** and **Stretch**. As of this century, the fastest supercomputers are **clusters**, the least expensive being **Beowulf** clusters whose nodes are **microcomputers**. But the fastest of all as of mid-2002 is a cluster of *vector computers* (*see* **Parallel computer**) rather than a cluster of microcomputers, the **Nippon Electric Corporation (NEC)** *Earth Simulator*. This machine has a peak performance of 40 Tflops (teraflops—trillions of floating-point operations per second), faster than all the rest of the world's supercomputers combined. The Earth Simulator, whose research and development cost of $400 million was funded by the Japanese government, is configured as 640 nodes of eight vector processors each, and the machine occupies an area of 150 × 200 feet.

Additional reading: Sterling, "How to Build a Hypercomputer."

Superconducting devices HARDWARE

The phenomenon of *superconductivity*, the disappearance of electrical resistance in certain materials at temperatures close to absolute zero, was discovered by the Dutch physicist Kamerlingh Onnes in 1911. The principal *superconducting device*, the *Josephson tunnel junction*, is based on an effect first predicted by Brian Josephson in 1962. It consists of a thin 30 Ångstrom insulating layer about ten atomic layers thick sandwiched between two superconducting (zero-resistance) films. When placed in a suitable cryogenic environment (such as liquid helium at 4.2° K ≈ −269° C), these junctions form the basis of ultrafast switching circuits with transitions times in the picosecond (ps) range and power dissipation of less than two microwatts. Such high-speed and low-power dissipation make the technology of Josephson junction devices a strong contender for use in future high-performance **supercomputers**. Josephson switching devices frequently consist of two or more junctions incorporated into a superconducting loop to form a *Superconducting Quantum Interference Device* (SQUID), invented by James Zimmerman and Arnold Silver in 1965. In logic applications, current diverted from one SQUID as it makes a transition to a nonzero voltage state can be used to induce switching in another SQUID. Complete SQUID logic families have been successfully designed and tested.

Additional reading: *Encyclopedia of Computer Science*, pp. 1723–1725.

SWAC COMPUTERS

SWAC (Standards Western Automatic Computer) was built under the direction of Harry Huskey for the U.S. National Bureau of Standards (NBS) field station, the *Institute for Numerical Analysis* at the *University of California at Los Angeles* (UCLA). At the time of its dedication in August 1950, SWAC was the fastest computer in existence. The machine was originally called the *Zephyr* but was later renamed SWAC to emphasize its relation to **SEAC**. The SWAC was a **parallel computer** using a Williams tube **electrostatic memory** with a **memory** cycle of 16 ms (milliseconds). A 4,096-word **magnetic drum** was used for **auxiliary storage**. The machine continued in useful operation until December 1967. Parts of the SWAC are now on exhibit in the *Museum of Science and Industry* in Los Angeles.

Additional reading: *Encyclopedia of Computer Science*, pp. 1725–1726.

Switching theory   THEORY

Switching theory is a theoretical basis for *logic design*, the sensible interconnection of **logic circuits** (*gates*) to achieve a particular result. Switching theory is concerned with logic design, **finite state machines**, combinational logic, asynchronous sequential circuits, fault detection, and circuit minimization. Applications of switching theory date back to the introduction of technology for the transmission, storage, and processing of digital information. Although its foundations were established earlier in mathematical logic and related areas, switching theory as a separate engineering discipline began in 1938 through the publications of Claude Shannon in the United States, Kyoichi Nakashima in Japan, and Eugeny Shestakov in Russia.

Additional reading: Petzold, *CODE*, pp. 102–128.

Systolic array   COMPUTER TYPES

A *systolic array* is a particular form of parallel **computer architecture** that uses multiple processors to solve computationally intensive scientific problems. All of the *systolic cells* in the **array** compute simultaneously as **data** is passed from cell to cell. A systolic array is analogous to the human circulatory system; data is pulsed through cells where it is processed. The cells can be connected linearly to form a one-dimensional **pipelined computer**, or they can be arranged as a two-dimensional array for problems that can be reduced to **matrix** operations. In a systolic array, each cell provides its **central processing unit (CPU)** with direct access to **input / output (I/O)** ports, in contrast to message-passing machines for which each processor can access only its local **memory**. Systolic architecture is well suited to **VLSI** (Very Large Scale Integration) implementation. The concept was proposed in 1978 in a paper by Hsiang-Tsung Kung and Charles Leiserson and first realized in 1986 as a 10-processor machine called *Warp* built by **General Electric (GE)** for Carnegie Mellon University. In 1990, Intel designed and built a successor called *i-Warp*. This machine was supported by a set of **software** tools called *Archimedes* that included mesh generators and a parallelizing **compiler** for use with **finite element** calculations.

Additional reading: *Encyclopedia of Computer Science*, pp. 1741–1743.

T

Table lookup
ALGORITHMS

The most powerful of all **algorithms** avoids the need for any calculation at all—just store all conceivably needed results in a certain problem domain in an **array** and, upon request, "look up" the specific answer that corresponds to one or more input **parameters**. To be applicable, the **digital computer** being used must have enough main **memory** to store a possibly very large table of values that span the range of interest (and may also require interpolation to estimate an answer that lies between two tabulated values). For example, reversing the bits in a stored bit sequence is an important intermediate step in implementation of the **Fast Fourier Transform**. The logic needed to reverse the bits of a 16-bit (**2-byte**) sequence is reasonably complex, but the reverse of *all* 16-bit patterns could be pre-stored in a one-dimensional array of 2^{16} elements, 128KB (kilobytes, where 1KB = 1,024 bytes). Then, using an original 16-bit number as an index into the array, its reversed bit pattern can be fetched in one machine cycle. Of course, to reverse a 32-bit (4-byte) sequence by *table lookup* would require more than 10GB (ten gigabytes, 10 billion bytes) of memory, which is currently beyond the capacity of the **random access memory (RAM)** of an affordable **personal computer**. (This is not beyond the capacity of current PC **hard disks**, but the time to access a disk would exceed the time to reverse the bits programmatically.) But all that need be done is to separate left and right 16-bit table lookups and combine the answers. As recently as 20 years ago, sales clerks often used table lookup in printed tables to determine applicable sales tax, but by now, of course, the process has been replaced by use of automated calculation. Charles Babbage foresaw the trade-off between calculation and table lookup while designing his **Analytical Engine** and predicted that the methods would alternate in efficiency over time in accord with the prevailing relative cost and performance of memory and logic. In 1949, Nathaniel Rochester of **IBM** proposed use of "memorized" rather than logically formed arithmetic, and this was used to a degree on the **IBM 1620**. The efficiency of table lookup is so obvious that its origin is not traceable; yet it is often overlooked. Beginning students are best advised: "If you're able, use a table."

TAC
COMPUTERS

TAC, the Tokyo Automatic Calculator, was a **vacuum tube** computer built in 1950 by a team led by Hideo Yamashita. The TAC became the basis for the country's first **transistor**-based computers. Earlier, Yamashita, the father of Japan's computer industry, had created an electrical statistical **calculator** that used relay-based binary logic. He also founded the *Information Processing Society of Japan*.

Additional reading: http://sextant.u-aizu.ac.jp/ftp/m3/c_p_text.html

Taylor series   MATHEMATICS

In 1671, James Gregory showed that most mathematical functions can be expanded in an infinite series that allows their computation using only addition, subtraction, and multiplication. But the result lay dormant for 44 years and is now attributed to Brook Taylor in 1715. The sine function, for example, can be written as

$$\sin(x) = x - x^3/6 + x^5/120 - x^7/5040 + \text{higher-order terms.}$$

When x is small—less than 0.5, say—the higher-order terms are so small that they can be neglected and the series becomes a *polynomial*. Discovery and rediscovery of the *Taylor series* are certainly milestones of mathematics but are also important milestones in computing since it is only through its arithmetic operations that a **digital computer** can compute anything.

Additional reading: Weisstein, *Concise Encyclopedia of Mathematics*, p. 1790.

TCP / IP   COMMUNICATIONS

TCP / IP, an abbreviation for two **protocols** within the *Internet Protocol Suite* (IPS), stands for Transmission Control Protocol / Internet Protocol. These protocols were developed to provide a connection between the **ARPAnet** and other **networks**, but they are now the heart and soul of the **Internet**. TCP was first described in general terms in a paper by Vinton Cerf and Robert Kahn in 1974. A year later, Stanford graduate student Yogen Dalal translated the proposal into specifications that could actually be implemented. The original TCP was intended to do everything that TCP and IP do now, but after a hallway discussion at a 1978 conference, Cerf, Jon Postel, and Danny Cohen decided that it would be better to create a division of labor between two separate protocols. The revised TCP would be responsible for breaking a message into separately transmitted packets at one end of the transmission and reassembling them and placing them in logical order at the other (*see* **Packet switching**). IP, in turn, would be responsible for routing, with anything not needed for that purpose left in TCP. Each system that uses IP has a numeric address and a domain name. IP addresses are 32 **bits** long and are normally written as four decimal integers (one per eight bits) separated by periods read as *dots* (e.g., "204.101.80.3"). But since such an address is not user-friendly, each has a corresponding textual *domain name* (e.g., www.ibm.com). The domain name is mapped into the IP address using the *Domain Name Server* (DNS) protocol.

Additional reading: Hafner, *Where Wizards Stay Up Late*, pp. 226–263.

Telegraph   COMMUNICATIONS

In 1825, British inventor William Sturgeon invented the electromagnet, a device that laid the foundations for long-range electronic communication. In 1830, American Joseph Henry demonstrated the potential of Sturgeon's device by sending an electric current over one mile of wire to activate an electromagnet that caused a bell to strike. On 24 May 1844, Samuel F. B. Morse demonstrated his *telegraph* by sending the message "What hath God wrought?"—a phrase from Numbers XXIII: 23—from the U.S. Supreme Court in Washington to his assistant Albert Vail at the Mount Clair Depot in Baltimore. This was certainly a technological milestone, and it is a computing milestone in that it inaugurated digital **data transmission**. His patent had been obtained in 1837, although there had been an earlier more rudimentary telegraph system developed by the English inventors Charles Wheatstone and William F. Cooke. And in 1833 the prolific mathematician Carl Friedrich Gauss built the first European telegraph, one that connected his home to his laboratory in Göttingen. But the Morse telegraph is notable for its use of his **Morse code**, a milestone in itself because it anticipated one of the major principles of **coding theory**.

Additional reading: Coe, The Telegraph: *A History of Morse's Invention and Its Predecessors in the United States.*

Telephone   COMMUNICATIONS

The invention of the *telephone* was a communications milestone that became a milestone in computing with the advent of transmission and receipt of **data** as well as voice signals over telephone lines. In 1837, Charles Grafton Page of Salem, Massachusetts, wrote in his notebook that "when a current flowing through an electromagnet is interrupted, a sound is emitted." This principle is fundamental to all of the many telephone systems ever invented. In 1871, Italian inventor Antonio Meucci obtained a provisional patent for a telephone but was financially unable to defend its claim to priority over the U.S. patent granted to Alexander Graham Bell on 7 March 1876. Bell, in turn, had filed his patent application only two hours ahead of another American, Elisha Gray. When Bell uttered the famous "Watson, come here, I want you!" on 10 March 1876, it was not the first voice transmission, and it was done with a liquid contact transmitter invented by Gray. But Bell did use his own electromagnetic transmitter three months later for his 25 June demonstration at the Philadelphia Centennial Exposition.

Others besides Meucci and Gray who claimed to have invented some form of telephone included Johann Philipp Reis, Emile Berliner, Innocenzo Manzetti, and Charles Bourseul in Europe, and Amos Dolbear, Sylvanus Cushman, Daniel Drawbaugh, Edward Farrar, and James McDonough in the United States. In 1878, David Edward Hughes invented the carbon microphone, which was essential to the development of telephony as we know it. In 1889, William Gray patented a coin-operated telephone. On 19 June 1900, the Yugoslav Michael Pupin obtained two related patents for the inductive loading of telephone lines, a technique that substantially improved the range and quality of telephone conversations.

The automatic exchange was patented in 1888 by the undertaker Almon Strowger of the United States, and the first dial telephone was patented in France on 18 May 1923 by Antoine Barnay. The first touch-tone dialing system was installed in Baltimore, Maryland, in 1941, but it proved too expensive for general use. By the early 1960s, **transistors** made use of touch-tone dialing possible. The first commercial touch-tone phones were previewed at the 1962

World's Fair in Seattle, Washington, and became widely available two years later. Erna Schneider Hoover received a software patent for a computerized telephone switching system in 1971. The first **cellular telephone** systems began operation in Tokyo in 1979 and in Chicago in 1983. Widespread use of **data transmission** via telephone began with the invention of the **modem**. While it was once thought that a 200 bps (**bits** per second) transmission rate was the best that could be achieved over twisted-pair copper wire, now millions of users regularly interact with the **Internet** with modems capable of 56,000 bps. Gradually, however, users are switching to even faster service using either **Digital Subscriber Lines (DSLs)** or a portion of the **bandwidth** of their cable **television** service.

Additional reading: Coe, *The Telephone and Its Several Inventors.*

Television COMMUNICATIONS

The development of *television*, certainly a milestone in communications, ultimately became a milestone in **information technology (IT)** when it evolved to become a digital interactive medium (*see* **Digital television**). A tiny fraction of the **bandwidth** of the **fiber optic** cable that brings television to many homes can now provide high-speed **Internet** access in parallel with normal TV service. Then fast **Internet** connections allow the downloading of TV clips that can be viewed on the **monitor** of a **personal computer (PC)**, or a cable-enabled TV set can be used in lieu of a PC to support *Web TV*. And, of course, both television and the electronic **digital computer** could not have happened without the prior invention of the **vacuum tube**, and both make use of the **cathode ray tube (CRT)**. And going further back, television would not be possible without the discovery of photoconductivity by Joseph May and Willoughby Smith in 1872 and the discovery of the visual persistence of images on the human retina by the French engineer Maurice LeBlanc in 1880, confirming an 1824 prediction of the English thesaurist Peter Mark Roget.

The term *television* was coined by Constantin Perskyi of France on 25 August 1900 at the International Electricity Congress, part of the

Paris Exposition of that year. The Congress voted to endorse the term over a competing proposal that transmission of moving images be called *radiovision*, but there was to be no such thing under either name for almost a quarter century. But as so often happens when technology reaches a certain threshold, a number of persons create similar innovative inventions at almost the same time. The Russian Vladimir Zworykin, working at **Radio Corporation of America (RCA)** in the United States, patented a television scanning camera called the *iconoscope* in 1923, and the American teenager Philo Farnsworth, whose work antedates that of Zworykin, patented a similar television tube a year later. RCA also had to contend with similar patents held by Kolomon Tihanyi of Hungary, François Henrouteau of Canada, and Kenjiro Takayanagi of Japan. The first commercially viable TV system, developed by Scottish inventor John Logie Baird of Great Britain, was a mechanical system based on a *scanning disk* invented in 1884 in Poland by Paul Nipkow. Baird's device produced an image of 30 lines at 12 frames per second. The receiver displayed a tiny, uneven orange-and-black image. The prolific American electrical engineer Charles Francis Jenkins invented a device similar to that of Baird consisting of an electric motor and prismatic rings; it displayed a cloudy 40-line image on a six-inch-square mirror. On 13 June 1925, he performed his first public wireless transmission of television images from Anacosta, Virginia, to Washington, D.C. On 2 July 1928, Jenkins commenced broadcasting on his commercial TV station, W3XK in Washington. The station broadcast programs several nights a week continuously for several years. Just a few months earlier, the U.S. government had issued the first television broadcasting license to station W2XB in Schenectady, New York. Also in 1928, the station telecast the first TV drama, *The Queen's Messenger*. In 1942, the call letters W2XB were changed to WRGB as a tribute to Walter R. G. Baker, a **General Electric (GE)** vice-president and pioneer in both television and **radio**.

Baird, who went on to invent radar, fiber optics, and an infrared device to enhance night vision, had also demonstrated his system publicly in 1925, as did Farnsworth. Farnsworth was honored on a 20-cent U.S. stamp in 1983, and two recent books, one by Evan Schwartz and a similar one by Daniel Stashower, argue persuasively that he should be considered "the father of electronic television" (*see* Figure 39). Another candidate for the title is the Swedish-born Ernst F. W. Alexanderson. Soon after reporting to work at General Electric on 23 February 1902, Alexanderson showed how to make high-frequency, high-power continuous electromagnetic waves for broadcasting. To do so, he invented the magnetic amplifier and a high-frequency alternator. Among his 344 patents were many that furthered the development of both radio and television. In 1928, Alexanderson broadcast a television signal from his laboratory at GE to his and two other homes in Schenectady. The "program" was very short, consisting of just a man taking his glasses off, putting them on again, and then blowing a smoke ring. On 22 May 1930, Alexanderson showed a television picture on a six-foot screen at Proctor's Theater in Schenectady, an event that *Life* magazine, in its millennium edition of 1998, ranked fourteenth in a list of the 100 most important events of the last thousand years. (The **transistor** made the list, at thirtieth, but, through an incredible oversight, the **digital computer** did not.)

The first color television system was invented in 1940 by Peter Goldmark of CBS. Eventually, however, the competing RCA color system developed by a team led by Edward Herold was adopted as the U.S. standard by the NTSB (National Television Standards Board). Herold perfected a system that was upward compatible (from black and white) invented by Henry Kasperowicz of the Allen B. Dumont Laboratories; RCA later bought the relevant patent.

On 1 July 1941, station WNBT, with facilities in the Empire State Building in New York City, transmitted the first TV commercial, an image of a Bulova clock that remained on the screen for 60 seconds. The cost to the sponsor was $4, the same as is now charged for 50 microseconds of the half-time break at the Super Bowl.

Additional reading: Fisher and Fisher, *TUBE: The invention of Television.*

Figure 39. Philo T. Farnsworth at his desk in his Philadelphia TV research lab, 1936. (Special Collections Department, J. Willard Marriott Library, University of Utah)

Telnet

NETWORKS

Telnet is a **protocol** devised in 1972 that allows someone using the **Internet** to open an interactive command-line session on a remote **digital computer**. By far the most common command-line operating system contacted is **Unix**. Once opened, an authorized user can sign in and run any **programs** that would ordinarily be available only to those having a direct connection to the remote computer.

Additional reading: Berners-Lee and Fischetti, *Weaving the Web*, pp. 47–48.

Ternary number system

NUMBER SYSTEMS

The **binary number system** is ideal for use in **digital computers** because of its close affinity to **Boolean algebra**. But it is actually the *ternary* (or *trinary*) *number system* (base 3 arithmetic) that is closer to the system of minimum information-theoretic construction cost. Assume, for the moment, that the cost of building a **logic circuit** that holds one digit is proportional to the number of digits in play, 16 for the **hexadecimal number system**, 10 for decimal, 8 for the **octal number system**, and so on, down to 2 for the minimum usable base, for which **flip-flops** can be used. But then note that the cost of representing a number is also pro-

portional to the number of digits—decimal or hex digits or *trits* (base 3 digits) or **bits**—needed to do so. For example, a binary number contains about 3.3 times as many bits as its decimal representation has digits. For what base b is the product of the two cost factors a minimum? It turns out that the optimum base would be e, the base of the natural logarithms—2.718281828459045 . . . —a fact first noted and published by engineers at **Engineering Research Associates (ERA)** in 1950. But since a number base must be an integer, the closest we can come is to use base 3 (although 2 is next best). In the ternary system, the straightforward symbol set for the three needed trits would be {0, 1, 2}, but Knuth has shown that using a *balanced ternary* number system based on {−1, 0, +1} has certain advantages, and Brian Hayes (*see* reference) has traced use of this system back to the days of Johannes Kepler. And with regard to normal ternary, Paul Erdös and Ronald L. Graham published a conjecture about the ternary representation of powers of 2 more than 20 years ago. They observed that 2^2 and 2^8, ternary 11 and 100111, respectively, can be written in ternary without any 2s. But every other positive power of 2 seems to have at least one 2 in its ternary expansion, which is the same as saying that no other power of 2 can be expressed as a simple sum of powers of 3. Ilan Vardi of the Institut des hautes études scientifiques in France has searched powers up to 26,973,568,802 without finding a counter example, but the conjecture remains open.

The first ternary computer must have been very slow since it was made entirely of wood! It was built in 1840 by the Englishman Thomas Fowler, 12 years after he invented and patented the *Thermosiphon*, the first central heating system. In the early 1950s, H. R. J. Grosch, of later **Grosch's law** fame, proposed that **Whirlwind** be a balanced ternary computer, but binary was chosen instead. The first balanced ternary computer, and the first electronic ternary computer of any kind, was built in 1958 by the Russian engineer Nikolay Brusentsov, who named it the *Setun* after the river than flows near Moscow University. About 50 Setuns were built between 1958 and 1965. But no other ternary computer has since been built in commercial quantity, nor is one ever likely to be.

Additional reading: Brian Hayes, "Third Base," *American Scientist,* **89**, *6* (November–December 2001), pp. 490–494.

Test-scoring machine

I/O DEVICES AND MEDIA

A *test-scoring machine* is a special-purpose machine that calculates scores for multiple-choice tests by detecting the marks used to select choices. The first such machine, the IBM Type 805 International *Test Scoring Machine,* was introduced in 1937. Based on the invention of Reynold B. Johnson while still a high school science teacher, this machine scored objective tests by sensing the conductivity of graphite pencil marks. **IBM** hired Johnson to perfect his machine, and he went on to develop several other inventions of value to the company. Optical successors to the IBM 805 are still in use today; hundreds of thousands of a small countertop version process tens of millions of lottery tickets every week.

Additional reading: Bashe et al., *IBM's Early Computers,* p. 275.

TEX

LANGUAGES

TEX, pronounced "tech," is a **computer**-controlled typesetting system designed by Donald Knuth. The name is the Greek root for English words such as "technique" and "technology." In 1977, Knuth became dissatisfied with the quality of the typesetting in proofs for a new edition of one of his **Knuth textbooks**. Having just learned about digital typesetters that used **bit-maps**, images made up of tiny black dots arranged on a grid, he realized that to make his books look good again, all he had to do was write a computer **program** to put tiny bits of ink in the right places. The language he created, TEX, is especially adept at mathematical typesetting that involves embedded formulas, subscripts, superscripts, a multitude of symbols, and tabular matter. The finished output can be directed to many devices, including video screens, impact printers, **laser printers**, and phototypesetters. TEX is **open source software** in the public domain. Its **algorithms** may be freely incorporated into other systems, but

Knuth requires that the name TEX be restricted to systems that are fully compatible with his program. A metaphor of boxes and glue is used to illustrate the way TEX assembles elements on a page. Each letter or **character** can be thought of as a small box containing an image. These boxes are glued together; a horizontal **string** of characters forms a bigger box to make up a word. Words, in turn, make sentences, which combine to make paragraphs, and so on. TEX programmers can create **macros** to remember commonly used sequences. Several dialects of TEX have been created, each using its own set of macros. LATEX, by Leslie Lamport, encourages its users to create documents with nested structure—chapters, sections, subsections, illustrations, equations. LATEX has automatic facilities for assigning numbers to itemized lists and for generating indexes and tables of contents. Many computer scientists and mathematicians use LATEX to prepare camera-ready manuscripts for submission to journals. A TEX Users Group (TUG) was formed in 1979 and numbers 1,500 members.

Additional reading: *Encyclopedia of Computer Science,* pp. 1756–1759.

Texas Instruments (TI)

INDUSTRY

Texas Instruments (TI) is the current name of the Texas firm founded in 1930 as *Geophysical Service, Inc.* (GSI) by physicists John Karchner and Eugene McDermott. The company's original business plan was to market equipment for seismic oil exploration. In 1939, GSI became a subsidiary of the Coronado Corporation and then made submarine detection devices for the U.S. Navy during World War II. On 6 December 1941, the day before the Japanese attack on Pearl Harbor, GSI was sold to McDermott, John Erik Jonsson, Cecil H. Green, and Henry Bates Peacock. In 1951, GSI was renamed Texas Instruments, with Jonsson as president and later chairman of the board through 1966. In 1952, TI entered the **semiconductor** business by purchasing a license to build **transistors** from Western Electric. In 1954, TI marketed the first commercial silicon transistor, one developed by Gordon Teal, and in 1958, TI engineer Jack St. Clair Kilby invented the **integrated circuit (IC)**, for which he received the Nobel Prize in Physics in 2000. In 1967, Kilby co-invented the first electronic handheld **calculator** with Jerry Merryman and Jim Van Tassel. In 1971, TI developed the first single-chip **microprocessor**. In 1972, TI introduced the TI-2500 portable calculator and the TI-3000 and TI-3500 desk calculators. Two years later, the **TI SR-50** became, with the **HP-35**, the most popular handheld calculators of that era. TI went on to market ever more powerful and less expensive calculators for some years, but eventually their product line succumbed to intense competition from Japan. Beginning in the late 1970s, TI began to concentrate on special-purpose integrated circuitry. In 1978, TI developed the first single-chip microprocessor for **speech synthesis**, and a year later both TI and Bell Labs developed a single-chip **digital signal processor (DSP)**. By 1990, TI had revenues of $6.6 billion and 70,000 employees. TI and **Motorola** are the current leaders in sales of high-performance DSPs.

Additional reading: Chandler, *Inventing the Electronic Century*, pp. 123–133.

Text editor

SOFTWARE

A *text editor* is a **program** used to prepare, edit, and display or print documents that consist entirely of text, **characters** encoded (most usually) in **ASCII**. All symbols in a text document are uniformly sized monospaced (uniform width) characters, the only kind that could be displayed on early **monitors**. Varied sizes and typefonts had to await the evolution of text editors into **word processing** systems used with graphics monitors and ink-jet or **laser printers**. The earliest text editors were **keyboard**-driven line-oriented editors. Multiline display screens with **cursor** addressability and local buffers made possible *full-screen* or *cursor editors*. An early example of a time-shared display editor is Stanford University's TVEDIT written by Brian Tolliver in 1963. Commands, represented by control character sequences, could be interspersed with the input of normal text. Users could move the cursor to the point of editing, rather than having to describe text arguments using a **programming language** syntax. The QED text editor was written in 1965 by Butler

Lampson and Peter Deutsch for the Berkeley **time sharing** system on the SDS 940. The **Unix** text editors *ed*, written by Ken Thompson for the PDP-7, and *vi*, done mostly by Bill Joy while at the University of California at Berkeley, are both descendants of QED.

In 1959, Douglas Engelbart at Stanford Research Institute introduced a major conceptual change with his NLS (oNLine System). It was implemented on the **Alto** in the late 1960s using display terminals, multicontext viewing, flexible **file** viewing, and a consistent user interface. NLS provided support for text structure and hierarchy: Users could manipulate documents in terms of their outline structure, not only their content (*see* **Hypertext**). Wilfred Hansen's EMILY text editor of 1969 extended the concept of the structure editor and developed the *syntax-directed editor*, which imposed on the program being edited the structure of the programming language itself, giving users the power to manipulate logical constructs such as do-while **loops** and their nested contents as single units. A product called *Volkswriter* written by Camilo Wilson enjoyed some success on early 1980s **personal computers (PCs)**. It was billed as a word processor, but it was actually an ASCII text editor.

XEmacs is a powerful, customizable text editor and development environment. It began as *Lucid Emacs*, which was in turn derived from *GNU Emacs*, a program written by Richard Stallman of the Free Software Foundation. GNU Emacs dates back to the 1970s and was modeled after a package called "Emacs" written in 1976 that was a set of **macro instructions** (macros) for TECO, a very early text editor written at MIT for the DEC PDP 10 (*see* **DEC PDP series**) under one of the earliest time sharing operating systems, ITS (Incompatible Timesharing System). TECO (Tape Editing COmmands) is famed for being a powerful but perversely complex text editor. By 1991, Emacs had replaced TECO and is still in use. Since the **source code** for a **high-level language** is normally stored as a text file, simple text editors are still useful despite the presence of word processors. **MS-DOS** has an *Edit* command, and **Microsoft Windows** contains both *Notepad*, which is strictly a text editor, and *Wordpad*, which can be told to save files in text format, among others.

Additional reading: *Encyclopedia of Computer Science*, pp. 1759–1773.

Theorem proving. *See* **Automatic theorem proving**.

3Com   INDUSTRY

The *3Com* corporation was founded in Santa Clara, California, on 4 June 1979 by former **Xerox PARC** employee Robert Metcalfe, co-inventor of **Ethernet** and enunciator of **Metcalfe's Law**. The company's name was derived from the three instances of the prefix "com" in *computing, communications,* and *compatibility.* Its business plan was to produce products and services related to the Ethernet **protocol** and other products related to the three disciplines of its name. The corporation has grown to be a Fortune 500 company mostly through 25 acquisitions, the largest being the 26 February 1997 purchase of **modem** manufacturer *U.S. Robotics* for $8.5 billion in 3Com stock. A supposedly incidental by-product of that transaction was the acquisition of *Palm Computing,* which developed and sold millions of *Palm Pilot* **personal digital assistants (PDAs)** before being spun off as an independent company in 1999.

Additional reading: http://americanhistory.si.edu/csr/comphist/montic/metcalfe.htm

TI SR-50   CALCULATORS

The *TI SR-50* handheld electronic **calculator** marketed by **Texas Instruments** in 1974 performed most classical mathematical functions: powers and roots, factorials, trigonometric and hyperbolic functions and their inverses, common and natural **logarithms**, and the exponential function. Its conventional algebraic notation attracted many users who were not fond of the reverse **Polish notation** (RPN) used with its competitor, the **HP-35**. The SR-50 featured a 14-character **light emitting diode (LED)** display and a **pi** key.

Time sharing   OPERATING SYSTEMS

Time sharing is an extension of **multiprogramming** in which all or most of the **programs** "in the mix" of jobs being alternately processed by an **operating system** originate at a remote terminal. Users at those terminals have the expectation that simple requests such as asking for display of a **file** or directory or compilation of a small program be satisfied without perceptible delay. When response time exceeds a few seconds or more, time sharing degenerates to mere **remote job entry (RJE)**. Christopher Strachey of England may have been the first to publish a description of a proposed time sharing system and applied, apparently unsuccessfully, for a patent on the idea in 1959. But in that same year, John McCarthy of MIT had written an internal memo that proposed certain modifications to an IBM 709 to facilitate time sharing. By the early 1960s, implementation of several time sharing systems began in earnest. At MIT, two systems were developed in 1961, the Compatible Time Sharing System (**CTSS**) for the **IBM 700 series** by Fernando Corbató and a system for the DEC PDP-1 by Jack Dennis. CTSS became the basis for **Project MAC**, an MIT research effort led by Robert Fano to explore the implications of time sharing and human-machine interactions. At **Bolt, Beranek, and Newman (BBN)** in Cambridge, another **DEC PDP series** time sharing system was developed by a team consisting of John McCarthy, Sheldon Boilen, Edward Fredkin, and J. C. R. Licklider. Other early systems were the 1964 Dartmouth College **Basic** System of John Kemeny and Thomas Kurtz for the GE 235 (*see* **GE 200 Series**); the **JOSS** System implemented at the **Rand Corporation** by Cliff Shaw; and a system developed at the System Development Corporation by Jacob Schwartz for the IBM AN/FSQ-32 military computer. The first commercial time-shared operating system was developed for the PDP-6 by the **Digital Equipment Corporation** in 1964. By the mid-1960s, development of extensive new time sharing systems had begun, the foremost being the development of **Multics** by Project Mac and Bell Labs. The original **IBM 360 series** architecture and operating systems of 1964 were not well suited for time sharing, and **IBM** and much of the computing community were shocked when **General Electric (GE)** was selected to provide the host computer for Multics, a GE 635 delivered in 1965 and replaced by a GE 645 in 1967 (*see* **GE 600 series**). In response, IBM began development of the TSS System for a newly designed IBM 360/67, but the project was not as successful as the (University of) Michigan Terminal System (MTS) developed for the same computer based on a 1966 paper by Bruce Arden, Bernard Galler, Frank Westervelt, and Tom O'Brien published in the January 1966 issue of the *Journal of the ACM*. In the late 1960s, SUNYATSEN (*State University of New York at Albany Time Shared Executive Network*) was conceived and named by Edwin D. Reilly and implemented for the CDC 3100 by John Watson. By the early 1970s, dozens of other time sharing systems were being implemented, and time sharing became a significant mode of computer interaction. By the mid-1980s, interest in localized time sharing systems began to wane in proportion to the proliferation of powerful **personal computers (PCs)** whose users did not have to share them with anyone else. Nonetheless, time sharing lives on through **client-server computing** and the largest time sharing system ever built, the **Internet**.

Additional reading: *Encyclopedia of Computer Science*, pp. 1778–1783.

Tomography. *See* Computerized Tomography (CT).

Toshiba INDUSTRY

The origins of the Japanese company *Toshiba* go back to 1885 with the founding of the Shibaura Electric company, which made **telegraph** equipment. In 1904, the company allied with **General Electric (GE)** to produce light bulbs under the name Tokyo Electric. GE's ties to Tokyo-Shibaura were necessarily terminated during World War II but were restored in 1946 when the company adopted the shortened form of its name, Toshiba. For almost 30 years, Toshiba concentrated on consumer electronic products, but it entered the **digital computer** industry in 1971 through an alliance with the **Nippon Electric Corporation (NEC)** to make **minicomputers**. Through the 1980s and 1990s, Toshiba obtained licenses to market **semiconductor** products and, by 1994, became a world leader in the production of **DRAM** (Dynamic Random Access Memory) chips and **liquid crystal displays (LCDs)**. The company now enjoys stable sales of over $50 billion per year.

Additional reading: Chandler, *Inventing the Electronic Century*, pp. 201–205.

Touch screen I/O DEVICES AND MEDIA

Certain display **monitors** have a *touch screen* that can be used to select menu options or to activate **icons** by touching them lightly with a finger or other stylus. Such monitors can thus be used for applications in which neither a **mouse** nor a **keyboard** is otherwise required, such as voting machines and library catalog access terminals. Touch screens use one of three physical principles for detecting the point of touch. Resistive screens, the first invented, work with any stylus. Capacitive screens must be touched by a finger or an electrically grounded stylus to conduct current. Wave screens that are based on surface acoustic waves must be touched by a finger or a soft stylus such as a pencil eraser to absorb energy; infrared wave screens work with any stylus. The first "touch sensor" was developed by Samuel Hurst, founder of the Elographics company, while he was an instructor at the University of Kentucky in 1971. His "Elograph" was not transparent as are touch screens but was nevertheless an important milestone for touch technology. The first true touch screen was developed by Hurst in 1977. In that same year, the most popular touch screen technology in use today, five-wire resistive technology, was developed by Bill Colwell of Elo TouchSystems, Inc., the current name of Elographics.

Additional reading: White, *How Computers Work*, p. 233.

Trackball I/O DEVICES AND MEDIA

A *trackball* is a pointing device that moves the **cursor** on a display **monitor** in response to the rotation of a palm-size ball. The ball is mounted in a housing such that it can be controlled by the palm without changing the ball's location on the desktop. A trackball is often described as an upside-down **mouse**. That is true in the sense that one can hold a mouse in one hand and manipulate its roller with the other, but the description is misleading in the sense that the trackball was invented first; the mouse is an upside-down trackball. The first U.S. patents for what are now called *trackballs* were awarded to William F. Alexander in 1958, Robert A. Koster in 1967, Norman J. Bose (the first to use the term "Track Ball") in 1971, and Tom R. Luque in 1983, the last of which is most similar to the trackball of today. All of these cite as antecedents the 1952 patent of Joseph Harrington, Jr., but judging by the drawing submitted with his patent application, his device was controlled by a sphere the size of a bowling ball.

Additional reading: White, *How Computers Work*, pp. 233–234.

TRADIC   COMPUTERS

TRADIC (Transistorized Airborne Digital Computer), the first general-purpose **digital computer** to be transistorized, was designed and built for the U.S. Air Force in 1954 at Bell Labs by a team led by J. H. Felker. TRADIC, the first completely solid state computer, used 700 point-contact **transistors** and 10,000 germanium diodes. The machine had a sustained speed of one MIPS (million **instructions** per second), which was competitive with the fastest **vacuum tube** computers of that time and was, of course, much more reliable.

Additional reading: Riordan and Hoddeson, *Crystal Fire*, p. 204.

Transistor   HARDWARE

Invention of the *transistor*, the successor to the **vacuum tube** as the essential **semiconductor** logic element of a modern **digital computer**, is officially credited to John Bardeen, Walter Brittain, and William Shockley of Bell Telephone Laboratories in 1948, work for which they received the Nobel Prize in Physics in 1956. The word *transistor* was coined at their request by their colleague John Pierce (who writes science fiction under the name *J. J. Coupling*, an angular momentum term recognizable to physicists). Working independently at a Westinghouse laboratory in Paris, the German physicists Herbert Mataré and Heinrich Welker developed a virtually identical device that they called a *transistron*, but their work was not an-

nounced until two months after that of the Bell Labs group. But like most important breakthroughs, there were antecedents. In 1930, Julius Lilienfeld of the University of Leipzig obtained a patent for what would now be called a *Metal Oxide Semiconductor Field Effect Transistor* (MOSFET). Although there is little evidence as to whether such a device was actually demonstrable, the patent prevented Shockley from obtaining a patent for what he did call a *Field Effect Transistor* (FET). What the Bell Lab team invented was a *point-contact transistor*, a type used in the **TRADIC** but which did not prove practical for use on a larger scale. An improved FET was built by Stanislaw Teszner of Poland in 1958, but the first transistors usable *en masse* were the planar transistors of the Swiss-born Jean Hoerni, one of the original eight scientists who left Shockley's group to found **Fairchild Semiconductor**. Mass-produced transistors on **microcomputer** chips have now been the dominant constituent of computer logic for three decades (*see* **Intel 80×86**; **Moore's Law**).

Additional reading: Riordan and Hoddeson, *Crystal Fire*.

Tree   DATA STRUCTURES

A *tree*, or more precisely a *rooted tree*, is a special form of directed **graph** that has a distinguished vertex called the *root*, which has no predecessor, and in which all other vertices have a unique predecessor. The term *tree* for such a restricted form of graph was first used by Arthur Cayley in 1857, but the concept had been explored by the Prussian electrical engineer Gustav Kirchhoff and the German mathematician Karl Georg Christian von Staudt about a decade earlier. Mathematical trees are normally drawn with their root at the top and their leaves toward the bottom, just the opposite of a living tree. Vertices (or *nodes*) of a tree that have successors are called *nonterminal vertices*, or *parent nodes*, while vertices that have no successors are called *terminal vertices* or *leaves*. Nodes that have a parent (all those other than the root) are called *child nodes*, and nodes that have the same parent are called *sibling nodes*. Trees in which each nonterminal vertex has at most *n* successors are called *n-ary* trees.

Trees in which each nonterminal vertex has at most two successors would then be 2-ary trees, but such structures have little or no application in **computer science** without the additional restriction that children are either *left child nodes* or *right child nodes*, in which case they are called *binary trees*.

Tree structures may be indicated by parentheses, nesting, or indentation. The representation (*a*)(*b*) (*c*(*d*) (*e*)) may be viewed as a list structure in which the successor nodes of *a* are represented by the sublists (*b*) and (*c* (*d*) (*e*)). This representation is used to represent trees as **linked lists** in languages such as **Lisp**. Of particular importance are *height-balanced trees*, in which the height (maximum distance from the root to a leaf) of the left subtree of any node differs from that of the right subtree by at most one. Such trees are called *AVL trees* after their inventors of 1962, G. M. Adel'son-Vel'skii and E. M. Landis. Keeping a tree balanced in this way provides far superior search time as compared to trees that become highly unbalanced.

Trees play an important role in **computer science**. The information in the *n* nodes of a properly organized binary tree can be searched in O(log *n*) time as compared to O(*n*) time needed to search an unordered linear list (*see* **Analysis of algorithms**). Binary trees are also a natural **data structure** for expressing the **operator**-operand structure of arithmetic **expressions**. There are three fundamentally different ways to list the nodes of a binary tree (other than simply listing them in breadth-first order, which is seldom of interest). When applied to an expression tree, the first three listings yield a recognizable variation of the original expression. (By highly improbable coincidence, the mathematician J. B. Listing studied trees in 1862.) *Preorder* (or *depth-first*) *traversal*, when applied recursively, yields the prefix form of **Polish notation**. *Inorder* (or *symmetric*) *traversal* yields the original *infix* form of the expression. *Postorder* (or *endorder*) *traversal* yields the *reverse Polish* or *Polish postfix* form of the expression. A computer's file *directory* is stored in the form of an *n*-ary tree, and early hierarchical **database management systems** stored their **data** as a tree.

Additional reading: *Encyclopedia of Computer Science*, pp. 1792–1795.

Turing machine
THEORY

A *Turing machine* is an abstract computing device invented by Alan M. Turing in 1936. A Turing machine consists of *a control unit*, which can assume any one of a finite number of possible states; a *tape*, marked off into discrete squares, each of which can store a single symbol taken from a finite **set** of possible symbols; and a *read/write* (R/W) *head*, which moves along the tape and transmits information to and from the control unit. A Turing machine computes via a sequence of discrete steps; its behavior at a given time is completely determined by the symbol currently being scanned by the read-write head and by the internal state of the control unit. On a given step, it writes a symbol on the tape, moves along the tape at most one square to the left or right, and enters a new internal state. The new symbol is permitted to be the same as the current symbol; similarly, it is permissible to stay on the same tape square on a given step or to reenter the same state. Certain symbol-state situations may cause the machine to halt (*but see* **Halting problem**). There is considerable evidence that the languages recognized by Turing machines are exactly those computed by effective procedures, that is, by **algorithms** (*see* **Church-Turing thesis**). Turing machines with multiple tapes (or, in general, *n*-dimensional tapes) and R/W heads can compute and recognize no more than can Turing machines with a single one-dimensional tape and a single R/W head, although they may compute and recognize more efficiently. Turing machines are the most capable class of machines of the four that comprise the **Chomsky hierarchy**.

Additional reading: *Encyclopedia of Computer Science*, pp. 1797–1801.

Turing test
ARTIFICIAL INTELLIGENCE

In a famous paper of 1950, Alan Turing addressed the philosophical question "Can Machines Think?" He thought the question too meaningless for serious discussion and proposed instead a test to ascertain whether a machine, a **digital computer** in particular, was exhibiting intelligent behavior. The *Turing test*, originally called "The Imitation Game," consisted of placing a human and a computer in a windowless room connected by teletype to a human examiner in another room. The examiner is challenged to converse with the unknown respondents through a series of typed questions and determine which is the computer and which is the human. The computer, of course, is programmed to display its **artificial intelligence** (a term not yet coined) as humanly as possible but not to give the game away by answering too quickly or perfectly accurately if asked to perform a computational feat beyond the capability of a human. The human respondent, presumably highly educated, is told to respond naturally. Turing claimed that if, after repeated testing, a group of examiners could not tell the computer from the human with significantly better than 50% accuracy, the computer must be deemed intelligent. Despite monetary prizes offered over six decades, no computer has passed the Turing test. Of course, a machine can be very intelligent when the domain of discourse is limited (*see* **Eliza**, for example) and when it is in engaged in **computer chess**. But if no topic is ruled out, computers do not fool anyone for very long.

In 1990, Hugh Loebner of the United States established the Loebner Prize (of up to $100,000 and a gold medal) to the first person whose program would meet his criteria for passage of a Turing test. Small sums have been awarded for progress each year since 1991, with Joseph Weintraub of *Thinking Software* winning in four of the first five years for the performance of his program, the *Talking PC Therapist*. The big prize remains unearned, so Turing's prediction that his test would be passed within 50 years has proved incorrect.

Additional reading: *Encyclopedia of Computer Science*, pp. 1801–1802.

Figure 40. The Glyn Hughes bronze statue of Alan Turing unveiled in 2000 at Sackville Park in Manchester, UK. The plaque in front of the bench reads: ALAN MATHISON TURING 1912–1954 FATHER OF COMPUTER SCIENCE, MATHEMATICIAN, LOGICIAN, WARTIME CODEBREAKER, VICTIM OF PREJUDICE. "Mathematics, rightly viewed, possesses not only truth but supreme beauty, a beauty cold and austere like that of sculpture."—Bertrand Russell.

On the bench, a second plaque reads: Alan Mathison Turing 1912–1954 AZLLSFTR PLIB ACIK TTRL AEF ("Founder of computer science" in Enigma-coded German). Turing is holding an apple (small "a," of course) that is not shown in the photo, and is draped with a Loebner medal (*see* **Turing test**). (Photo courtesy of Dr. Richard Wallace)

gle weekend. The machine used the Flexowriter printer first developed for the **SAGE** project. Clark was also the designer of the *TX-2* of 1959, the first computer with an extensive **computer graphics** facility (*see* **Sketchpad**). On 26 October 1960, CBS-TV aired "The Thinking Machine," which featured a series of short plays with a western theme produced by a TX-2 program called SAGA II, written by Douglas Ross and Harrison Morse of MIT's Electronics Systems Laboratory.

Additional reading: Ceruzzi, *A History of Modern Computing*, pp. 127–128.

TX-0 / TX-2   COMPUTERS

The *TX-0*, built at the MIT Lincoln Laboratory in 1953, was the first general-purpose **digital computer** to use **transistors** instead of **vacuum tubes** and was the first to incorporate an **interrupt**. The **computer architecture** of the TX-0 was designed by Wesley Clark over a sin-

Typewriter   HISTORIC DEVICES

Since the *typewriter* was the first device capable of **word processing**, however rudimentary, its invention deserves inclusion as a milestone of **Information Technology (IT)**. The first attempts to build typewriting machines were made early in the nineteenth century, but all were cumbersome and some were as big as a piano. But William Burt's *Typographer* of 1830 worked reasonably well. In 1867, the American Carlos Glidden read an article in *Scientific American* about the 24-letter typing machine of John Pratt and called it to the attention of Christopher Latham Sholes. Sholes then proceeded with help from Glidden and Samuel Soule to invent what became the first practical "typewriter," a term coined in the article. The trio sold their patent to entrepreneurs who, in 1873, signed a contract with Eliphalet Remington and Sons, gunsmiths, of Ilion, New York, to manufacture the Sholes machine in quantity. The first model was marketed in 1874 and sold under the name of Remington. Among its features that were still standard in machines of a century later were the cylinder, with its line-spacing and carriage-return mechanism; the escapement, which causes the letter spacing by carriage movement; the scalloped arrangement of typebars so as to strike the paper at a common center; use of an inked ribbon; and the QWERTY **keyboard** layout that is virtually identical to the arrangement that is now universal. The American Samuel Langhorne Clemens (Mark Twain) purchased a Remington and was the first author to submit a typewritten book manuscript.

In 1873, teenager John Thomas Underwood emigrated from London to the United States to join his father in business as John Underwood & Company. Beginning in that year, they pioneered the manufacture of typewriter supplies, carbon paper, and other accessories to support the early typewriters being manufactured by Remington. In 1895, Underwood the son bought the patent to Franz Wagner's invention of the "front-stroke" machine, which, unlike previous models, enabled the typist to see what was being typed. The new Underwood typewriter revolutionized the industry. By 1915, the Underwood factory in Hartford, Connecticut, was the largest of its kind in the world. An Underwood typewriter was featured in the 1939 New York World's Fair to represent the "World of Tomorrow." That world came and went, but many homes and offices continue to house a vintage typewriter if only to fill out the few remaining forms that are not accessible on the **Internet**.

Additional reading: Beeching, *Century of the Typewriter*.

Figure 41. A hagiographic image of Christopher Latham Sholes, apparently in Paradise. In the reference listed for the article, Beeching writes: "Christopher Latham Sholes has been described as 'the most unselfish, kind-hearted, and companionable man who ever lived.' . . . He had the light blue far-away eyes of a visionary, and with his long flowing hair was a familiar figure in the streets of Milwaukee in 1866. He was the fifty-second man to 'invent' the typewriter." Despite the ranking, the modern typewriter is a direct descendent of his design. (Corbis/Bettmann)

Typography   GENERAL

[The] association of die, matrix, and lead in the production of durable typefaces in large numbers and with each letter strictly identical, was one of the two necessary elements in the invention of typographic printing in Europe. The second necessary element was the concept of the printing press. Johannes Gutenberg [of Germany] is generally credited with the simultaneous discovery of both these elements, though there is some uncertainty about it, and disputes arose early to cloud the honour.

—Online *Encyclopedia Britannica*

Figure 42. The Applegath press, circa 1848. Augustus Applegath's eight-feeder vertical rotary "type-revolving" press designed for *The Times* of London printed at the rate of 8,000 impressions an hour. It consisted of one central type cylinder surrounded by eight smaller cylinders known as impression cylinders. (Compare to the description of the architecture of the **CDC 6600**.) Each of these cylinders had a feeding apparatus with two boys in attendance. Sheets of paper were fed down by hand through tapes and then passed sideways between the impression cylinder and the type cylinder, thus producing sheets printed first on one side and then, through a second pass at another cylinder, on the other. (National Museum of Photography, Film & Television/Science & Society Picture Library)

Certainly the invention of *typography* in general and the printing press in particular is a major milestone of **information technology (IT)**. The cloud that *Britannica* refers to is that Gutenberg's financial backer Johann Fust and Fust's son-in-law, Peter Schöffer, founders of the first commercially successful printing firm, later claimed they had played a role in these inventions. Whatever, the liberation of the general public from restricted access to a small number of hand-copied books was a major ad-vance in civilization. The first practical photo-typesetting machine was developed in 1949 by Louis Marius Moyroud and René Alphonse Higonnet. An online printer is perhaps the most indispensable computer **peripheral**, and interest in the design of readable and artistic typefonts has continued unabated from Gutenberg in the fifteenth century to Donald Knuth, among others, in the twenty-first (*see* TEX).

Additional reading: http://www.britannica.com/eb/article?eu=117314&hook=417279#417279.hook

U

Ubiquitous computing
GENERAL

Ubiquitous computing, a term coined by Mark Weiser of **Xerox PARC** in 1988, implies the pervasive use of wireless **handheld computers** so small, perhaps even wearable, that their users are never out of range of the **Internet**. No other term was needed, but the same concept is now sometimes called *pervasive computing*. *See also* **Augmented reality**; **Mobile computing**.

Additional reading: Mark Weiser, "The Computer for the 21st Century," *Scientific American*, **265**, 3 (September 1991), pp. 94–104.

Ultrasonic memory
HARDWARE

Ultrasonic memory, particularly the mercury *acoustic delay line*, played an important role in the early development of **digital computers** but is now only of historical interest. The mercury delay line was invented at Bell Labs in the early 1940s by William Shockley, the later co-inventor of the **transistor**. The report on the **EDVAC** drafted by John von Neumann in June 1945 clearly envisaged use of this type of **memory**, although it did not describe its physical principles. Of the early machines, the initial versions of **ENIAC**, **EDSAC**, **SEAC**, **ACE / Pilot ACE**, EDVAC, **LEO**, and **UNIVAC** all had ultrasonic memories, as did Zuse's Z4 computer of 1950 in Germany (*see* **Zuse computers**). In a typical memory of this kind, a train of pulses representing the number to be stored is modulated onto a carrier and applied to a piezoelectric crystal in contact with a column of mercury. The ultrasonic pulses generated travel along the column until they reach a second crystal at the far end that converts them back into electric signals, which are then amplified and rectified. The resulting pulses are applied to a gate together with pulses from a continuously running clock pulse generator. This gating operation provides regeneration and synchronization. The emerging pulses, which are exact replicas of the original pulses, are reapplied to the modulator and continue to circulate. The operations of reading, clearing, and writing can be performed by gating suitable waveforms synchronized with the clock. A main memory usually consisted of a group of 32 *tanks*, each between 0.5 and 1.5 meters long, giving a delay of between 0.33 and 1 millisecond.

Additional reading: Williams, *A History of Computing Technology*, pp. 308–313.

Unbundling
SOFTWARE

One of the most significant milestones in computing was not a theoretical discovery or clever invention but rather a simple management decision. Prior to 1969, the then-dominant **computer** vendor **IBM** had provided all **software** written for a particular system free of additional cost beyond purchase or rental charges. But on 23 June of that year, IBM stunned the computing industry by announcing that henceforth each of its software packages other than a basic **operating system** would be priced separately from the **hardware** needed to run it. The consequence, whether intended or not, was the birth of a vigorous third-party software industry, leading to the eventual founding of **Microsoft** and many other companies who specialize in writing and marketing software for a multitude of computers including, of course, those of IBM.

Additional reading: Pugh, Humphrey, and Grad, "Origins of Software [Un]Bundling."

Unicode   CODING THEORY

Unicode, developed in 1988 by a consortium of computer companies, is a 16-bit-character **code** that will ultimately be able to represent 2^{16} or 65,536 **characters** as compared to the 256 of **EBCDIC** or extended **ASCII**. As of 4 September 1998, Version 2.1 of Unicode had encoded 38,887 characters of the world's most common alphabets. For upward compatibility, the rightmost seven bits of the first 128 Unicode characters are identical to their ASCII counterparts. Unicode is now standard 10646 of the International Organization for Standardization.

Additional reading: Petzold, *CODE*, p. 300.

Unisys   INDUSTRY

Unisys was formed in 1986 through the merger of **Burroughs Corporation** and **Sperry Rand**. The long history of the company's progenitors is summarized on the Website cited as a reference or can be traced through the articles highlighted above and the ones on **Eckert-Mauchly Computer Corporation (EMCC)**, **Engineering Research Associates (ERA)**, **Remington Rand**, **UNIVAC**, and **Univac 1100 Series**. Unisys is now concentrating on support for **electronic commerce**. Its major **hardware** product, introduced in 2000, is the Cellular Multi-Processing (CMP) server.

Additional reading: http://www.unisys.com/about_unisys/history/index.htm

UNIVAC   COMPUTERS

UNIVAC (UNIVersal Automatic Computer) was the first **digital computer** sold commercially in the United States. Work on the prototype was begun by the **Eckert–Mauchly Computer Corporation (EMCC)** in 1948, and

it was completed and delivered to the U.S. Bureau of the Census in 1951. A total of 46 UNIVACs were delivered during the period 1951–1958. The UNIVAC was the first computer that could handle numeric and literal character **strings** equally well. Information from its **ultrasonic memory** could be accessed at a speed of 40 to 400 microseconds. New **data** was transcribed to metallic **magnetic tape** by a key-to-tape device, or data already on **punched cards** was transcribed to magnetic tape with a card-to-tape converter. Input could also be fed to a running program from the **keyboard** of a control console. The first program for the UNIVAC was written by Ida Rhodes. Buffered storage **registers** permitted the **central processing unit (CPU)** to operate in parallel with data being read from magnetic tape. UNIVAC was a one-address decimal computer with a **word length** of 12 characters. Its circuitry operated at a 2.25 MHz **bit** rate. Memory capacity was 1,000 **words** (12,000 characters). Addition or subtraction took 0.525 milliseconds (ms), multiplication 2.150 ms, and division 3.890 ms. The UNIVAC came to worldwide attention on 4 November 1952 when CBS used one to predict Eisenhower's margin of victory over Adlai Stevenson to an accuracy of 1% early in the evening with only 7% of the vote in. In the 1957 Spencer Tracy–Katharine Hepburn movie *Desk Set*, the computer looked suspiciously like a UNIVAC. Until its sales lead was overtaken by **IBM** in the late 1950s, "UNIVAC" came very close to becoming a generic word for **digital computer** in the way "Kleenex" and "Xerox" are so often used to refer to competitive products.

Additional reading: Stern, *From ENIAC to UNIVAC: An Appraisal of the Eckert-Mauchly Computers*.

Univac 1100 series   COMPUTERS

Over a 20-year period, the Univac division of **Engineering Research Associates** and **Sperry Rand** made two **digital computer** series of different and hence incompatible **computer architecture** but numbered both in the range 1100 to 1110. The first consisted of the **ERA 1101** and the **Univac 1103** (*see* next article). The second consisted of the Univac 1107 of 1967, the Univac 1108 of 1968, and the 1110

Figure 43. UNIVAC. Grace Murray Hopper at the UNIVAC keyboard, c. 1960. (Smithsonian Institution Photo No. 83–14878)

of 1970, and it is these computers that are referred to as the *Univac 1100 Series*. All three used a 36-bit **word**, but the three computers did not quite form either an upward- or downward-compatible series because each computer's **instruction set** contained **instructions** not present in the others. But a certain degree of compatibility was maintained through use of the *Sleuth* **assembler**, the Exec-8 **operating system**, and standardized **compilers** for **Fortran** and **Cobol**. The 1100 was an especially powerful system whose performance was comparable to **Control Data**'s scaled-down **CDC 6600**, the CDC 6400. But the 1100 series did not enjoy a long or profitable life, and the series was phased out long before Sperry Rand and the **Burroughs Corporation** merged to form **Unisys** in 1986.

Additional reading   http://www.fourmilab.ch/documents/univac/

Univac 1103   COMPUTERS

The *Univac 1103*, the first "scientific" **digital computer**, was not a part of the **Univac 1100 series** of the 1970s but rather was an improved version of the **ERA 1101** of 1953. Seymour Cray, who went on to design many computers for the **Control Data Corporation (CDC)**, **Cray Research**, and the *Cray Computer Corporation* (CCC), was a member of the 1103 design team. The 1103, originally the *ERA 1103*, had a 12-microsecond cycle, an **electrostatic memory** of 1,024 36-bit **words**, and stored two

18-bit one-address **instructions** per word. A field change made in 1954 to the Univac 1103 installed at the U.S. National Advisory Committee for Aeronautics made that computer the first to have an **interrupt** facility, and that feature was then made standard on the 1103A of a year later. The 1103A was the first commercial computer to use **magnetic core** as **memory**.

Additional reading: *Encyclopedia of Computer Science*, p. 556.

Universal Product Code (UPC)

APPLICATIONS

The *Universal Product Code (UPC)* now in use was invented by George J. Laurer in 1973, but the first patent for a *bar code* was issued to inventors Joseph Woodland and Bernard Silver on 7 October 1952. UPC-encoded bar codes now appear on almost all retail products and library books. Bar codes are designed to be read by an **optical scanner**. In June 1974, the first UPC scanner was installed at a Marsh's supermarket in Troy, Ohio. The first product to bear a bar code was Wrigley's Gum. Bar codes are obviously nonsecret because their ten-digit codes are interpreted (displayed numerically) just below the bars that encode them. The five leftmost digits identify the manufacturer through a code assigned by the Uniform Code Council. The five rightmost digits are assigned by the manufacturer to identify various individual products; thus, the price itself is not encoded, but instead a product identification number, from which a **digital computer** to which the scanner is attached can obtain the price by **table lookup**. Each digit of the code has a unique sequence of bars or, more precisely, a pair of such sequences, since the pattern of a digit on the right-hand side is the encoded 1's **complement** value of the corresponding pattern on the left. This is done so that the **program** processing the scanner's input can detect whether the product was passed over the reading aperture right to left or left to right. In the early 1970s, the first scanners to be widely used by retailers were called *pen* (or *wand*) *scanners*. They used a **light emitting diode (LED)** at their tip and a light detector in their barrel and had to actually touch the bar code in order to read it. But **laser** scanners do not need contact; some can even read a bar code from several feet away rather than the few inches needed for a supermarket scanner, and they are much better able to read a bar code imprinted on a curved surface. Bar code readers have applications other than for retail checkout, inventory control and library circulation being obvious candidates. For this purpose, a pen scanner or a small handheld laser scanner is a very effective tool.

Additional reading: Petzold, *CODE*, pp. 79–83.

Unix
OPERATING SYSTEMS

Unix, the first highly interactive **operating system**, was developed in the early 1970s at Bell Laboratories by Ken Thompson and Dennis Ritchie, who received the **Association for Computing Machinery (ACM)** Turing Award for their contributions to computing. Unix has grown, evolved, and endured as one of the most influential **software** systems in computing history. Unix had its roots in the 1960s **Multics** project, in which Thompson and Ritchie participated. Unix first ran on a DEC PDP-9 in 1971. In 1973, Unix became the first operating system to be written in a **high-level language** when most of its **code** was rewritten in **C**, a **programming language** specifically developed to implement Unix. Bell Labs awarded Unix License Number 1 to the University of Illinois in 1975. Unix is a **multitasking**, multiuser, **time sharing** operating system. Unix carefully distinguishes between the operating system *kernel* and user applications. Each user **program** runs as a separate *process*, making system calls into the kernel to perform such tasks as accessing **files** or allocating additional **memory**. The kernel itself was deliberately kept very small. The original user interface for Unix was a command-line **interpreter**, called a *shell*. A shell processes **keyboard** input and performs user-requested tasks. Users can modify the shell as easily as any other program. Several shells, including the *C shell* and *Korn shell*, have become popular in addition to the original *Bourne shell*. Having shell commands run as regular processes makes it possible for ordinary users to write *command scripts*, files containing arbitrary commands. The shell can process commands contained in a file as easily as those entered at the **keyboard**.

Unix is essentially a textual command-line operating system, but with the help of the **X window system** developed at MIT, **graphical user interfaces (GUIs)** have been developed to offer a more user-friendly interface for simple operations. Unix provides a hierarchical file system in which **directories** hold files and other (sub)directories. The resulting file **tree** is shared by all users, making it straightforward to name and find files. Unix introduced many tools now taken for granted in computing, including document formatters and **spelling checkers**. One of the main goals of the **open source software** movement, particularly the Free Software Foundation (FSF), is the creation of a public domain Unix system. The GNU **Linux** operating system is a popular **freeware** system for **personal computers (PCs)**. Other popular PC systems are FreeBSD and Net-BSD based on the BSD 4.4 **source code**.

Additional reading: *Encyclopedia of Computer Science*, pp. 1816–1819.

User group
GENERAL

A *user group* is an association of users of a particular **digital computer** who meet regularly to share experience and to try to influence the manufacturer with regard to its technological and **software** support. The first user group, SHARE, was formed in 1955 by users of the IBM 701 in Los Angeles (*see* **IBM 700 Series**). Their first formal meeting was held in August of that year. Participating were representatives from the U.S. National Security Agency (NSA), the **Rand Corporation**, the Los Alamos and Livermore national laboratories, **General Electric (GE)**, General Motors, Standard Oil of California, eight aerospace organizations, and of course, **IBM** itself. A library of public domain software written by programmers from member organizations was established and operated by Ben Faden of North American Aviation. A few months after the founding of SHARE, a group of users of IBM commercial computers (the IBM 702 and 705) founded GUIDE. Other user groups were founded including VIM for the **CDC 6600**, TUG for the **Philco Transac S-2000** family, and HUG for users of **Honeywell Equipment**. The user group concept has broadened beyond the computer itself. There now exist user groups that support copiers, printers, **programming languages**, **operating systems** such as **Unix** and **Linux**, and many commercial **spreadsheet**, **da-**tabase, and **word processing** products. With attention turned to the **IBM 360 series** and its successors, SHARE still exists and is approaching its golden anniversary.

Additional reading: *Encyclopedia of Computer Science*, pp. 1819–1821.

V

Vacuum tube

The electronic *vacuum tube* as we came to know it in early **radio** and **television** sets was invented in 1904 by the English electrical engineer John Fleming. Fleming's electron tube was a two-element device called a *diode* that could convert alternating current (AC) into direct current (DC). In 1907 the American Lee De Forest added a third element called a *grid* and thus invented the *triode*, a tube that could amplify signals rather than merely transmit them. Vacuum tubes made it possible to build electronic **digital computers** rather than just much slower computers that used electromagnetic relays, and vacuum tubes reigned as the technology of choice in the field until a few years after the invention of the **transistor** in 1947.

Additional reading: Mount and List, "Electron Tube," *Milestones in Science and Technology*, p. 44.

Variable
PROGRAMMING

Neither mathematics nor **computer science** could have progressed very far without invention of the concept of a *variable* by the French lawyer François Viète, and this did not happen until surprisingly recently. In 1591, Viète, an accomplished amateur mathematician, published his book *In Artem Analyticam Isagoge* in which, for the first time, letters were used in **expressions** to represent variable quantities, the plus and minus signs were used in the way we

do now, and also for the first time, $A \times A$ was written as A^2. In computing, a *variable* is a constituent of a **high-level language** program whose value may literally be varied, that is, modified during the course of running the **program**. A variable is named during program creation by writing a **declaration** that associates the chosen name with the **data type** of the variable—real, integer, complex, and so on. The **compiler** associated with the language being used will assign the variable to a particular storage location in **random access memory (RAM)**. At the **machine language** level, all locations in RAM are variable, and all locations in **read-only memory (ROM)** are not. But some languages allow the declaration of *constants* placed in RAM but protected from change through **software**, such as a compiler that reports a compilation error if a store-to-memory **statement** is misdirected to a **constant**.

Additional reading: Struik, *A Concise History of Mathematics*, p. 94.

VAX series. *See* DEC VAX series.

Vector computer. *See* Parallel computer.

Very Long Instruction Word (VLIW)
COMPUTER ARCHITECTURE

Use of a *Very Long Instruction Word (VLIW)* in conjunction with specialized circuitry is an alternative to **instruction lookahead** as a way to achieve *instruction-level parallelism* in a Multiple **Instruction** Single Data (MISD) **parallel computer** (*see* **Flynn taxonomy**). The concept of a VLIW computer and its nomenclature was developed by Joseph "Josh" Fisher in 1979 while a professor at Yale. Later research in this area was pioneered by Fisher and the late Bob Rau while both were working as senior fellows at **Hewlett-Packard (hp)**, and by Glen Culler of the University of California at Santa Barbara. The VLIW idea was inspired by the use of horizontal (i.e., parallel) *microcode* on the first **supercomputers** such as the **CDC 6600** and IBM 360/91 (*see* **IBM 360 series**). Such computers tried to identify sections of linear **code** that could be executed in parallel at execution time. As an alternative, VLIW depends for its success on use of specialized **compilers** that try to identify such parallelizable segments during compilation of **programs** expressed in a **high-level language**. Each identified segment is then coded as a sequence of individual *operations* that are packed into a single very-long instruction, an instruction having a **word length** of 256 **bits** up to, in some implementations, 1,024 bits. At execution time, the running **microprocessor** would then execute packed operations in parallel. In the 1970s, many **digital signal processors (DSPs)** used VLIW-like wide instructions in **read-only memory (ROM)** to implement the **Fast Fourier Transforms (FFT)** and other **algorithms**. The first true VLIW machines were small supercomputers of the early 1980s built by Fisher's company, *Multiflow*, Rau's company, *Cydrome*, and Culler's company, *Culler Scientific*, but neither the machines nor those companies have endured. Some of the notable and more successful VLIW processors of recent years are IA-64 or *Itanium* from **Intel**, the Crusoe processor from Transmeta, the Trimedia processor from Philips, and TMS320C62× DSPs from **Texas Instruments (TI)**. Advocates of the VLIW **computer architecture** see it as a successor to that of the **reduced instruction set computer (RISC)**.

Additional reading: B. Rau and J. A. Fisher, "Instruction-Level Parallel Processing: History, Overview, and Perspective," *Journal of Supercomputing* 7, *1/2* (1993), pp. 9–50.

Virtual memory   OPERATING SYSTEMS

Virtual memory is an **operating system** technique that allows execution of a **program** whose combined **object code** and data space exceeds the portion of main **memory** allocated to it. The concept is normally used in conjunction with **multiprogramming** that allows alternative attention to several different programs that are "in the mix" simultaneously. Running programs are subdivided into either relatively large variable-length *segments* or small fixed-length *pages* that are swapped between main and **auxiliary storage** as needed. In a paging system, several pages are loaded into main memory at once, and additional pages are fetched from auxiliary storage—once a **magnetic drum** but now usually a **hard disk**—only when the operating system detects a page fault. A *page fault* is the detection that the next page needed to continue execution is not already in main memory, in which case it is swapped with a page believed to be inactive. The first virtual memory operating system was used on the Ferranti **Atlas** in 1958. The first virtual memory systems commercially available in the United States were **Multics** for the GE 635 (*see* **GE 600 series**) and the Michigan Terminal System (MTS) for the IBM 360/67 (*see* **IBM 360 Series**), both of which became operational in the late 1960s. Important theoretical work relative to virtual memory management was done in 1969 by David Sayre of **IBM** and in 1970 by Peter Denning. Denning devised the *working set model*, which capitalizes on *locality of reference*, the experimental observation that during any given short time interval, running programs tend to be referencing data stored over a small range of memory addresses. Although the most common elements of virtual memory are fixed-length pages, Ted Kaehler of **Xerox PARC** made an interesting virtual memory operating system in 1975 called *OOZE* (Object-Oriented Zoned Environment) that allocated main memory in terms of variable-length *objects*, the storage units of **Smalltalk**, the first **object-oriented programming** system.

Additional reading: *Encyclopedia of Computer Science*, pp. 1832–1835.

Virtual reality   COMPUTER GRAPHICS

Virtual reality is the use of **information technology** that provides humans with the perception that they are immersed in an environment other than their own. Perhaps the first visual reality device was the Link trainer invented by Edwin Link and used by the United States to train pilots in World War II. The flight simulator **programs** on current **personal computers** provide a visual experience that is similar, but not the physical movement that the Link could provide. Many arcade and videogames give some sense of virtual reality as they allow users to steer along race courses or speed downhill on a ski slope. By now, however, *virtual reality*, a term coined by Jaron Lanier of VPL Research in the late 1980s, connotes the use of a *head-mounted display* (HMD) with special goggles that permit viewing a projected three-dimensional scene and a haptic device such as an electronic *data glove* connected to a **digital computer**. *Haptic*, from the Greek *haptesthai*, means "to touch" or "to feel." Perhaps the most flexible haptic device is Thomas Massie's PHANToM, which has unique, pivoting thimblelike receptacles mounted at the ends of computerized arms, into which a person can insert his or her fingers and then virtually "feel" the shape, texture, and weight of objects on the computer screen. The experience, envisioned in some of the early **cyberspace** novels of William Gibson, gives the user a sense of being physically present in a distant and perhaps unfamiliar environment and being able to interact with it. Virtual reality has caught the public's attention as a form of entertainment, but its serious applications continue to provide, as the famous Link did, ways to train people to operate complex devices in seemingly threatening environments without danger. *See also* **Augmented reality**.

Additional reading: *Encyclopedia of Computer Science*, pp. 1835–1839.

Visual language   LANGUAGE TYPES

A *visual language* is a **programming language** whose features allow the creation of **programs** that interact with a **graphical user interface (GUI)** during execution. Programs generated by a visual language may display **icons** (small images) and *radio buttons* or *menus* from which program options may be selected by click of a **mouse** or **trackball**. The first visual language is considered to have been Ivan Sutherland's **Sketchpad** for the TX-2 at MIT in 1963, and his brother William also used the TX-2 to develop a simple visual **dataflow machine** language in 1965 (*see* **TX-0 / TX-2**). The visual languages currently in widespread use are the latest versions of **Microsoft** *Visual Basic* of 1987, Microsoft *Visual C++* written by Stanley Lippman in 1993, Borland *Visual C++* of 1994, and a version of **Pascal** included with the Borland *Delphi* development package first released in 1995.

Additional reading: http://www.cs.orst.edu/~burnett/vpl.html

Visualization. *See* **Data visualization**.

VLSI   HARDWARE

VLSI (very large scale integration) is a level of **logic circuit** miniaturization corresponding to **microprocessors** containing hundreds of thousands of **transistors** on a single chip, a level attained in 1975. The potential of VLSI technology was vigorously brought to world attention by California Institute of Technology (CalTech) professor Lynn Conway and Xerox systems designer Carver Mead. The impact of Mead and Conway on the design of VLSI circuits brought about a fundamental reassessment of how **integrated circuits (ICs)** are used. They optimized the VLSI process by melding the concepts of fabrication at the device level and **computer architecture** at the system level to produce truly integrated systems, not just circuits. For their work on VLSI in general and most particularly for their highly influential textbook *Introduction to VLSI Systems* that took university **computer science** and engineering departments by storm, Carver Mead and Lynn Conway earned *Electronics* magazine's annual Achievement Award in 1981. Densities of millions of transistors per chip are called ULSI, *ultra large-scale integration*, but at the current rate of progress, a new term will soon be needed.

Additional reading: http://ai.eecs.umich.edu/people/conway/Awards/Electronics/ElectAchiev.html

Von Neumann machine

COMPUTER TYPES

A **stored program computer** whose organization is similar to that of the **EDVAC** is called a *von Neumann machine*. Such machines are single **instruction** stream, single **data** stream (SISD) **digital computers** (uniprocessors) according to the **Flynn taxonomy** of **computer architecture**. The name *von Neumann machine* honors the senior author of the most famous paper in **computer science**, "Preliminary Discussion of the Logical Design of an Electronic Computing Instrument," a 1946 report of the Institute for Advanced Study (IAS) in Princeton, New Jersey, by John von Neumann, Arthur W. Burks, and Herman H. Goldstine. The eminent computer pioneer Sir Maurice Wilkes believes that a more appropriate name would be *Eckert–von Neumann machine*, giving equal recognition to J. Presper Eckert, the lead designer of the **ENIAC**. All "non-von" computers are **parallel computers** of one type or another.

The early von Neumann computers other than EDVAC that are the subjects of separate articles are the **AVIDAC, BINAC, EDSAC, IAS computer, ILLIAC, JOHNNIAC, MANIAC, MISTIC, ORDVAC, SEAC, SILLIAC, SWAC, UNIVAC,** and **WEIZAC.**

Additional reading: *Encyclopedia of Computer Science*, pp. 1841–1842.

Figure 44. With the possible exception of Albert Einstein, John von Neumann was the most brilliant scientist of the twentieth century. His prodigious memory and the incredible speed of his incisive reasoning led his colleagues to speculate that he must be an extraterrestrial but was so clever that he could give a flawless imitation of a human. (From the collection of the author)

W

WEIZAC

COMPUTERS

The *WEIZAC* (WEIZmann Automatic Computer), a copy of the **IAS computer**, was built for the Applied Mathematics Department at the Weizmann Institute of Science in Rehovot, Israel, during 1954–1955. The project leader for WEIZAC, which operated until the early 1960s, was Gerald Estrin. The machine was succeeded by two incompatible successors, GOLEM I and II in 1964 and 1972, respectively. The GOLEMs were designed and constructed at the Institute and used for research there and at other Israeli research centers. The 75-bit **word length** of GOLEM, equivalent to 22 decimal digits, is believed to be a world record.

Whirlwind

COMPUTERS

Project *Whirlwind* was sponsored at MIT by the U.S. Navy. It was originally started in 1944 to investigate the solution of aircraft stability and control problems through construction of an **analog computer**. But by 1946, the decision was made that a 16-bit-**word** general-purpose **digital computer** using 256 words of **electrostatic memory** and a 1 MHz pulse rate be used instead. Whirlwind was constructed under the leadership of Jay Wright Forrester. When first put in service in August 1949, the computer had 3,300 **vacuum tubes** and 8,900 crystal diodes. By June of 1950, one hour of error-free operation had been achieved, and by March of 1951, Whirlwind was operational on a 35-hours-per-week schedule. During 1953, a **magnetic tape** system and a **magnetic drum** system were installed, and electrostatic storage was replaced by two banks of **magnetic core** memory consisting of 1,024 16-bit words each. By December 1954, the computer had grown to 12,500 tubes and 23,800 diodes. Whirlwind occupied a two-story building. Its **central processing unit (CPU)**, control console, and **cathode ray tube (CRT)** displays occupied the second floor. Each **bit** of the **arithmetic-logic unit (ALU)** was a bay of equipment 2 feet wide and 12 feet high. The drum storage system and data communications interface occupied the ground floor. Whirlwind **instructions** consisted of a 5-bit **operation code (opcode)** and a single 11-bit address; all 32 possible opcodes were used. Its **loader**, implemented in **hardware** rather than the more usual **software**, was a bank of 32 toggle switches. Whirlwind operated until 1959. Parts of it are now in the Smithsonian Institution in Washington, D.C. and the Computer Museum History Center in Mountain View, California.

Additional reading: Robert R. Everett, "Whirlwind," in Metropolis, Howlett, and Rota, *A History of Computing in the Twentieth Century*, pp. 365–384.

Williams tube. *See* Electrostatic memory.

Windows. *See* Microsoft Windows.

Wireless connectivity COMMUNICATIONS

While *wireless connectivity* could certainly imply conventional **radio**, it is used here to describe the most recent milestone cited in this book, the availability as of 2002 to replace the snake pit of wires used to connect a desktop **personal computer (PC)** to its **peripheral** devices with connectivity via short-range high-frequency radio waves. Using **spread-spectrum communications** in the 2.4 GHz range, suitably equipped devices such as printers and **optical scanners** can communicate with a nearby desktop PC just as if they had a conventionally cabled connection. A **personal digital assistant (PDA)** such as a *Palm Pilot* and a PC can now communicate via a wireless HotSync operation. Wireless connectivity may finally advance **mobile computing** to the level of **ubiquitous computing**.

Additional reading: *Computing Encyclopedia*, vol. 4, p. 208.

Word DEFINITIONS

On the earliest **digital computers**, the length in **bits** or digits of a single-precision integer, the length of an **instruction**, and the length of a unit of information retrieved through reference to a specific address were all the same. That unit of information, regardless of how it was to be interpreted, became known as a *word*, a term of very specific meaning that let us speak of the **word length** of the **IBM 700 series**, say, as being 36 bits and that of the **CDC 1604** as being 48 bits. "Word," though of unknown parentage, seemed like an appropriate name since 36 or 48 bits was just the right size to hold the 6 or 8 encoded **characters** of a word of average size in a **natural language** such as English or French. On the newest **byte**-addressable computers, however, *word length* and hence *word* are somewhat ambiguous. To most writers, a single byte is too short to serve as a satisfactory "word," integers occupy 4 or 8 bytes, and instructions have a variable byte length depending on their function. Nonetheless, *word* is still useful in narrative, and we trust that the reader will keep the ambiguity in mind when it is encountered in other articles.

Word length DEFINITIONS

The *word length* of a **digital computer** is the length of its **words** as measured in digits—**bits** on a computer that uses the **binary number system**. But see the prior article, which discusses why *word* is not necessarily a well-defined entity on modern **byte**-addressable computers.

Word processing APPLICATIONS

Word processing, the online preparation, editing, and formatting of documents for later printing, is a term introduced by **IBM** in 1964 in conjunction with the introduction of a *Selectric typewriter* augmented with cassette tape **memory**. "Word processing" is a translation of the German word *Textverarbeitung*, coined in the late 1950s by IBM engineer and former Luftwaffe pilot Ulrich Steinhilper. As free-standing devices, the earliest word processors were more **hardware** than **software**. Before the **time sharing** and **personal computer (PC)** eras, the principal firms making **word processing** machines were IBM and Wang, a company started by An Wang, the co-inventor of **magnetic core** memory. Despite a cost of up to $15,000, including training, for a stand-alone system, word processing became a bustling industry by the early 1970s. The rudiments of computer-based word processing can be traced to work in the late 1950s for the **TX-0 / TX-2** at MIT, where Jack Gilmore wrote *Scope Writer* and Stephen Piner wrote a **program** called *Expensive Typewriter*, both supported by an input device called the *Lincoln Writer*. Throughout the 1960s, a number of **text editors** whose capabilities approached those of a word processor were written for use with time sharing systems, so they required access to large **mainframes**, or at least **minicomputers**. By the mid-1970s, however, things began to change. The *Bravo* editor written by Charles Simonyi and the **Smalltalk** environment developed at **Xerox PARC** for the **Alto** brought major innovations in text handling and use of a **graphical user interface (GUI)**. Larry Tesler and Tim Mott, also of Xerox PARC, developed the *Gypsy* word processor in 1974, a program that became the basis for *Microsoft Word* when Mott and some of his colleagues later went to work with Richard Brodie at **Microsoft**. Tesler went on to work on the **Apple Lisa** project. These systems blended text and graphics on a high-resolution raster graphics screen (*see* **Bit map**), using a dynamic graphical interface provided by a dedicated PC that used the **mouse** as a pointing device. Bravo and Gypsy displayed the text on a bit-mapped screen in a facsimile of the document's final appearance, the approach called What You See Is What You Get (WYSIWYG), and were thus the first *interactive editor / formatters*.

In early 1975, filmmaker Michael Shrayer bought an **MITS Altair** kit and quickly developed *Electric Pencil*, the first word processor for a commercial PC, and for two years he had the market to himself. In 1978, Seymour Rubinstein formed the *Micropro International* software company and hired Rob Barnaby to write a new word processor called *WordStar*, a product that dominated the market for six years. The first word processor for the **Apple II** was *EasyWriter*, a takeoff on the title of the movie *Easy Rider*, written in 1982 by John Draper, a.k.a. "Captain Crunch." New computers, exemplified by the **Apple Macintosh** of 1984, put a priority on ease of use, consistency of user interface, and interactive editing throughout their operating environment. The earliest full-featured word processor for the Macintosh was called *FullWrite Professional*, marketed by Ashton-Tate. The original product did not survive, although an attempt to resurrect it was made in the mid-1990s. The most widely used word processors in both the **Apple** and **IBM PC** environments are now *Microsoft Word*, whose first version was written by Richard Brodie, and *Word Perfect*, written by Bruce Bastian and originally marketed by a company of that name founded by Bastian and Alan Ashton. These products cost hundreds of dollars, but there are excellent **open source software** equivalents that run under **Linux**, as does the **Sun Microsystems** *StarOS* office suite that includes a word processor. With the proliferation of word processing technology, content and editing have now become so intertwined that very little computer-assisted text processing is distinct from formatting. It is now common for a single piece of software to be used for everything from word processing to typographical style selection, page placement, and updating of the table of contents, index, and bibliography. Thus word processing is gradually becoming indistinguishable from **desktop publishing**.

Additional reading: http://www.stanford.edu/~bkunde/fb-press/articles/wdprhist.html

Workstation COMPUTER TYPES

A *workstation* is a single-user **digital computer** capable of high-performance **computer graphics**. In the early 1980s, workstations were distinguished by their good-sized **memory**, large **bit-map** image **monitors**, and their use of a **mouse**-driven **graphical user interface (GUI)**, at a time when most office and laboratory workers used relatively small **ASCII** terminals connected to time-shared **mainframes** or to **personal computers (PCs)** of limited memory and an **operating system** with a **keyboard**-driven command-line interface such as **MS-DOS**. By 2000, every new PC had a powerful processor, ample memory, and a high-resolution display, and there is now a great overlap in performance among a top-of-the line PC, a workstation, and a small **minicomputer**, which were once quite distinctive classes of computer. The concept of a personal workstation originated at **Xerox PARC** during the 1960s. It is characteristic of a workstation that its **software** run under **Unix** on a **Reduced Instruction Set Computer (RISC)** microprocessor. The most advanced workstations are made by **Sun Microsystems**, using its own Sparc chip, and **Silicon Graphics, Inc.** (SGI), which uses a MIPS chip made by its subsidiary, MIPS Computer Systems, Inc., a company that SGI bought in 1993. **Clusters** of workstations (COWs) now achieve **supercomputer** performance levels (*see* **Beowulf**).

Additional reading: Goldberg, *A History of Personal Workstations*.

World Wide Web (WWW) NETWORKS

The *World Wide Web (WWW)*, a major milestone that brought access to the **Internet** to millions, was conceived and implemented by Tim Berners-Lee of England. The glimmer of his idea of the Web began in 1980 while Berners-Lee was in Geneva, Switzerland, consulting at CERN, which once meant *Conseil Européen pour la Recherche Nucléaire* but is no longer an acronym because of a broadening of mission. There he wrote a program called *Enquire* for Norsk Data machines. He considered Enquire to be short for "Enquire Within upon Everything," the title of a book of Victorian advice that he had read as a child. The **program** organized informational items important to Berners-Lee in **hypertext** form similar to that of Vannevar Bush's hypothetical **memex**. Berners-Lee quickly began to realize that Enquire's links to local information could just as well be made links to **files** on remote computers of a **network**, the essential idea of a World Wide Web, the name he soon coined. But it took ten years for him to convince CERN that implementation of the Web concept was worthwhile. In 1990, he wrote a prototype Web browser / editor for a **personal computer** marketed by NeXT, Inc., a company founded by Steve Jobs during a break in his **Apple** employment. In the expansion of his work to other machines, Berners-Lee acknowledges programming assistance from Nicola Pellow, a British student intern at CERN, and French visitor Jean-François Groff. The subsequent worldwide fascination with the Web rose in proportion to the development of ever more powerful browsers, a story told in that article.

Additional reading: Gillies and Cailliau, *How the Web Was Born.*

X

X window system COMPUTER GRAPHICS

The *X window system*, sometimes called just *X* but not to be called *X Windows*, is a **Graphical User Interface (GUI)** used with various flavors of **Unix**, including **Linux**. X was developed at MIT with support from the **Digital Equipment Corporation (DEC)** in order to produce a distributed, **hardware**-independent user interface for the Athena Project of the MIT Computer Science Laboratory. Development began in May 1984 by Robert Scheifler and Ron Newman from MIT and Jim Gettys from DEC. The Unix-based X window system was the basis for several early 1990s **browsers** for the **World Wide Web**. In X, every independently managed rectangle displayed on the **monitor** screen is called a *window*. The windows are managed using a **tree**, a hierarchical **data structure** that represents nested rectangles of screen area and associated information that specifies which rectangles are entirely or partially visible and which are partially obscured by rectangles activated more recently.

In 1989, the German student Thomas Roell began to modify X for use on **personal computers** that used the **Intel 80×86** chip. Roell then worked with Stephen Gildea and Mark Snitily to produce X386, which was released in 1990. Refinements were then made by the "gang of four" consisting of Jim Tsillas, Glenn Lai, David Wexelblat, and David Dawes. From 1992 to the present, this group, augmented by Dirk Hohndel, Richard Murphey, and Jonathan Tombs, has released a series of versions of XFree86, X386 renamed to emphasize that it may be downloaded free of charge. XFree86, Linux, and **Apache** are widely considered to be the three most successful examples of **open source software**.

Additional reading: Michael J. Hammel, "The History of Xfree86," *Linux Magazine*, December 2001, pp. 26–31.

Xerography HARDWARE

Xerography, a combination of the Greek words for "dry" and "writing," was invented by Chester F. Carlson in 1938, although he called his process *electrophotography*. The process is used with plain-paper **facsimile transmission (fax)** machines, **laser printers**, and photocopiers. Copying begins by placing a uniform static charge on the photoconducting surface of a drum. When a **laser** discharges just those parts of the surface that correspond to the black portions of the item being copied and "toner" is applied, the image on the drum is transferred to paper and fused with heat. Carlson had difficulty marketing his invention until 1944 when he was able to contract with the Batelle Memorial Institute in Columbus, Ohio, for further development. In 1947, Batelle licensed the Haloid Corporation of Rochester, New York, to market a xerographic copier. The Xerox 914, introduced in 1959, was so successful that Haloid changed its name to Xerox in 1961.

Additional reading: *Computing Encyclopedia*, vol. 2, pp. 21–22.

Xerox PARC INDUSTRY

Xerox PARC (Palo Alto Research Center) was founded by Xerox chief scientist Jacob Goldman in 1970, the year after his company gained a foothold in the **digital computer** business through purchase of Scientific Data Systems (SDS) for $920 million. Goldman named George Pake as first director, and he in turn hired Robert Taylor from the U.S. Defense Advanced Projects Agency (DARPA) to head recruiting (*see* **ARPAnet**). Among the distinguished scientists that served at Xerox PARC over the years were Alan Kay (*see* **Alto**), Robert Metcalfe (see **Ethernet**; **Metcalfe's law**), Lynn Conway (*see* **VLSI**), and James Clark (*see* **Browser; Netscape Communications; Silicon Graphics, Inc.**). Over the more than three decades since its inception, scientists at Xerox PARC invented personal **distributed systems**, the **graphical user interface (GUI)**, the first commercial **mouse**, **bit-map**-image **computer graphics** displays, Ethernet, **client-server computing**, **object-oriented programming**, **laser** diodes, the **laser printer**, **Smalltalk**, and many of the basic **protocols** of the **Internet**. Xerox scientists have also made major contributions to **network** architecture, **data visualization**, **liquid crystal displays**, **Page Description Languages (PDLs)**, **expert systems**, and **data compression**. In January 2002, Xerox PARC became the Palo Alto Research Center Incorporated.

Additional reading: Hiltzik, *Dealers of Lightning: Xerox PARC and the Dawn of the Computer Age.*

Xerox Star COMPUTERS

The *Xerox Star* 8010, introduced in May 1981, exhibited many features that were developed for the **Alto** at **Xerox PARC**, including a **laser printer** and **word processing** software for combining text and graphics in the same laser-printed document. Development of the Star was an important event in the history of **personal computers** because it changed notions of how interactive systems should be designed. The Star influenced the design of both the **Apple Lisa** and the **Apple Macintosh**. Over a relatively brief lifetime of three years, 100,000 Stars were sold.

Additional reading: Jeff Johnson and Teresa L. Roberts, "The Xerox Star: A Retrospective," *IEEE Computer*, **22**, 9 (September 1989), pp. 11–29.

XML LANGUAGES

XML (eXtensible Markup Language) is an enhancement of **HTML**, the **page description language (PDL)** used to define Web pages. XML is an **extensible language** because it allows users to define their own specialized *tags*, commands delineated $< \ldots >$ that specify links to music **files**, images, and new entities not envisioned in the language definition itself. XML is upward compatible from HTML; that is, it can be used to write HTML **code** that will still work with older **browsers**. XML specifications, edited by Tim Bray, were developed in 1997 by W3C, the **World Wide Web** Consortium led by Web inventor Tim Berners-Lee. The working group that produced XML was headed by Jon Bosak. *Xpath*, a language for addressing parts of an XML document, was created in 1999 by James Clark and Steven DeRose of Brown University (*see* **Graph**).

Additional reading: Berners-Lee and Fischetti, *Weaving the Web*, pp. 160–166, 181–188.

Y

Y2K problem

In the last few years before the year 2000, alarms were sounded throughout the information processing world that many crucial computer **programs** would not be able to handle dates beyond 12/31/99 because it had been a (careless) tradition to use only two digits to represent a year. The false economy stemmed from the early days of computing when every column of an 80-column **punched card** and every **byte** of **memory** was precious. The problem was given the colorful name *Y2K problem* (Y2000 problem) even though *K* usually means 1,024 (2^{10}), not decimal 1,000, to computer people. Credit card companies had to cope with this well before the new millennium and did so almost without incident. As 2000 approached, there were pessimists who began to hoard supplies and feared that no planes would be able to fly on New Year's Day and that the world's banking system would collapse. But, due to the diligence with which companies updated their **software**, that day came and went with no serious disruption of services. Nonetheless, those that did express grave concern should be credited with spurring the necessary and largely successful preventive measures. The real Y2K problem turned out to hinge on whether *chad*, the small pieces of cardboard that should be cleanly punched out and removed from a punched card by a card punch, are actually detached or whether they are left hanging.

Z

Zero

NUMBER SYSTEMS

The invention of **positional number systems** was a vitally important computing milestone, but the concept was not complete until *zero* had been accepted as a digit on a par with all others that comprise a particular system. The earliest positional number systems had nothing like a zero at all, making it impossible to distinguish, say, 379 from 307009. A breakthrough of sorts occurred in about 300 B.C. when the Babylonians invented a "placeholder," so that the latter number could be expressed as 3☐7☐☐9 (though they used a more complicated symbol than ☐). But since, unlike a true digit, the ☐ had no value of its own, its use in the units place was not allowed, lest a lone ☐ stand for "nothingness," a religiously forbidden concept. Finally, in about the ninth century in India, which had adopted the Babylonian's base 10 system four centuries earlier, a true zero was allowed in all decimal places, a truly significant milestone in the history of computing. Conquering Muslims then brought the now complete decimal system back west as far as Africa, and some historians believe that the zero was known to the French mathematician Gerbert of Aurillac, Pope Sylvester II, in time to solve the Y1K problem by writing it as 1000 rather than M. There are various theories as to why zero was given the same shape as the letter "O," but its most likely origin was that a round stone removed from counters in the dust left such an impression. Whatever, the concept of zero as a self-respecting digit in its own right left a profound impression on Leonardo of Pisa, better known as Fibonacci, he of the famous Fibonacci sequence (*see* **Discrete mathematics**). When he learned of it while traveling in Africa in about 1225, he hurried home and helped spread the news throughout Western Europe.

Additional reading: Seife, *Zero: The Biography of a Dangerous Idea*

Zilog Z-80

COMPUTERS

The *Zilog Z-80*, an upward compatible competitor to the **Intel 8080**, was first marketed in 1978. Zilog was founded in 1974 by former **Intel** employees Federico Faggin and Ralph Ungermann. The Z-80 was faster than the 8080 and contained ten additional **instructions** in its **instruction set**. The Z-80 **microprocessor** was used in a number of early **personal computers**, including the **Radio Shack TRS-80**, and was also used in some **Texas Instruments** electronic **calculators** and in early versions of Nintendo's *Game Boy*. The Z-80 was followed rather quickly by the Z8000, which used a 16-bit rather than an 8-bit **data path** and featured two modes, one for the **operating system** and one for user programs in order to prevent interference with **interrupt** handling. Despite early success, Zilog was sold in 1997 to the Texas Pacific Group, an investment company based in Fort Worth, Texas.

Additional reading: Freiberger and Swaine, *Fire in the Valley*, pp. 130–134.

Zuse computers

COMPUTERS

Konrad Zuse designed and built a number of so-called *Zuse computers* between 1938 and 1969. These **digital computers** served as the inspiration for much theoretical work in programming and **programming languages** by Zuse's colleague Heinz Rutishauser and by Zuse himself (*see* **Plankalkül**). The power and design of the first four machines, the Z1 through Z4, are particularly remarkable because they were built during the prelude to and duration of World War II when Zuse was isolated from computing developments in the United States and Great Britain.

The Z1, begun in 1936 and completed in 1938, used an arrangement of slotted metal plates through which pins passed to store, read, and write **bits** of information. Zuse was familiar with the **binary number system** before he started to build computers and discovered for himself its relation to **Boolean algebra** while working on the **arithmetic-logic unit (ALU)** of the Z1. The Z1 never worked properly, however, due to technological difficulties. For his Z2, Zuse used surplus **telephone** relays instead of mechanical devices to perform arithmetic. The machine, built in a workshop at his parents' Berlin apartment, was completed in 1939 and was demonstrated at the German Aerodynamics Research Institute that same year. Zuse's Z3 of 1941 was the first of his machines to work reliably. The Z3 used 1,800 telephone relays for **memory**, 600 for calculation, and 200 for logical control. Input was encoded in an 8-bit **code** by manually punching holes onto strips of discarded 35mm movie film. Although Zuse envisioned a **stored program computer** in a paper he wrote in 1937, the concept was not used in his early machines, probably because of their very limited memory; the Z3 could store only 64 22-bit **floating-point** numbers consisting of 7 bits for the base 2 exponent, 1 bit for the sign, and 14 for the mantissa. Thus the computer was externally programmed, executing linear sequences of incoming **instructions** with no provision for conditional jumps. Operations included ordinary arithmetic, square root, store and recall from memory, and binary to decimal conversion. Clock speed was about 4 to 5 Hz, with one multiplication taking about 3 to 5 seconds. The Z3 was completed in early December 1941 and ran until it was destroyed in an Allied bombing raid on Berlin in 1944.

From 1942 to 1945, Zuse built the Z4, a much more capable machine comparable in sophistication to the **Harvard Mark I** and the **Bell Labs relay computers**. Like the Z3, its memory consisted of only 64 locations, but its **word length** was increased to 32 bits, and its **instruction set** was much more extensive. After World War II, Zuse ran a successful computer company and, upon retirement, became a proficient artist whose oil paintings now hang in museums. In 1950, the only Z4 that had survived was leased by Eduard Stiefel for the *Eidgenossische Technische Hochschule* (ETH) in Zurich.

Additional reading: Rojas and Hashagen, *The First Computers: History and Architectures*, pp. 237–275.

Figure 45. Konrad Zuse in his laboratory with his Z4 in the background. Working in isolation in Germany during World War II, Zuse independently invented all major concepts of a stored program computer and designed the highly sophisticated high-level language Plankalkül. (Photo courtesy of Horst Zuse; *see* http://home.t-online.de/home/horst.zuse/show_en.html)

Listing of Cited References

The following are references to sources that were referred to in the narrative of article entries or recommended as additional reading.

Abramowitz, Milton; Stegun, Irene A. *Handbook of Mathematical Functions*. New York: Dover; 9th printing, 1970. 1,046p.

This classic, first published in 1964, contains very little history, but its introduction contains material on interpolation and recurrence relations that are pertinent to this book. And the great bulk of the remaining pages continue to provide a valuable resource for comparing the output of newly written mathematical routines to previously tabulated values.

Anders, George. *Perfect Enough: Carly Fiorina and the Reinvention of Hewlett-Packard*. New York: Penguin Putnam; 2003. 248p.

A fascinating profile of a historic company and its current colorful leader.

Aspray, William. *John von Neumann and the Origins of Modern Computing*. Cambridge, MA: MIT Press; 1990. 394p.

About the stored program concept and much that followed.

Aspray, William (Ed.). *Computing Before Computers*. Ames: Iowa University Press; 1990. 266p.

A well-illustrated book that covers many historic mechanical calculators and some early electronic calculators that were not quite "computers" in the modern sense.

Austrian, Geoffrey D. *Herman Hollerith: Forgotten Giant of Information Processing*. New York: Columbia University Press; 1982. 418p.

Hollerith has gained in stature over the years since this fascinating biography was published, and he is no longer "forgotten."

Babbage, Charles. *Passages from the Life of a Philosopher*. Piscataway, NJ: IEEE Press; 1994. 383p.

A reprint of a classic, highly entertaining autobiographical work first published in 1864. Edited with a new introduction by historian Martin Campbell-Kelly. Aspiring computer scientists should read this cover to cover as a first priority.

Bachtold, Adrian; Hadley, Peter; Nakanishi, Takeshi; Dekker, Cees. "Logic Circuits with Carbon Nanotube Transitors." *Science*, **294**, *5545* (9 November 2001), pp. 1317–1320.

See Huang et al. for the other of two breakthrough papers on nanoelectronics.

Bachtold, Adrian; Hadley, Peter; Nakanishi, Takeshi; Dekker, Cees. "Logic Gates and Computation from Assembled Nanowire Building Blocks." *Science*, **294**, *5545* (9 November 2001), pp. 1313–1317.

See also Bachtold et al. in the same issue for a report of a similar milestone in nanoelectronics.

Backus, John. "The History of Fortran I, II, and III." *IEEE Annals of the History of Computing*, **20**, *4* (October–December 1998), pp. 68–78.

Bashe, Charles J.; Johnson, Lyle R.; Palmer, John H.; Pugh, Emerson W. *IBM's Early Computers*. Cambridge, MA: MIT Press; 1986. 716p.

In addition to material reflecting its title, this work is also an excellent history of the punched-card era that preceded it.

Bauer, Roy A.; Collar, Emilio; Tang, Victor. *The Silverlake Project: Transformation at IBM*. London: Oxford University Press; 1992. 219p.

The suspenseful story of the IBM AS/400, a highly successful minicomputer that almost never saw the light of day.

Baumann, R.; Feliciano, M.; Bauer, F. L.; Samelson, K. *Introduction to Algol*. Englewood Cliffs, NJ: Prentice-Hall; 1964. 142p.

The best of the very few Algol books ever written.

Beeching, Wilfred A. *Century of the Typewriter*. New York: St. Martin's Press; 1974. 276p.

A comprehensive history, profusely illustrated with sketches and pictures of historic machines and layouts of 289 different keyboards.

Bergin, Thomas J., Jr.; Gibson, Richard G., Jr. (Eds.). *The History of Programming Languages*. New York: Academic Press; 1996. 864p.

A successor to the Wexelblat book of the same name.

Berkeley, Edmund Callis. *Giant Brains, or Machines That Think*. New York: Wiley; 1949, 270p.

An early but still widely quoted history of the origins of digital computing.

Berlinski, David. *A Tour of the Calculus*. New York: Vintage Books; 1997. 331p.

A very accessible survey.

Berlinski, David. *The Advent of the Algorithm: The 300-Year Journey from an Idea to the Computer*. San Diego, CA: Harcourt; 2001. 368p.

A decent history of a key milestone.

Berners-Lee, Tim; with Fischetti, Mark. *Weaving the Web: The Original Design and Ultimate Destiny of the World Wide Web*. San Francisco: HarperCollins; 1999. 226p.

A firsthand account by the creator of the Web. Gillies and Cailliau was used as the main reference for the Web article because it is more densely packed with information, but this book is interesting more for what Berners-Lee thinks should happen in the future.

Biggs, N. L.; Lloyd, E. K.; Wilson, R. J. *Graph Theory: 1736–1936*. New York: Oxford University Press; 1976. 239p.

A fascinating and very readable history of the first 200 years of classical graph theory.

Blaauw, Gerrit A.; Brooks, Frederick P., Jr. *Computer Architecture: Concepts and Evolution*. 2 vols. Reading, MA: Addison-Wesley; 1997. 1,264p.

An excellent exposition of the subject.

Blatner, David. *The Joy of π*. New York: Walker & Co; 1997. 130p.

All you might ever want to know about π and how to compute it.

Bowden, B(ertram) V(ivian). *Faster Than Thought*. London: Pitman; 1953. 416p.

An early classic whose appendix contains Ada Augusta's famous translation of Menabrea's paper describing the Analytical Engine.

Brooks, Frederick P., Jr. *The Mythical Man-Month: Essays on Software Engineering*. Reading, MA: Addison-Wesley; 1975. 195p.

One of the top ten classics of computer science.

Buchholz, Werner. *Planning a Computer System: Project Stretch*. New York: McGraw-Hill; 1962. 322p.

Another one of the ten books that every young computer scientist should read.

Burks, Alice R.; Burks, Arthur W. *The First Electronic Computer: The Atanasoff Story*. Ann Arbor: University of Michigan Press; 1988. 387p.

Includes extensive discussion of the Honeywell versus Sperry Rand suit that established Atanasoff's priority of invention of the electronic digital computer.

Burks, Arthur W. (Ed.) *Essays on Cellular Automata*. Urbana: University of Illinois Press; 1970. 375p.

Over 30 years later, this is still the essential reference. Contains classic contributions by Burks, John Holland, Edward Moore, John Myhill, J. W. Thatcher, and Stanislaw Ulam.

Bush, Vannevar. *Pieces of the Action*. New York: William Morrow; 1970. 366p.

An autobiography of the inventor of the Differential Analyzer.

Campbell-Kelly, Martin; Aspray, William.

Computer: A History of the Information Machine. New York: Basic Books; 1995. 342p.

A scholarly yet entertaining history through the mid-1990s. Contains a generous sampling of quotes from the work of the people who made it happen. Includes an extensive bibliography.

Casti, John L. *The Cambridge Quintet: A Work of Scientific Speculation.* Reading, MA: Addison-Wesley; 1998. 181p.

A must read. The author describes the lively discussion of five of the greatest thinkers of the twentieth century at a fictional dinner party on a stormy evening in 1949. The diners invited to converse with C. P. Snow at his home are Erwin Schrödinger, Ludwig Wittgenstein, J. B. S. Haldane, and Alan Turing. Although reticent in real life, Turing does most of the talking and manages to convey the essence of his three greatest achievements: the Turing machine, the Turing Test for machine intelligence, and the undecidability of the halting problem.

Ceruzzi, Paul E. *A History of Modern Computing.* Cambridge, MA: MIT Press; 1998. 398p.

A comprehensive and well-written history of computing from its punched-card origins in the mid-1940s to the end of the last decade. Includes an extensive bibliography of historical work.

Ceruzzi, Paul E. *Reckoners: The Prehistory of the Digital Computer, from Relays to the Stored Program Concept, 1935–1945.* Westport, CT: Greenwood Press; 1983.

Covers the period of the subtitle in fascinating detail.

Chandler, Alfred D., Jr. *Inventing the Electronic Century: The Epic Story of the Consumer Electronics and Computer Industries.* New York: Free Press; 2001. 321p.

Covers Japanese computer companies as extensively as those of the United States.

Clarke, Arthur C. *2001: A Space Odyssey.* New York: New American Library; 1968. 256p.

Perhaps the most influential science fiction book (and movie) of all time.

Coe, Lewis. *The Telegraph: A History of Morse's Invention and Its Predecessors in the United States.* Jefferson, NC: McFarland & Co.; 1993. 192p.

Morse, re-Morse, and a host of others.

Coe, Lewis. *The Telephone and Its Several Inventors: A History.* Jefferson, NC: McFarland & Co.; 1995. 230p.

"Several" is something of an understatement.

Coe, Lewis. *Wireless Radio.* Jefferson, NC: McFarland & Co.; 1996. 204p.

Yet another invention of multiple parentage.

Cohen, I. Bernard. *Howard Aiken: Portrait of a Computer Pioneer.* Cambridge, MA: MIT Press; 1999. 329p.

The definitive biography of the designer of the Harvard Mark I (IBM ASCC) by a leading historian of computing.

Computing Encyclopedia. Lincoln, NE: Sandhills Publishing Co.; 2002. 5 vols. of 240p. each.

Simply written for popular consumption, this work contains much historical material but concentrates on terminology, hardware, and software currently in use that may or may not ultimately qualify as "milestones." The five volumes are printed on newspaper stock but are nonetheless profusely and attractively illustrated in color. The fifth volume contains biographical profiles of many noteworthy persons who are now or once were working in some aspect of computing.

Cormen, Thomas H.; Leiserson, Charles E.; Rivest, Ronald L. *Algorithms.* Cambridge, MA: MIT Press; 1990. 1,028p.

The definitive textbook to date.

Davis, Martin. *The Universal Computer: The Road from Leibniz to Turing.* New York: Norton; 2000. 256p.

The title describes the era covered.

de Carvalho, Maria Pires. "Chaotic Newton's Sequences." *The Mathematical Intelligencer,* **24,** *1* (Winter 2002), pp. 31–35.

See also the related article "Graphic and Numerical Comparison between Iterative Methods" by Juan L. Varona on pages 37–46 of the same issue.

de Latil, Pierre. *Thinking by Machine: A Study of Cybernetics.* Boston: Houghton Mifflin; 1957. 355p.

An early popular account of cybernetics with an interesting forward by Isaac Asimov.

Denning, Peter J. (Ed.). *The Invisible Future: The Seamless Integration of Technology into Everyday Life.* New York: McGraw-Hill; 2001. 256p.

The editor is the author of chapter 23, an extensive definition and ringing endorsement of the new field called *information technology* (IT).

Dewdney, A. K. *The Turing Omnibus: 61 Excursions in Computer Science.* Rockville, MD: Computer Science Press; 1989. 415p.

A delightful collection of articles by a writer noted for extreme clarity of exposition.

Encyclopedia of Computer Science. 4th ed. Edited by Anthony Ralston, Edwin D. Reilly, and David Hemmendinger. London: Nature Publishing Group; 2000. 2,034p.

Contains 642 articles on computer science and technology, with special emphasis on the origins and history of the subject. Includes a 16-page color centerfold and biographical profiles of 38 computer pioneers.

Evans, Christopher. *The Making of the Micro*. New York: Van Nostrand Reinhold; 1981. 118p.

I love this thin little book, if only for its exquisitely detailed drawings of several early calculators. Computers were just verging on being "micro" in 1981, but the book presents a very good history of the calculators and earlier technology that preceded the first early microchips.

Federighi, F. D.; Reilly, E. D. *Weighting for Baudot and other problems for you and your computer*. Wayne, NJ: Avery Publishing Group; 1978. 240p.

Fisher, David E.; Fisher, Marshall Jon. *TUBE: The Invention of Television*. Washington, DC: Counterpoint; 1996. 427p.

An engrossing history of the multitude of physicists, engineers, and inventors who contributed to the evolution of television.

Flannery, Sarah. *In Code: A Mathematical Journey*. New York: Workman Publishing; 2001. 341p.

With just a little help from her dad, this teenage author produced a fascinating and inspirational book.

Flower, Derek Adie; Zahran, Mohsen. *The Shores of Wisdom: The Story of the Ancient Library of Alexandria*. Philadelphia: Xlibris; 1999. 196p.

The sad story of the loss of an irreplaceable treasure.

Foster, Caxton C. *Real Time Programming—Neglected Topics*. Reading, MA: Addison-Wesley; 1981. 190p.

Freiberger, Paul; Swaine, Michael. *Fire in the Valley: The Making of the Personal Computer*. 2nd ed. New York: McGraw-Hill; 1999. 463p.

A comprehensive and credible history of the personal computer.

Garfinkel, Simson. *Architects of the Information Society: Thirty-five Years of the Laboratory for Computer Science at MIT*. Cambridge, MA: MIT Press; 1999. 86p.

Does for the MIT lab what Hiltzik's book does for Xerox PARC.

Garliński, Jósef. *The Enigma War*. New York: Charles Scribner's Sons; 1979. 219p.

Contains excellent descriptions of both the Enigma cipher machine and the "Bombe" that was devised by the Allies to break its messages.

Gay, Martin K. *Recent Advances and Issues in Computers*. Phoenix, AZ: Oryx Press; 2000. 261p.

Contains, among several other topics, biographical sketches of 14 people who have advanced the state of the computing art and science.

Gillies, James; Cailliau, Robert. *How the Web Was Born*. New York: Oxford University Press; 2000. 372p.

This book does for the World Wide Web what the Hafner and Lyon history does for the Internet that supports it. In addition to the main narrative, contains a list of over 180 persons and (briefly) their contribution to the development of the Web.

Glossbrenner, Alfred; Glossbrenner, Emily. *Search Engines for the World Wide Web*. Berkeley, CA: Peachpit Press; 2001. 348p.

Discusses the strong and weak points of many leading search engines.

Golambos, Louis; Abrahamson, Eric John. *Anytime, Anywhere: Entrepreneurship and the Creation of a Wireless World*. New York: Cambridge University Press; 2002. 300p.

The birth and explosion of cellular and wireless communication.

Goldberg, Adele (Ed.). *A History of Personal Workstations*. Reading, MA: Addison-Wesley; 1988. 537p.

History as seen from the perspective of those who pioneered their development.

Goldstine, Herman H. *The Computer from Pascal to von Neumann*. Princeton, NJ: Princeton University Press; 1972. 378p.

An indispensable history up to the late 1950s.

Hafner, Katie; Lyon, Matthew. *Where Wizards Stay Up Late: The Origins of the Internet*. New York: Simon and Shuster; 1996. 304p.

A very readable history of the Internet through 1994, just before the impact of the World Wide Web. Among the wizards are Baran, Cerf,

Crocker, Heart, Kahn, Kleinrock, Licklider, Postel, Roberts, Taylor, and Tomlinson.

Hecht, Jeff. *City of Light: The Story of Fiber Optics.* New York: Oxford University Press; 1999. 316p.
> Why fast Internet access is possible.

Hiltzik, Michael. *Dealers of Lightning: Xerox PARC and the Dawn of the Computer Age.* New York: HarperCollins; 1999. 448p.
> The story of Xerox PARC, which has been home to an astonishing number of researchers who have done so much to shape the last three decades of computing.

Hoare, C.A.R. *Communicating Sequential Processes.* Englewood Cliffs, NJ: Prentice-Hall; 1985. 256p.
> A formal treatment but one presented with extreme clarity by a master teacher.

Hodges, Andrew. *Alan Turing: The Enigma.* New York: Simon and Schuster; 1983. 587p.
> The Enigma is both Turing himself and the World War II German cipher machine. The bestselling biography that evolved into Hugh Whitemore's Broadway play entitled (also with double meaning) *Breaking the Code.*

Hogan, James P. *Mind Matters: Exploring the World of Artificial Intelligence.* New York: Ballantine; 1997. 381p.

Holland, John H. *Emergence: From Chaos to Order.* Reading, MA: Perseus; 1998. 258p.
> Despite the subtitle, only two pages mention chaos theory specifically, but the book gives excellent coverage to game theory and the emergent behavior of neural nets.

Huang, Yu; Duan, Xiangfeng; Cui, Yi; Lauhon, Lincoln J.; Kim, Kyoung-Ha; Lieber, Charles M. "Logic Gates and Computation from Assembled Nanowire Building Blocks." *Science,* **294**, 5545 (9 November 2001), pp. 1313–1317.
> *See also* Bachtold in the same issue for a report of a similar milestone in nanoelectronics.

Ifrah, Georges. *The Universal History of Computing from the Abacus to the Quantum Computer.* Translated from the French and with notes by E. F. Harding. New York: Wiley; 2001. 410p.
> An excellent survey. Contains a remarkably coherent four-page flowchart detailing the history of computation in which each box is roughly comparable to a milestone of this book.

Inmon, William H. *Data Architecture: The Information Paradigm.* 2nd ed. New York: Wiley; 1993. 272p.
> An influential work by the man who coined the term "data warehouse."

Iverson, Kenneth. *A Programming Language.* New York: Wiley; 1962. 286p.
> The classic APL notation and later language as described by its inventor.

Jackson, Tim. *Inside Intel: Andy Grove and the Rise of the World's Most Powerful Chip Company.* New York: Dutton; 1997. 424p.
> The subtitle is aptly descriptive.

Jager, Rama D.; Ortiz, Rafael. *In the Company of Giants.* New York: McGraw-Hill; 1997. 232p.
> Conversations with 16 "Visionaries of the Digital World": Jobs, Rodgers, Eubanks, Case, Cook, Kurtzig, Warnock, Geschke, Dell, Wang, Gates, Grove, Hawkins, McCracken, Olsen, and Hewlett (in the order presented).

Johnson, George. *A Shortcut Through Time: The Path to the Quantum Computer.* New York: Knopf; 2003. 204p.
> An excellent and accessible summary from a contributing science writer for the *New York Times.*

Kaehler, Ted; Patterson, Dave. *A Taste of Smalltalk.* New York: Norton; 1986. 136p.
> A very readable introduction to the first object-oriented language.

Kahn, David. *The Codebreakers: The Story of Secret Writing.* London: Weidenfeld and Nicolson; 1967. 1164p. New York: Scribner; 1996, rev. ed. 1,181p.
> The definitive history of the subject, so thorough as originally published that only a 14-page update was deemed necessary 31 years later.

Kidwell, Peggy Aldrich; Ceruzzi, Paul E. *Landmarks in Digital Computing.* Washington, DC: Smithsonian Institution Press; 1994. 148p.
> Landmarks are presented chronologically rather than alphabetically, with both overlap and underlap with milestones of this book. The average length of a landmark article is a bit longer than that of a milestone, and all are accurate and well written.

Kruth, Donald. *The Art of Computer Programming.* Vol. 1: *Fundamental Algorithms,* 3rd ed.; 1997. 700p. Vol. 2: *Seminumerical Algorithms,* 3rd ed.; 1997. 775p. Vol. 3: *Sorting and Searching,* 2nd ed.; 1998. 780p. Reading, MA: Addison-Wesley.
> *The* fundamental reference to computer science.

Lavington, Simon. *Early British Computers.* Bedford, MA: Digital Press; 1980. 139p.

Contains some valuable tables that intercompare several of the early British computers. Especially intriguing is its reproduction of a machine language program of Tom Kilburn that is alleged to have been the first program to run successfully on any digital computer (*see* Figure 32).

Lee, J.A.N. *Computer Pioneers.* Los Alamitos, CA: IEEE Computer Society Press; 1995. 816p.

Biographies of over 250 computer pioneers, ranging in length from one line to several pages.

Levy, Steven. *Artificial Life: The Quest for a New Creation.* New York: Pantheon; 1992. 390p.

Contains some beautiful color plates that help show the excitement of this relatively new field.

Levy, Steven. *Crypto: How the Code Rebels Beat the Government, Saving Privacy in the Digital Age.* New York: Viking; 2001. 356p.

Contains fascinating detail about the British cryptologists who invented public-key cryptosystems but couldn't admit it for many years. A great read, as are all of Levy's books.

Levy, Steven. *Hackers: Heroes of the Compute Revolution.* Garden City, NY: Anchor Press/ Doubleday; 1984. 458p.

Levy's hackers are good guys, not the bad hackers ("crackers") that arose more recently. This book is a real page-turner, one that captures the spirit of early machine language programming better than any other known to this author.

Levy, Steven. *Insanely Great: The Life and Times of the Macintosh, the Computer That Changed Everything.* New York: Viking; 1994. 328p.

An overstatement, perhaps, but not by much.

Lewis, F. L. *Applied Optimal Control and Estimation.* Englewood Cliffs, NJ: Prentice-Hall; 1992. 656p.

Chapter 1 provides an interesting and extensive history of automatic control theory.

Lewis, Philip M., II; Rosenkrantz, Daniel J.; Stearns, Richard E. *Compiler Design Theory.* Reading, MA: Addison-Wesley; 1976. 647p.

This book and their earlier technical papers published relative to its subject changed compiler construction from a mixture of art and lore to something much more of a science.

Lindgren, Michael. *Glory and Failure: The Difference Engines of Johann Müller, Charles Babbage, and Georg and Edvard Scheutz.* Cambridge, MA: MIT Press; 1990. 415p.

An interesting contrast of designs and personalities.

Lohr, Steve. *Go To: The Story of the Programmers Who Created the Software Revolution.* New York: Basic Books; 2001. 250p.

Literary colorful coverage of Basic, Cobol, C++, Fortran, Java, and the open source software movement.

Lukoff, Herman. *From Dits to Bits: A Personal History of the Electronic Computer.* Portland, OR: Robotics Press; 1979. 219p.

An amateur radio buff used to the dits and dahs of Morse code becomes an engineer who worked on bits of BINAC and UNIVAC for Eckert and Mauchly. Contains some rarely seen pictures of these early computers.

Maddux, Cleborne; Johnson, D. Lamont (Eds.). *Logo: A Retrospective.* Binghamton, NY: Haworth Press; 1997. 212p.

A series of nostalgic essays that speculate on why Logo failed to have the expected impact on elementary education.

Malone, Michael S. *Infinite Loop: How Apple, the World's Most Insanely Great Computer Company, Went Insane.* New York: Doubleday; 1999. 597p.

Well researched by an author willing to risk use of a provocative subtitle.

Malone, Michael S. *The Microprocessor: A Biography.* New York: Springer-Verlag; 1995. 333p.

A very readable history.

Martin, Ernst. *The Calculating Machines* (Die Rechenmaschinen)—*Their History and Development.* Translated from the 1925 German edition and edited by Peggy Aldrich Kidwell and Michael R. Williams. Vol. 16 in the Charles Babbage Institute Reprint Series for the History of Computing. Cambridge, MA: MIT Press; 1992. 392p.

Describes many less well-known vintage calculators that are not covered in this book. The final 25 pages contain delightful reproductions of German advertisements for these and other calculators.

McCartney, Scott. *ENIAC: The Triumphs and Tragedies of the World's First Computer.* New York: Walker & Co.; 1999. 262p.

A very complete account, although a more accurate subtitle would be "world's first *programmable, general purpose, electronic, decimal, digital* computer." Omitting any of the italicized adjectives places some other computer "first."

McCorduck, Pamela. *Machines Who Think*. San Francisco: W. H. Freeman; 1979. 375p.
Sparkling profiles of the principal early workers in Artificial Intelligence.

Metropolis, Nicholas; Howlett, J.; Rota, Gian-Carlo (Eds.). *A History of Computing in the Twentieth Century: A Collection of Essays*. New York: Academic Press; 1980. 659p.
Historical essays by 39 famous computer pioneers, including Backus, Dijkstra, Hamming, Knuth, Mauchly, Stibitz, Wilkes, and Zuse.

Moreau, René. *The Computer Comes of Age*. Cambridge, MA: MIT Press; 1984. 227p.
An appendix contains an interesting table that gives the architectural characteristics of several early Soviet computers.

Mount, Ellis; List, Barbara A. *Milestones in Science and Technology*. 2nd ed. Phoenix, AZ: Oryx Press; 1994. 206p.
In proportion to how much older science in general is than computer science, Mount and List contains about three times as many articles as this book, but they average one-third as long in length. For this reason, the 10% or so of articles that overlap the present work are only occasionally cited as "additional reading," but the articles contain many useful leads to inventors whose accomplishments in Information Technology I might have overlooked.

Nasar, Sylvia. *A Beautiful Mind*. New York: Simon & Schuster; 1998. 460p.
The biography of John Forbes Nash, winner of the 1994 Nobel Prize in Economics. This book, made into the award-winning movie of the same name, contains a more accessible exposition of Game Theory than most technical books on the subject.

Nash, Stephen G. (Ed.). *A History of Scientific Computing*. Reading, MA: Addison-Wesley; New York: ACM Press; 1990. 359p.
Contains James Cooley's historical essay "How the FFT Gained Acceptance."

Newborn, Monty. *Automated Theorem Proving*. New York: Springer-Verlag; 2001. 231p.
An excellent survey of a difficult subject.

Newborn, Monty. *Kasparov versus Deep Blue: Computer Chess Comes of Age*. New York: Springer-Verlag; 1997. 322p.
This definitive history of computer chess unfortunately went to press just before Deep Blue's victory in the 1997 rematch.

Nyce, James M.; Kahn, Paul (Eds.). *From Memex to Hypertext: Vannevar Bush and the Mind's Machine*. San Diego, CA: Academic Press; 1991. 380p.
Two great ideas from one great mind. Contains the full text of Bush's famous *Atlantic Monthly* essay "As We May Think."

Petzold, Charles. *CODE: The Hidden Language of Computer Hardware and Software*. Redmond, WA: Microsoft Press; 1999. 393p.
A very clearly written and illustrated exposition of applied coding theory, covering the Baudot, Braille, and Morse codes and the ASCII and EBCDIC character codes. The title also embraces the alternative meaning of "code" as a sequence of program instructions or statements.

Pugh, Emerson W. *Building IBM: Shaping an Industry and Its Technology*. Cambridge, MA: MIT Press; 1995. 405p.
The inside story.

Pugh, Emerson; Humphrey, Watts S.; Grad, Burton. "Origins of Software [Un]Bundling." *IEEE Annals of the History of Computing*, **24**, *1* (January–March 2002), pp. 57–71.
These are actually three independent but related papers, one by each author.

Pugh, Emerson; Johnson, Lyle R.; Palmer, John H. *IBM's 360 and Early 370 Systems*. Cambridge, MA: MIT Press; 1991. 819p.
The definitive work on the architectural series that has spanned almost four decades.

Ralston, Anthony; Wilf, Herbert S. *Mathematical Methods for Digital Computers*. Vol. 1. New York: Wiley; 1960. 293p.

Randell, Brian. *The Origins of Digital Computers: Selected Papers*. 3rd ed. New York: Springer-Verlag; 1983. 580p.
A basic resource.

Redmond, Kent C.; Smith, Thomas M. *From Whirlwind to Mitre: The R & D Story of the SAGE Air Defense Computer*. Bedford, MA: Digital Press; 2000. 566p.
In essence, a sequel to their earlier book. Here, the emphasis is on how the SAGE project influenced and substantially changed how R & D is done in the United States.

Redmond, Kent C.; Smith, Thomas M. *Project Whirlwind: The History of a Pioneer Com-*

puter. Bedford, MA: Digital Press; 1980. 280p.

The excitement shows through.

Reid, Robert H. *Architects of the Web*. New York: John Wiley; 1997. 370p.

Chapter-length profiles of eight people who were instrumental in the development of the World Wide Web: Andreessen, Glaser, Polese, Pesce, Poler, Yang, Anker, and Minor.

Reid, T. R. *The Chip: How Two Americans Invented the Microchip and Launched a Revolution*. New York: Simon & Schuster; 1984. 243p.

Despite the chauvinistic title, this is very readable story about the relative merits of the achievements of Jack St. Clair Kilby of Texas Instruments and Robert Noyce of Fairchild Semiconductor.

Reilly, Edwin D.; Federighi, Francis D. *Pascalgorithms: A Pascal-Based Introduction to Computer Science*. Boston, MA: Houghton-Mifflin; 1989. 800p.

Contains the only published explanation of the Federighi gerund function.

Ribenboim, Paulo. *The Little Book of Big Primes*. New York: Springer-Verlag; 1991. 237p.

Just about all that is known and much that is unknown about prime numbers.

Rifkin, Glenn; Harrar, George. *The Ultimate Entrepreneur: The Story of Ken Olsen and Digital Equipment Corporation*. Chicago: Contemporary Books; 1988. 332p.

Much about Olsen, of course, but also contains a nice summary of each computer in the DEC PDP series.

Riordan, Michael; Hoddeson, Lillian. *Crystal Fire: The Invention of the Transistor and the Birth of the Information Age*. New York: Norton; 1998. 368p.

All about the foundation for the computer's third generation.

Roberts, Eric S. *Thinking Recursively*. New York: John Wiley; 1986. 179p.

The most readable book-length exposition of the subject.

Robertson, James E. "The ORDVAC and the ILLIAC." In Metropolis, Howlett, and Rota pp. 347–364.

A brief sketch of two early von Neumann computers.

Rojas, Raúl; Hashagen, Ulf (Eds.). *The First Computers: History and Architectures*. Cambridge, MA: MIT Press; 2000. 455p.

In addition to discussion of the historic American and British computers, this book provides a rare look at some early German and Japanese computers.

Sangalli, Arturo. *The Importance of Being Fuzzy*. Princeton, NJ: Princeton University Press; 1998. 173p.

The title continues "and other insights from the border between math and computers," announcing correctly that the book covers much more than just fuzzy sets. Included are very readable sections on neural nets and on the limits of classical computing and formal reasoning.

Schroeder, Manfred R. *Number Theory in Science and Communication With Applications in Cryptography, Physics, Biology, Digital Information, and Computing*. Berlin: Springer-Verlag; 1984. 324p.

A readable discussion of the applicability of number theory to several scientific disciplines.

Seife, Charles. *Zero: The Biography of a Dangerous Idea*. New York: Penguin Putnam; 2000. 248p.

The fascinating story of how it took centuries for zero to become a self-respecting digit.

Shurkin, Joel. *Engines of the Mind*. New York: Norton; 1984. 352p.

A quite complete history to the date of publication, one that emphasizes the personalities of the people behind the machines.

Slater, Robert. *Portraits in Silicon*. Cambridge, MA: MIT Press; 1987. 374p.

Articles about or interviews with 34 computer scientists from Charles Babbage to Donald Knuth, the last possibly being the best profile of Knuth in print.

Smith, Douglas K; Alexander, Robert C. *Fumbling the Future*. New York: Morrow; 1988. 274p.

The subtitle "How Xerox invented, then ignored, the first personal computer" is a bit harsh on Xerox management, though the authors' opinion is somewhat widely shared.

Stephenson, Neal. *In the Beginning Was the Command Line*. New York: Avon; 1999. 151p.

The author's irreverent philosophy of how an operating system ought to be designed.

Sterling, Thomas. "How to Build a Hypercomputer." *Scientific American*, **285**, *1* (July 2001), pp. 38–45.

Stern, Nancy. *From ENIAC to UNIVAC: An Appraisal of Eckert-Mauchly Computers.* Bedford, MA: Digital Press; 1981. 286p.

Strathern, Paul. *The Big Idea: Turing and the Computer.* New York: Doubleday; 1999. 105p.
A simply written primer on computing up through and including Turing.

Struik, Dirk J. *A Concise History of Mathematics.* 4th rev. ed. New York: Dover; 1987. 195p.

Swade, Doron. *The Difference Engine: Charles Babbage and the Quest to Build the First Computer.* New York: Viking; 2001. 342p.
Covers both the Difference Engine and the Analytical Engine.

Torvalds, Linus; Diamond, David. *Just for Fun: The Story of an Accidental Revolutionary.* New York: HarperCollins; 2001. 262p.
How Linux came to be as told anecdotally by its creator.

Tsang, Cheryl. *Microsoft: First Generation.* New York: John Wiley; 2000. 253p.
Profiles of 12 people who helped shape Microsoft in its early years: O'Rear, Oki, Brodie, Borland, Evans, Neir, Cole, Yee, Harding, Sribhibhadh, Steele, and Dziko.

Understanding Computers. Alexandria, VA: Time-Life Books; 1989. 3,072p.
A 24-volume series, profusely illustrated in color, of the principles and history of computing from antiquity through the year of publication. Each 128-page unnumbered volume has a unique title that is used as a series subtitle when used as additional reading. One volume, an extensive index, also contains an illustrated timeline.

Waldrop, M. Mitchell. *The Dream Machine: J. C. R. Licklider and the Revolution That Made Computing Personal.* New York: Viking; 2001. 512p.
This book may finally bring Licklider the recognition that his legacy deserves.

Weisstein, Eric W. *Concise Encyclopedia of Mathematics.* Boca Raton, FL: CRC Press; 1999. 1,969p.
Everything you always wanted to know about applied mathematics but were afraid to ask.

Weizenbaum, Joseph. *Computer Power and Human Reason.* San Francisco: W. H. Freeman; 1976. 300p.
A classic philosophy of computing that is still worth reading, if only for his description of experience with his Eliza program.

Wexelblat, Richard L. (Ed.). *History of Programming Languages.* New York: Academic Press; 1981. 758p.
For the successor volume, see Bergin and Gibson.

White, Ron. *How Computers Work.* Millennium edition. Indianapolis, IN: Que; 1999. 421p.
Not a historical work but an invaluable and beautifully illustrated book whose title could have been broadened to *How Computers and Their Peripheral Devices Work.* Covers the mouse, trackball, scanner, monitor, disk drives, and many other devices.

Wiener, Norbert. *Cybernetics, or Control and Communication in the Animal and the Machine.* 2nd ed. Cambridge, MA: MIT Press; 1961. 212p.
The classic work by the giant of the field.

Williams, Michael R. *A History of Computing Technology.* Englewood Cliffs, NJ: Prentice-Hall; 1985. 432p.
A basic resource.

Wilson, John F. *Ferranti: A History: Building a Family Business, 1882–1975.* London: Carnegie Publishing; 2000. 640p.

Wirth, Niklaus. *Algorithms + Data Structures = Programs.* Englewood Cliffs, NJ: Prentice-Hall; 1976. 366p.
The inventor of Pascal shows off the beauty and utility of the language.

Wood, Gaby. *Edison's Eve: A Magical History of the Quest for Mechanical Life.* New York: Knopf; 2002. 304p.
According to Joyce Carol Oates, "A treasure trove of marvels and information. Wittily and cogently written, this unusual cultural analysis provides us with unsettling insights into our historic fascination with human simulacra through the centuries."

Listing of References
of Interest

The following citations are to references that were consulted in preparing this book but which, though not cited as "Additional Reading" at the end of an article, are nevertheless significant sources worth further examination. Some were mentioned in the body of an article.

Åström, Karl. *Control of Complex Systems.* New York: Springer-Verlag; 2000. 482p.
May well become the standard reference to the subject.

Augarten, Stan. *Bit by Bit: An Illustrated History of Computers.* New York: Ticknor and Fields; 1984. 324p.

Baer, Robert M. *The Digital Villain: Notes on the Numerology, Parapsychology, and Metaphysics of the Computer.* Reading, MA: Addison-Wesley; 1972. 187p.
A little gem. Thirty years later, Baer's insights hold up well. Contains excerpts from eight works of fiction that depict computers in society.

Bardini, Thierry. *Bootstrapping: Douglas Engelbart, Coevolution, and the Origins of Personal Computing.* Palo Alto, CA: Stanford University Press; 2001. 328p.
An engrossing story of a key founder of personal computing.

Cohen, I. Bernard; Welch, Gregory W. *Makin' Numbers: Howard Aiken and the Computer.* Cambridge, MA: MIT Press; 1999. 329p.
Includes recollections and reminiscences of Aiken by Brooks, Calingaert, Hopper, Bloch, Oettinger, Wilkes, and Tropp.

Cortada, James W. *An Annotated Bibliography on the History of Data Processing.* West-port, CT: Greenwood Press; 1983. 215p.
An invaluable resource, but an updated version is sorely needed.

Dummer, G. W. A. *Electronic Inventions 1745–1976.* New York: Pergamon; 1977. 158p.
Describes hundreds of electronic inventions listed by name, name of inventor, and chronology.

Dyson, Esther. *Release 2.1: A Design for Living in the Digital Age.* New York: Broadway Books; 1998. 370p.
Perceptive opinions about the role of education and government in the digital age.

Encyclopedia of Personal Computing. Edited by Stan Gibilisco. New York: McGraw-Hill; 1995. 1,216p.
Useful, but a lot has happened since 1995.

Feynman, Richard P. *Feynman Lectures on Computation.* Edited by Anthony J. G. Hey and Robin W. Allen. Reading, MA: Addison-Wesley; 1996. 303p.
The first half of this book is essentially a primer on computing that is quite accessible to the beginning student. The second half is more technically challenging as the master teacher discusses the limitations of computation imposed by the laws of physics.

Greenia, Mark W. *History of Computing: An Encyclopedia of the People and Machines That Made Computer History*. Antelope, CA: Lexikon; 2002. An e-book on one CD-ROM.

In addition to almost 2,000 well-researched short articles, 1,600 photos make this the best single source of pictures of computers and people.

Hillis, Daniel. *The Pattern on the Stone: The Simple Ideas That Make Computers Work*. Los Angeles: Perseus Press; 1999. 164p.

Even simple ideas are sometimes hard to explain, but Hillis does so magnificently.

Hofstadter, Douglas R. *Gödel, Escher, Bach: An Eternal Golden Braid*. New York: Basic Books; 1979. 777p.

This is a must read for every student of mathematics and computer science.

Hyman, Anthony. *Charles Babbage: Pioneer of the Computer*. Princeton, NJ: Princeton University Press; 1982. 287p.

Perhaps the best short biography of Babbage. Contains some beautiful plates of portions of both the Difference Engine and the Analytical Engine.

Ifrah, Georges. *The Universal History of Numbers from Prehistory to the Invention of the Computer*. New York: Wiley; 2000. 633p.

Translated from the French by David Bellos, E. F. Harding, Sophie Wood, and Ian Monk. The definitive work on the subject.

Johnson, R. Colin; Brown, Chappell. *Cognizers: Neural Networks and Machines That Think*. New York: Wiley; 1988. 260p.

A popular treatise on the history of neural net research.

Kidder, Tracy. *The Soul of a New Machine*. New York: Modern Library; 1997. 336p.

A nicely reprinted edition of the 1981 Pulitzer Prize–winning story of the dedicated team of engineers who were charged with designing and testing the Data General MV-8000 Eagle minicomputer in the short space of one year.

Kurzweil, Ray. *The Age of Intelligent Machines*. Cambridge, MA: MIT Press; 1990. 552p.

Informed speculation by a noted inventor of optical character recognition (OCR) machines.

Kurzweil, Ray. *The Age of Spiritual Machines: When Computers Exceed Human Intelligence*. New York: Viking; 1999. 388p.

A somewhat more philosophical sequel to his *Age of Intelligent Machines* of nine years earlier. Valuable for its very extensive bibliography as much as for its insights.

Lundstrom, David E. *A Few Good Men from Univac*. Cambridge, MA: MIT Press; 1987. 227p.

A jarring but accurate title.

Macrae, Norman. *John von Neumann: The Scientific Genius Who Pioneered the Modern Computer, Game Theory, Nuclear Deterrence, and Much More*. New York: Pantheon Books; 1992. 405p.

The only full-length biography of the preeminent computer scientist of the twentieth century. The subtitle needs no further embellishment.

Mead, Carver; Conway, Lynn. *Introduction to VLSI Systems*. Reading, MA: Addison-Wesley; 1980. 396p.

The now-classic text that inspired thousands of students to embrace VLSI design.

Millhauser, Steven. *In the Penny Arcade*. New York: Knopf; 1986. 165p.

Includes the Pulitzer Prize–winning novelist's short story "August Eschenburg," in which the author uses his usual meticulously precise and readable prose to describe automata of similarly meticulous precision. Not a single word can ever be cut from one of his exquisite sentences.

Mollenhoff, Clark R. *Atanasoff: Forgotten Father of the Computer*. Ames: Iowa State University Press; 1988. 274p.

Details of the famous Honeywell versus Sperry Rand suit that established Atanasoff as the inventor of the first electronic digital computer. Covers material similar to that of Burks and Burks.

Moschovitis, Christos J. P.; Poole, Hilary; Schuyler, Tami; Senft, Theresa M. *History of the Internet: A Chronology, 1843 to the Present*. Santa Barbara, CA: ABC-CLIO; 1999. 312p.

Milestones of the Internet.

Murray, Charles J. *The Supermen: The Story of Seymour Cray and the Technical Wizards behind the Supercomputer*. New York: John Wiley & Sons; 1997. 232p.

The subtitle is an apt description.

Negroponte, Nicholas. *Being Digital*. New York: Knopf; 1995. 243p.

Insights into a conceivable future by the founding director of the Media Lab at MIT.

Penzias, Arno. *Ideas and Information*. New York: Simon & Schuster; 1989. 224p.

Musings on the history and conceivable future of computing by a Nobel Prize–winning physicist.

Ritchie, David. *The Computer Pioneers: The Making of the Modern Computer.* New York: Simon & Schuster; 1986. 238p.
People plus thumbnail sketches of major computers up to the 1970s.

Rochester, Jack B.; Gantz, John. *The Naked Computer.* New York: William Morrow; 1983, 335p.
Seriously outdated by now, but its collection of amusing anecdotes about computers, programmers, and befuddled computer users is fun reading.

Schwartz, Evan I. *The Last Lone Inventor: A Tale of Genius, Deceit, and the Birth of Television.* New York: HarperCollins; 2002. 322p.

Shasha, Dennis; Lazere, Cathy. *Out of Their Minds: The Lives and Discoveries of 15 Great Computer Scientists.* New York: Springer-Verlag; 1995. 291p.
The 15 are, in the order presented: Backus, McCarthy, Kay, Dijkstra, Rabin, Knuth, Tarjan, Lamport, Stephen Cook, Levin, Brooks, Burton Smith, Hillis, Feigenbaum, and Lenat. Almost all are associated with a milestone in this book.

Stashower, Daniel. *The Boy Genius and the Mogul: The Untold Story of Television.* New York: Broadway Books; 2002. 277p.
One of two recent books, the other being Evan Schwartz's *The Last Lone Inventor* (cited above), that proclaim that Philo Farnsworth should be regarded as the true "father of television." But the story is hardly "untold"; see the Fishers' *TUBE* among the primary references.

Stephenson, Neal. *Cryptonomicon.* New York: Avon Books; 1999; 918p.
Though this 900-pound—oops, page—*magnum opus* is a fascinating work of fiction, it alludes, accurately, to a great number of milestones or terms relating to milestones. Among them are abacus, Baudot code, C and C++, Colossus, Euclidean algorithm, Fibonacci sequence (*see* Discrete mathematics), the File Transfer Protocol (FTP), genetic algorithm, the Tera supercomputer, Turing Test, ultrasonic memory, and the X window system.

Turing, Alan. "Computing Machines and Intelligence." *MIND: A Quarterly Review of Psychology and Philosophy*, **59**, *236* (October 1950), pp. 433–460.
An image of the cover appears at http://theworldsgreatbooks.com/turing.htm, the Website of the Manhattan Rare Book Company. Their advertisement reads, "The complete issue, in original gray wrappers. First printing. Light wear to wrappers with crease to front wrapper, front joint very tender. Ink markings, not to Turing's article. A very good copy, rare in original wrappers. $2000." Since Turing's paper is the single most influential publication in the history of computer science, the asking price can only increase as time goes by.

Weinberg, Gerald. *The Psychology of Computer Programming.* New York: Van Nostrand Reinhold; 1971. 285p.
The classic by the leading advocate of "egoless programming."

Wilkes, Maurice V. *Memoirs of a Computer Pioneer.* Cambridge, MA: MIT Press; 1985. 250p.
The invaluable insight of the computer patriarch who is now Sir Maurice Wilkes.

Wilkes, Maurice V.; Wheeler, David; Gill, Stanley. *The Preparation of Programs for an Electronic Digital Computer.* Reading, MA: Addison-Wesley; 1951. 167p.
This first book on the subject of programming pertains to the EDSAC.

Wolfram, Stephen. *A New Kind of Science.* Champaign, IL: Wolfram Media, Inc.; 2002. 1,258p.
In which the creator of the *Mathematica* computer algebra program speculates that the universe is a cellular automaton and why viewing it as such leads to the title of the book. Aside from the prospects that physical scientists will agree, the book—the *Gödel, Escher, Bach* of 2002—is a valuable storehouse of information about computation and the mathematics pertaining to it.

Wooley, Benjamin. *The Bride of Science: Romance, Reason, and Byron's Daughter.* New York: McGraw-Hill; 1999. 416p.
A definitive biography, warts and all, of Babbage's assistant Augusta Ada Byron King, Countess of Lovelace.

Classification of Articles

All articles appear in at least one category, and some are listed more than once. For example, **Machine translation** is both an APPLICATION and an example of ARTIFICIAL INTELLIGENCE.

ALGORITHMS
Buffon's needle
Data Encryption Standard (DES)
Difference calculation
Euclidean algorithm
Fast Fourier Transform (FFT)
Finite element method
Genetic algorithm
Horner's rule
Kalman filter
Least squares method
Monte Carlo method
MP3
Newton's method
Pattern matching
Quicksort
Random number generation
RSA algorithm
Sieve of Eratosthenes
Simplex method
Simpson's rule
Sorting
Table lookup

APPLICATIONS
Automatic Teller Machine (ATM)
Automation
Bulletin Board System (BBS)
CAD / CAM
Computer algebra
Computer animation

Computer-Assisted Instruction (CAI)
Computer-assisted proof
Computer music
Computer vision
Computerized Tomography (CT)
Data compression
Data encryption
Data Encryption Standard (DES)
Data mining
Data recognition
Data reduction
Data visualization
Data warehouse
Database Management System (DBMS)
Desktop publishing
Digital photography
Electronic book (e-book)
Electronic commerce (e-commerce)
Electronic Funds Transfer (EFT)
Electronic mail (e-mail)
Embedded system
Finite element method
Geographic Information System (GIS)
Global Positioning System (GPS)
Groupware
Image processing

Machine translation
Magnetic Ink Character Recognition (MICR)
Management Information System (MIS)
Optical Character Recognition (OCR)
Pattern recognition
Pretty Good Privacy (PGP)
Public-Key Cryptosystem (PKC)
Robotics
Simulation
Smart card
Speech recognition
Speech synthesis
Spelling checker
Universal Product Code (UPC)
Word processing

ARTIFICIAL INTELLIGENCE
Artificial Intelligence (AI)
Artificial Life (AL)
Automatic theorem proving
Computer algebra
Computer animation
Computer chess
Computer game
Computer vision
Eliza
Expert system
Genetic algorithm

Classification of Articles

Machine learning
Machine translation
Neural net
Pattern recognition
Perceptron
Robotics
Speech recognition
Speech synthesis
Turing test
Virtual reality

CALCULATORS

Abacus
Arithmometer
Baldwin / Odhner calculator
Burroughs adding machine
Card Programmed Calculator
  (CPC)
Comptometer
Curta calculator
Dalton adding machine
Difference Engine
Electronic calculator
Friden calculator
Grant calculators
HP-35
Leibniz calculator
Napier's bones
Pascal calculator
Schickard calculator
Slide rule
Stanhope calculator
Steiger calculator
TI SR-50

CODING THEORY

ASCII
Baudot code
Binary-Coded Decimal (BCD)
Coding theory
EBCDIC
Error-correcting code
Huffman encoding
Morse code
Parity
Unicode
Universal Product Code (UPC)

COMMUNICATIONS

Bandwidth
Baudot code
Bus
Cellular telephone
Channel
Cybernetics
Data transmission

Digital Subscriber Line (DSL)
Digital television
Electronic mail (e-mail)
Ethernet
Facsimile transmission (fax)
Fiber optics
FireWire
Modem
Morse code
MP3
Protocol
Public-Key Cryptosystem (PKC)
Radio
Router
Spread-spectrum communications
TCP / IP
Telegraph
Telephone
Television
Wireless connectivity

COMPUTER ARCHITECTURE

Bit
Byte
Complex Instruction Set
  Computer (CISC)
Computer architecture
Emulation
Flynn taxonomy
Index register
Instruction
Instruction lookahead
Instruction set
Machine language
Microinstruction
Multiple-address computer
Multiprocessing
Open architecture
Operation code (opcode)
Parallel computer
Pipelined computer
Reduced Instruction Set
  Computer (RISC)
Register
Shift register
Very Long Instruction Word
  (VLIW)
Word

COMPUTER GRAPHICS

Artificial Life (AL)
Bit map
Computer animation
Computer graphics
Data visualization

Digital photography
Digitizing tablet
Fractal
Geographic Information System
  (GIS)
Graphical User Interface (GUI)
Holography
Icon (symbol)
Monitor
Sketchpad
Virtual reality
X window system

COMPUTER TYPES

Analog computer
Automatic Teller Machine (ATM)
Beowulf
Calculator
Cellular automaton
Cluster
Complex Instruction Set
  Computer (CISC)
Computer
Connection Machine
Cosmic Cube
Dataflow machine
Digital computer
DNA computer
Electronic calculator
Fault-tolerant computer
Handheld computer
Hybrid computer
Interactive computer
Laptop computer
Mainframe
Microcomputer
Microprocessor
Minicomputer
Mobile computing
Multiple-address computer
Multiprocessing
Optical computing
Parallel computer
Perceptron
Personal computer (PC)
Personal Digital Assistant (PDA)
Pipelined computer
Portable computer
Programmable calculator
Quantum computing
Real-time system
Reduced Instruction Set
  Computer (RISC)
Server
Soft computing

Stored program computer
Supercomputer
Systolic array
Ubiquitous computing
Very Long Instruction Word
  (VLIW)
Von Neumann machine
Workstation

COMPUTERS
ABC
ACE / Pilot ACE
Alto
Alwac III-E
Analytical Engine
Apple II
Apple Lisa
Apple Macintosh
Atlas
AVIDAC
Bell Labs relay computers
Bendix G-15
BESM
BINAC
Burroughs B5000 series
CDC 1604
CDC 6600
Colossus
Commodore Amiga
Commodore PET
Commodore 64
Cosmic Cube
Cray-1
CUBA
Cyclone
DEC PDP series
DEC VAX series
Digital Signal Processor (DSP)
EDSAC
EDVAC
Elliott 803
English Electric KDF9
ENIAC
ERA 1101
GE 200 series
GE 600 series
HAL 9000
Harvard Mark I (ASCC)
IAS computer
IBM 360 series
IBM 650
IBM 700 series
IBM 1400 series
IBM 1620
IBM AS/400

IBM PC
ILLIAC
ILLIAC IV
IMSAI 8080
Intel 4004
Intel 8008
Intel 8080
Intel 80×86 series
Interface Message Processor
  (IMP)
JOHNNIAC
LARC
LEO
Librascope LGP-30
LINC
MADDIDA
Manchester Mark I
MANIAC
MISTIC
MITS Altair 8800
MOS 6502
Motorola 68×××series
NORC
ORACLE
ORDVAC
Osborne portable
Philco Transac S-2000
Radio Shack TRS-80
RAMAC
RCA Spectra series
SEAC
SILLIAC
Simon
SOL
SSEC
Stretch
SWAC
TAC
TRADIC
TX-0 / TX-2
UNIVAC
Univac 1100 Series
Univac 1103
WEIZAC
Whirlwind
Xerox Star
Zilog Z-80
Zuse computers

DATA STRUCTURES
Abstract data type
Array
Bit map
Database
Deque

File
Graph
Hypertext
Linked list
Pushdown stack
Queue
Record
Set
String
Tree

DEFINITIONS
Accuracy
Applet
Augmented reality
Auxiliary storage
Bandwidth
Bit
Bit map
Bug
Byte
Calculator
Channel
Character
Code
Computer
Constant
Cursor
Data
Data path
Data recognition
Data transmission
Declaration
Encapsulation
Expression
Firmware
Hardware
Heuristic
Icon (symbol)
Identifier
Input / Output (I/O)
Instruction
Interactive computer
Iteration
Loop
Machine learning
Machine-readable form
Microcomputer
Multiplexing
Multitasking
Multithreading
Natural language
Object code
Operand
Operation code (opcode)

Microsoft
Mitsubishi
Motorola
National Cash Register (NCR)
National Semiconductor
Netscape Communications
Nippon Electric Corporation
 (NEC)
Olivetti
Oracle Corporation
Packard Bell
Radio Corporation of America
 (RCA)
RAND Corporation
Remington Rand
Sharp Corporation
Siemens-Nixdorf
Silicon Graphics, Inc.
Sony
Sperry Rand
Sun Microsystems
Texas Instruments (TI)
3Com
Toshiba
Unisys
Xerox PARC

INFORMATION PROCESSING
Batch processing
Data compression
Data encryption
Data mining
Data processing
Data reduction
Data transmission
Data warehouse
Database
Database Management System
 (DBMS)
Distributed system
Geographic Information System
 (GIS)
Hypertext
Information retrieval
List processing
Management Information System
 (MIS)
Mobile computing
Relational database
Remote Job Entry (RJE)
String processing
Time sharing

I/O DEVICES AND MEDIA
Cathode Ray Tube (CRT)
Digital camera

Digitizing tablet
Keyboard
Laser printer
Light pen
Liquid Crystal Display (LCD)
Magnetic tape
Monitor
Mouse
MP3
Optical scanner
Optical storage
Punched card
Punched paper tape
Test-scoring machine
Touch screen
Trackball

LANGUAGE TYPES
Assembly language
Block-structured language
Extensible language
Functional language
Hardware Description Language
 (HDL)
High-level language
Intermediate language
Machine language
Natural language
Page Description Language
 (PDL)
Problem-Oriented Language
 (POL)
Procedure-Oriented Language
 (POL)
Programming language
Visual language

LANGUAGES
Ada
Algol 60
Algol 68
APL
APT
Autocode
Basic
C
C++
CGI (Common Gateway
 Interface)
CLU
Cobol
Comit
Dynamo
Forth
Fortran

GPSS
Haskell
HTML
Icon
IPL-V
Java
JOSS
Jovial
Lisp
LOGO
M (Mumps)
Modula-2
Pascal
Perl
Plankalkül
PL / I
PostScript
Prolog
RPG
SETL
Short Order Code
Simscript
Simula 67
Smalltalk
Snobol
SQL
T$_E$X
XML

MATHEMATICS
Analytic geometry
Automatic theorem proving
Boolean algebra
Calculus
Computer-assisted proof
Discrete mathematics
Fast Fourier Transform (FFT)
Fractal
Fuzzy logic
Gödel incompleteness theorem
Horner's rule
Kalman filter
Logarithm
Matrix
Newton's method
Number theory
Numerical analysis
Pi (π)
Polish notation
Prime number
Recurrence relation
Simpson's rule
Stirling's approximation
Taylor series

Classification of Articles

MEMORY
Associative memory
Auxiliary storage
Bit
Bubble memory
Byte
Cache
DRAM
Electrostatic memory
Firmware
Flash memory
Floppy disk
Hard disk
Interleaved memory
Magnetic core
Magnetic drum
Magnetic tape
Optical storage
Random Access Memory (RAM)
Read-Only Memory (ROM)
Smart card
Solid state memory
Ultrasonic memory
Virtual memory

NETWORKS
Apache
ARPAnet
Browser
Client-server computing
Common Gateway Interface
  (CGI)
Cooperative computing
Cyberspace
Distributed system
Electronic commerce
  (e-commerce)
Electronic mail (e-mail)
Ethernet
File Transfer Protocol (FTP)
Firewall
Gateway
Interface Message Processor
  (IMP)
Internet
Intranet
Local Area Network (LAN)
Metcalfe's law
MP3
Network
Neural net
Open Systems Interconnection
  (OSI)
Packet switching

Project MAC
Protocol
Router
SABRE
SAGE
Telnet
Wireless connectivity
World Wide Web (WWW)

NUMBER SYSTEMS
Binary number system
Complement
Floating-point arithmetic
Hexadecimal number system
Modular arithmetic
Octal number system
Positional number system
Ternary number system
Zero

OPERATING SYSTEMS
Concurrent programming
CP / M
CTSS
Dartmouth Time-Sharing System
  (DTSS)
Directory
Graphical User Interface (GUI)
JOSS
Linux
Microsoft Windows
MS-DOS
Multics
Multiplexing
Multiprogramming
Multitasking
Multithreading
Operating system
Time sharing
Unix
Virtual memory

PROGRAMMING
Assembler
Bug
Class
Code
Compiler
Constant
Control structure
Coroutine
Data type
Declaration
Encapsulation

Expression
Flowchart
Function
Identifier
Information hiding
Initial orders
Instruction set
Interpreter
Iteration
Macro instruction
Object code
Pointer
Procedure
Recursion
Source code
Statement
Subroutine
Variable

PROGRAMMING
METHODOLOGY
Automatic programming
Concurrent programming
Event-driven programming
Functional [programming]
  language
Literate programming
Logic programming
Microprogramming
Object-Oriented Programming
  (OOP)
Structured programming

SOFTWARE
Applet
Assembler
Browser
Compiler
Computer algebra
Database Management System
  (DBMS)
Desktop publishing
Free software
Freeware
Interpreter
Loader
Microsoft Windows
Open source software
Program
Programming language
Search engine
Shareware
Software
Software engineering

Spreadsheet
Text editor
Unbundling
Word processing
X window system

THEORY
Analysis of algorithms
Automata theory
Automatic control theory
Automatic theorem proving
Cellular automaton

Chaos theory
Chomsky hierarchy
Church-Turing thesis
Coding theory
Computational complexity
Computer-assisted proof
Finite State Machine (FSM)
Fuzzy logic
Game theory
Graph theory
Halting problem

Information theory
Lambda calculus
Linear Bounded Automaton
  (LBA)
NP-complete problem
Number theory
Numerical analysis
Program verification
Queueing theory
Switching theory
Turing machine

The Top Ten Consolidated Milestones

in approximate order of development

1. The **algorithm**, without which nothing can be computed.

2. The **positional number system**, without which computation as we know it is inconceivable.

3. The **Church-Turing thesis**, which reassures us that anything that can be algorithmically computed can be computed by a **Turing machine** and thus by a suitably equipped **stored program computer**.

4. The **stored program computer** concept, which makes the **digital computer** automatic and general-purpose.

5. Full-featured **procedure-oriented languages (POLs)**, such as **Algol 60**, **Cobol**, and **Fortran**, up through the more recent **Pascal** and **C++**.

6. *Solid-state electronics*, as exemplified by the **transistor**, **solid state memory**, and **inte-**grated circuits (ICs) that reduced the size and cost of the **digital computer** to manageable proportions.

7. **Operating systems** that supported remote-access **time sharing**, which freed us from the confines of on-site **batch processing**.

8. The **personal computer (PC)**, which brought computing into the school and the home.

9. The **Graphical User Interface (GUI)**. Recall the subtitle of the Levy book about the **Apple Macintosh**: *"The computer that changed everything."*

10. *"The Net,"* the **World Wide Web (WWW)** running on the **Internet** creating **cyberspace**, the technological equivalent of the more ethereal *noösphere* of the Jesuit anthropologist and philosopher Pierre Teilhard de Chardin.

Personal Name Index

Abramson, Norman (U.S.) [1932–]
Network

Ackermann, Wilhelm (Germany) [1896–1962]
Recursion

Adel'son-Vel'skii, Georgii Maksimovich
Tree

Adleman, Leonard Max (U.S.) [1947–]
DNA computer
Graph theory
Public-key Cryptosystem (PKC)
RSA algorithm

Adler, Allen (U.S.) [1916–1964]
Robotics

Ahl, David Howard (U.S.) [1939–]
Personal Computer (PC)

Ahmes the Scribe (Egypt) [~1680–1620 B.C.]
Pi (π)

Aiken, Howard Hathaway (U.S.) [1900–1973]
Bug
Harvard Mark I
SSEC

Airy, Sir George Biddell (England) [1801–1892]
Automatic control theory

Albrecht, Robert (U.S.)
Basic

Alcorn, Allan (U.S.) [1950–]
Atari

Alexander, Samuel Nathan (U.S.) [1910–1967]
SEAC

Alexander, William F. (U.S.)
Trackball

Alexanderson, Ernst F(redik) W(erner) (Sweden) [1878–1975]
General Electric (GE)
Radio
Television

al-Kashi, Ghiyath al-Din Jamshid Mas'ud (Persia) [~1380–1429]
Horner's rule

al-Khalifa, Hamad bin I'sa (Bahrain) [1950–]
Computer chess

al-Khayyami, Ghiyath al-Din Abu'l-Fath Umar ibn Ibrahim Al-Nisaburi / Omar Khayyam (Persia) [1048–1131]
Discrete mathematics

al-Khowarizmi, abu Ja'far Mohammed ibn Musa (Egypt) [~783–850]
Algorithm

Allen, Frances Elizabeth (U.S.) [1932–]
Fortran

Allen, James Ward (England) [1941–]
Light Emitting Diode (LED)

Allen, Paul G. (U.S.) [1953–]
Basic
Interpreter
Microsoft
MITS Altair 8800

Allison, Dennis (U.S.)
Basic

Allmark, Reg(inald) H. (England)
English Electric KDF9

Amarel, Saul (Greece/Israel/U.S.) [1928–2002]
Expert system

Amdahl, Gene Myron (U.S.) [1922–]
Amdahl Corporation
Amdahl's law
IBM 360 series

Ammann, Urs (Switzerland)
Intermediate language
Pascal

Amsler, Jacob (Switzerland) [1823–1912]
Planimeter

Anderson, Harlan E. (U.S.) [1930–]
Digital Equipment Corporation (DEC)

Boyer, Robert S. (U.S.)
Automatic theorem proving
Pattern matching

Boyle, Willard Sterling (Canada) [1924–]
Charge-coupled device (CCD)

Bradley, David J. (U.S.) [1949–]
IBM PC

Brahmagupta (India) [c. 598– 665]
Logarithm

Brainerd, John Grist (U.S.) [1904–1988]
ENIAC

Brainerd, Paul (U.S.) [1947–]
Desktop publishing

Brand, Stewart (U.S.) [1938–]
Open source software

Brattain, Walter Houser (U.S.) [1902–1987]
Transistor

Braun, (Karl) Ferdinand (Germany) [1850–1918]
Cathode Ray Tube (CRT)
Radio
Semiconductor

Bray, Tim (Canada) [1960–]
XML

Breiman, Leo (U.S.) [1928–]
Data mining

Bricklin, Daniel S. (U.S.) [1951–]
Spreadsheet

Briggs, Henry (England) [1561– 1631]
Difference calculation
Logarithm

Bright, Herbert Samuel (U.S.) [1919–1986]
Fortran

Brin, Sergey (Russia) [1974–]
Search engine

Brinch Hansen, Per (Denmark / U.S.) [1938–]
Concurrent programming

Brodie, Richard (U.S.) [1961–]
Word processing

Bromberg, Howard (U.S.)
Cobol

Brooker, Ralph Anthony "Tony" (England) [1925–]
Autocode

Brooks, Frederick Phillips, Jr. (U.S.) [1931–]
Byte
Computer architecture
IBM
IBM 360 series
Software engineering

Browne, Sir Thomas (England) [1605–1682]
Computer

Bruce, Thomas (U.S.)
Browser

Brusentsov, Nikolay Petrovich (Russia) [1925–]
Ternary number system

Bryce, James Wares (U.S.) [1880–1949]
Harvard Mark I

Buchanan, Bruce G. (U.S.) [1940–]
Expert system

Buchholz, Werner W. (Germany / U.S.) [1922–]
Byte
Stretch

Buck, Dudley (U.S.) [1927– 1959]
Magnetic core

Buffon. See Leclerc
Buffon's needle

Bull, Fredrik Rosen (Norway) [1882–1925]
Machines Bull

Burack, Benjamin (U.S.) [1914– 2001]
Boolean algebra

Burdette, Earl (U.S.)
ORACLE

Bürgi, Jobst (Switzerland) [1552–1632]
Logarithm
Sector

Burks, Arthur Walter (U.S.) [1915–]
EDVAC
ENIAC
IAS computer
Von Neumann machine

Burroughs, William Seward (U.S.) [1857–1898]
Burroughs adding machine
Burroughs Corporation

Burt, William Austin (U.S.) [1792–1858]
Typewriter

Bush, Vannevar E. (U.S.) [1890– 1974]
Analog computer
Differential Analyzer
Hypertext
Memex
Personal Computer (PC)

Bushinsky, Shay (Israel)
Computer chess

Bushnell, Nolan Kay (U.S.) [1943–]
Atari
Computer game

Butler, Samuel (England) [1835– 1902]
Artificial Intelligence (AI)

Butz, Albert M. (U.S.) [~1860–?]
Honeywell Corporation

Bybee, Jim (U.S.)
MITS Altair 8800

Byron, Lord George Gordon Noel, 6th Baron (England) [1788–1824]
Program

Byron-King, Augusta Ada / Countess of Lovelace (England) [1815–1852]
Program

Cadwell, J. H. (U.S.)
Finite element method

Cailliau, Robert (U.S.) [1947–]
Browser

Caldwell, Samuel Hawks (U.S.) [1904–1960]
Differential Analyzer

Descartes, René du Perron (France / Netherlands) [1596–1650]
Analytic geometry
Array
Data structure

Dethloff, Jurgen (Germany)
Smart card

Deutsch, David (England) [1953–   ]
Quantum computing

Deutsch, L(aurence) Peter (U.S.) [1946–   ]
PostScript
Text editor

Dickson, W(illiam) K(ennedy) L(aurie) (France / U.S.) [1860–1935]
Digital television

Diebold, John (U.S.) [1926–   ]
Automation

Dieckmann, Max (Germany) [1882–1960]
Cathode Ray Tube (CRT)
Facsimile transmission (fax)

Dietzgen, Eugene (Germany / U.S.) [~1862–1929]
Slide rule

Diffie, (Bailey) Whitfield (U.S.) [1944–   ]
Public-key Cryptosystem (PKC)

Dijkstra, Edsger Wybe (Netherlands) [1930–2002]
Algol 60
Algol 68
Concurrent programming
Control structure
Graph theory
Program verification
Pushdown stack
Structured programming

Dimond, Tom L. (U.S.)
Digitizing tablet

Dionysius Exiguus (Italy) [6th c.]
Positional number system

Discount, Norman (U.S.)
Cobol

Disney, Walter Elias "Walt" (U.S.) [1901–1966]
Computer animation

Dodgson, Charles Lutwidge / Lewis Carroll (England) [1855–1881]
Electronic book (e-book)

Dolbear, Amos Emerson (U.S.) [1837–1910]
Telephone

Dolby, Ray M. (U.S.) [1933–   ]
Digital television

Dos Passos, John Roderigo (U.S.) [1896–1970]
General electric (G.E.)

Draper, Charles Stark (U.S.) [1901–1987]
Automatic control theory

Draper, Hal (U.S.) [1915–1990]
Directory

Draper, John T. / "Captain Crunch" (U.S.) [1945–   ]
Word processing

Drawbaugh, Daniel (U.S.) [1827–1911]
Telephone

Drebble, Cornelius Jacobszoon (Netherlands) [1572–1634]
Automatic control theory

Drexler, K. Eric (U.S.) [1955–   ]
Nanoelectronics

Dreyer, John F. (U.S.)
Liquid Crystal Display (LCD)

Dreyfus, Philippe L. (France)
Computer science

Driscoll, James R. (U.S.)
Data structure

Drude, Paul W. (Germany) [1863–1906]
Semiconductor

Dubinsky, Donna (U.S.) [1956–   ]
Personal Digital Assistant (PDA)

Dudley, Homer Walter (U.S.) [1896–1980]
Speech recognition
Speech synthesis

Dummer, G(eoffrey) W. A. (England) [1909–2002]
Integrated Circuit (IC)

DuMont, Allen Balcom (U.S.) [1901–1965]
Television

Dunwell, Stephen W. (U.S.) [1913–1994]
Stretch

Dvorak, August (U.S.) [1894–1975]
Keyboard

Earnest, Les (U.S.) [1930–   ]
Desktop publishing
Spelling checker

Eastlake, Donald E. III (U.S.)
Computer chess

Eccles, William Henry (U.S.) [1875–1966]
Flip-flop
Logic circuit

Eckdahl, Donald E. (U.S.) [1924–2001]
MADDIDA

Eckert, J(ohn) Presper Eckert (U.S.) [1919–1995]
ABC
Eckert-Mauchly Computer Corporation (EMCC)
EDVAC
ENIAC
LARC
Remington Rand
Stored program computer
Von Neumann machine

Edison, Thomas Alva (U.S.) [1847–1931]
Automaton
Bug
Digital television
General Electric (GE)
Radio

Edmonds, Jack (U.S.)
Computational complexity

Edwards, D(avid) B. G. "Dai" (England) [1928–   ]
Manchester Mark I

Efron, Richard (U.S.)
GPSS

Egli, Hans W., Jr. (Switzerland / England) [1899–   ]
Steiger calculator

Hazen, Harold Locke (U.S.) [1901–1980]
Differential Analyzer

Hearn, Anthony Clem (Australia / U.S.) [1937–]
Computer algebra

Heart, Frank (U.S.) [1930–]
Interface Message Processor (IMP)
Internet

Heatherington, Dale (U.S.)
Modem

Heaviside, Oliver (England) [1850–1925]
Radio

Hebb, Donald O. (Canada) [1904–1985]
Perceptron

Hebern, Edward Hugh (U.S.) [1869–1952]
Hagelin machine

Hedwig, Eva Maria Kiesler / Hedy Lamarr (Austria / U.S.) [1913–2000]
Spread-spectrum communication

Heisenberg, Werner Karl (Germany) [1901–1976]
Matrix

Hejlsberg, Anders (Denmark / U.S.)
Java
Pascal

Hell, Rudolph (Germany) [1901–2002]
Facsimile machine (fax)

Hellman, Martin E. (U.S.) [1945–]
Public-Key Cryptosystem (PKC)

Hennessy, John Leroy (U.S.) [1952–]
Reduced Instruction Set Computer (RISC)

Henrouteau, François (Canada)
Television

Henry, Joseph (U.S.) [1797–1878]
Telegraph

Hepburn, Katharine Houghton (U.S.) [1907–]
UNIVAC

Hermann, J. H. (Germany) [19th c.]
Planimeter

Herold, Edward William (U.S.) [1907–1993]
Television

Herrick, Harlan (U.S.) [1922–1997]
Fortran

Hertz, Heinrich Rudolf (Germany) [1857–1894]
Radio

Hertzfeld, Andrew (U.S.) [1954–]
Apple Macintosh
Graphical User Interface (GUI)

Herzstark, Carl Albert (Austria) [1902–1988]
Curta calculator

Hewlett, Walter B. (U.S.) [1944–]
Hewlett-Packard (*hp*)

Hewlett, William Redington (U.S.) [1913–2001]
Hewlett-Packard (*hp*)

Higonnet, René Alphonse (France) [1902–1983]
Typography

Hilbert, David (Germany) [1862–1943]
Halting problem

Hildebrandt, Paul (U.S.)
Sorting

Hiller, Lejaren Arthur (U.S.) [1924–1994]
Computer music

Hillis, W(illiam) Daniel "Danny" (U.S.) [1956–]
Connection machine
Supercomputer

Hilsenrath, Joseph (U.S.) [1912–]
Spreadsheet

Hinton, Geoffrey E. (England / U.S. / Canada) [1947–]
Perceptron

Hitchcock, Frank L. (U.S.)
Simplex method

Hoane, A. Joseph, Jr. (U.S.) [1963–]
Computer chess

Hoare, Sir C(harles) A(ntony) R(ichard) "Tony"(England) [1934–]
Algol 68
Concurrent programming
Control structure
Elliott 803
Pascal
Program verification
Quicksort
Sorting

Hocquenghem, Alexis (France) [1908–]
Error-correcting code

Hoelzer, Helmut (Germany) [1912–1996]
Analog computer

Hoerni, Jean A. (Switzerland / U.S.) [1924–1997]
Fairchild Semiconductor
Integrated Circuit (IC)
Transistor

Hoff, Marcian Edward, Jr. "Ted" (U.S.) [1937–]
DRAM
Intel 4004
Intel 8080
Microprocessor

Hogan, Sharon Anne (U.S.) [1945–2002]
Information retrieval

Hohndel, Dirk (U.S.) [1968–]
X window system

Holberton, (Frances) Elizabeth ("Betty") Snyder (U.S.) [1917–2001]
ENIAC

Holland, John Henry (U.S.) [1929–]
Artificial Life (AL)
Cellular automaton
Neural net

Hollerith, Herman (U.S.) [1860–1929]
Hollerith machine
IBM
Punched card
Sorting

Kettering, Charles Franklin (U.S.) [1876–1958]
National Cash Register (NCR)

Keuffel, Wilhelm J. D. (Germany / U.S.) [1838–1908]
Planimeter
Slide rule

Khosla, Vinod (India / U.S.) [1948–]
Sun Microsystems

Kilburn, Tom (England) [1921–2001]
ACE / Pilot ACE
Atlas
Index register
Manchester Mark I
Multiprogramming
Virtual memory

Kilby, Jack St. Clair (U.S.) [1923–]
Fairchild Semiconductor
Flip-flop
Integrated circuit (IC)
Texas Instruments (TI)

Kildall, Gary A. (U.S.) [1942–1994]
CP / M
IBM PC
Operating system

King, Frank (U.S.)
Database Management System (DBMS)
Relational database

Kircher, Athanasius (Germany) [1602–1680]
Logarithm

Kirchhoff, Gustav Robert (Prussia) [1824–1887]
Graph
Tree

Kister, James Milton (U.S.) [1930–]
Computer chess

Kleene, Stephen Cole (U.S.) [1909–1994]
Recursion

Klein, Rudolph (U.S.)
ORACLE

Kleiner, Eugene (Austria / U.S.) [1922–]
Fairchild Semiconductor

Kleinrock, Leonard (U.S.) [1934–]
Internet
Packet switching
Queueing theory

Kleme, Virginia Carlock (U.S.)
ORACLE

Knopf, James / Jim Button (U.S.)
Shareware

Knuth, Donald Ervin (U.S.) [1938–]
Analysis of algorithms
Data structure
Euclidean algorithm
IBM 650
Knuth textbooks
Literate programming
Pattern matching
Simulation
TEX

Koch, Hugo Alexander (Netherlands) [1870–1928]
Hagelin machine

Koelbel, Charles (U.S.)
Fortran

Kohonen, Teuvo Kalevi (Finland) [1934–]
Speech recognition

Kolmogorov, Andrei Nikolaevich (Russia) [1903–1987]
Automatic control theory
Random number generation

Konigsberger, J. (Germany)
Semiconductor

Korfhage, Robert Roy (U.S.) [1931–1999]
Information retrieval

Korn, Arthur (Germany) [1870–1945]
Facsimile transmission (fax)

Korn, David (U.S.)
Unix

Koster, Robert Allen (U.S.) [1941–]
Trackball

Kotok, Alan (U.S.) [1941–]
Computer chess

Kovalevskaya, Sofia Vasilyevna (Russia / Sweden) [1850–1891]
Numerical analysis

Koza, John R. (U.S.) [1943–]
Artificial Life (AL)
Genetic algorithm

Kramnik, Vladimir (Russia) [1975–]
Computer chess

Kron, Gabriel (U.S.) [1901–1968]
Finite element method

Kruesen, Knut (Norway)
Machines Bull

Ktesibios of Alexandria (Greece) [3rd c. B.C.]
Automatic control theory

Kubrick, Stanley (U.S.) [1928–1999]
HAL 9000

Kuck, David Jerome (U.S.) [1937–]
Parallel computer

Kumar, Sanjay (Sri Lanka / U.S.) [1962–]
Computer Associates

Kung, H(siang) T(sung) (China / U.S.) [1945–]
Systolic array

Kurtz, Thomas Eugene (U.S.) [1928–]
Basic
Dartmouth Time-Sharing System (DTSS)
Time sharing

Kurzweil, Raymond (U.S.) [1948–]
Optical Character Recognition (OCR)
Speech synthesis

Lagrange, Joseph Louis (France) [1736–1813]
Logarithm
Numerical analysis

Lai, Glenn G. (U.S.)
X window system

Lai, Stefan K. (Japan) [1952–]
Flash memory

Murphey, Richard (U.S.)
X window system

Murray, Donald (U.S.) [1865–1945]
Baudot code

Murtha, John C. (U.S.)
Digital Signal Processor (DSP)

Murto, William (U.S.)
Compaq

Mutch, Eric N. [England]
EDSAC

Myers, Jack D. (U.S.) [1913–1998]
Expert system

Myhill, John R. (U.S.) [1923–1987]
Cellular automaton

Myhrvold, Nathan (U.S.) [1959–]
Microsoft Windows

Nakamura, Shuji (Japan) [1954–]
Light emitting diode (LED)

Nakashima, Kyoichi (Japan)
Switching theory

Napier, Laird John of Merchiston (Scotland) [1550–1617]
Logarithm
Napier's Bones

Napoleon Bonaparte (France) [1769–1821]
Computer chess
Data visualization

Nasar, Sylvia (Germany / Turkey / U.S.) [1947–]
Game theory

Nash, John Forbes, Jr. (U.S.) [1928–]
Game theory

Nassi, Isaac Robert "Ike" (U.S.) [1949–]
Flowchart

Naughton, Patrick (U.S.) [1965–]
Java

Naur, Peter (Denmark) [1928–]
Algol 60
Algol 68
Backus-Naur Form (BNF)

Nelson, Harry L. (U.S.) [1931–]
Computer chess

Nelson, Robert A. (U.S.)
Fortran

Nelson, Theodor Holm "Ted" (U.S.) [1937–]
Hypertext
Memex

Neustadter, Alfred (U.S.) [1910–1996]
Information retrieval

Newell, Allen C. (U.S.) [1927–1992]
Artificial Intelligence (AI)
Automatic programming
Automatic theorem proving
Computer chess
Hardware description language (HDL)
IPL-V
Linked list
Pushdown stack

Newell, Martin (U.S.) [1950–]
PostScript

Newhall, Edwin E. (U.S.)
Local Area Network (LAN)

Newman, Edward Arthur (England) [1918–1993]
ACE / Pilot ACE

Newman, Maxwell Herman Alexander "Max" (England) [1897–1984]
Index register
Manchester Mark I

Newman, Robert Bradford (U.S.) [1917–1983]
Bolt, Beranek, and Newman (BBN)

Newman, Ron (U.S.) [1957–]
X window system

Newton, Sir Isaac (England) [1642–1727]
Calculus
Difference calculation
Discrete mathematics
Horner's rule
Newton's method
Numerical analysis
Simpson's rule

Nipkow, Paul Gottlieb (Poland) [1860–1940]
Television

Nixdorf, Heinz (Germany) [1926–1986]
Electronic calculator
Siemens-Nixdorf

Nixon, Richard Milhous (U.S.) [1913–1994]
Bolt, Beranek, and Newman (BBN)

Noble, Benjamin (England)
EDSAC

Noble, David L. (U.S.) [1928–]
Floppy disk

Noe, Jerre (U.S.) [1923–]
General Electric (GE)
Magnetic Ink Character Recognition

Noll, Laura White (U.S.)
Snobol

Norris, William C. (U.S.) [1911–]
Control Data Corporation (CDC)
Engineering Research Associates (ERA)
Remington Rand

Novikoff, Albert Boris J. (U.S.)
Perceptron

Noyce, Robert Norton (U.S.) [1927–1990]
Fairchild Semiconductor
Integrated Circuit (IC)
Intel

Nutt, Roy (U.S.) [1931–1990]
Assembler
Fortran

Nutting, William (U.S.)
Computer game

Nyberg, Kim (Finland)
Browser

Nygaard, Kristen (Norway) [1926–2002]
Simula 67

Nyquist, Harry (Sweden / U.S.) [1889–1976]
Automatic control theory

Oates, Edward (U.S.) [1947–]
Database Management System
  (DBMS)
Oracle Corporation
Relational database

O'Brien, Thomas (U.S.)
Time sharing

**Odaira, Namihei (Japan)
[1874–1951]**
Hitachi

**Odhner, Willgodt Theophil
(Russia) [1845–1905]**
Baldwin / Odhner calculator

**Oettinger, Anthony Gervin
(Germany / U.S.) [1929–]**
Machine translation

Oikarinen, Jarkko (Finland)
Electronic mail (e-mail)

**Oliver, Bernard M. "Barney"
(U.S.) [1916–1995]**
HP-35

**Olivetti, Adriano (Italy) [1901–
1960]**
Olivetti

**Olivetti, Camillo (Italy) [1868–
1943]**
Olivetti

**Olsen, Kenneth Harry (U.S.)
[1926–]**
Digital Equipment Corporation
  (DEC)
Magnetic core

Omar Khayyam. *See* **al-
Khayyami**
Discrete mathematics

**Omidyar, Pierre (France / U.S.)
[1967–]**
Electronic commerce (e-
  commerce)

Opler, Ascher (U.S.)
Firmware

**Oppenheimer, J(ulius) Robert
(U.S.) [1904–1967]**
IAS computer

**Oppikofer, Johannes
(Germany) [19th c.]**
Planimeter

**O'Rear, Robert (U.S.)
[1944–]**
Microsoft
MS-DOS
Word processing

**Ornstein, Severo (U.S.)
[1930–]**
Interface Message Processor
  (IMP)

Orr, Kenneth (U.S.) [1945–]
Flowchart

**Osborne, Adam (Thailand /
India / England / U.S.) [1939–
2003]**
Homebrew Computer Club
Osborne portable
Portable computer

**Otis, Elisha Graves (U.S.) [1811–
1861]**
Automatic control theory

**Oughtred, William (England)
[1575–1660]**
Logarithm
Slide rule

**Packard, David (U.S.) [1912–
1996]**
Hewlett-Packard (*hp*)

**Page, Charles Grafton (U.S.)
[1812–1868]**
Telephone

**Page, Lawrence "Larry" (U.S.)
[1973–]**
Search engine

Page, Richard (U.S.)
Apple Lisa

**Paik, Woo (Korea / U.S.)
[1957–]**
Digital television

**Pake, George Edward (U.S.)
[1924–]**
Xerox PARC

Palevsky, Max (U.S.) [1927–]
Machines Bull
Personal Computer (PC)

**Papadimitriou, Christos H.
(Greece / U.S.) [1949–]**
NP-complete problem

**Papert, Seymour Aubrey
(South Africa / U.S.) [1928–]**
Artificial Intelligence (AI)
LOGO

Neural net
Perceptron

**Papian, William Nathaniel
(U.S.) [1916–]**
Magnetic core

**Pappalardo, A. Neil (U.S.)
[1943–]**
M (Mumps)

**Parker, John E. (U.S.) [1900–
1989]**
Engineering Research Associates
  (ERA)

**Parnas, David Lorge (U.S.)
[1941–]**
Information hiding

Parsons, Keith (U.S.)
Basic

**Pascal, Blaise (France) [1623–
1662]**
Discrete mathematics
Pascal
Pascal calculator

**Pask, Gordon (U.S.) [1928–
1996]**
Cybernetics
Perceptron

**Patterson, David Andrew (U.S.)
[1947–]**
Complex Instruction Set
  Computer (CISC)
RAID
Reduced Instruction Set
  Computer (RISC)

**Patterson, John Henry (U.S.)
[1844–1922]**
National Cash Register (NCR)

**Peacock, Henry Bates (U.S.)
[1894–1985]**
Texas Instruments (TI)

**Peano, Giuseppe (Italy) [1858–
1932]**
Gödel incompleteness theorem

**Peddle, Charles "Chuck" (U.S.)
[~1945–]**
Apple II
MOS 6502

**Peirce, Charles Sanders (U.S.)
[1839–1914]**
Boolean algebra

Pell, John (England) [1610–1685]
Logarithm

Pellow, Nicola (England) [1970–]
Browser

Pepper, Thomas (U.S.) [1975–]
MP3

Perkins, Alfred (England)
Information retrieval

Perlis, Alan Jay (U.S.) [1922–1990]
Computer science

Perot, H(enry) Ross (U.S.) [1930–]
Electronic Data Systems (EDS)

Perskyi, Constantin (Russia / France)
Television

Phelps, Byron E. (U.S.) [1914–]
SSEC

Philbrick, George A. (U.S.) [1911–1992]
Analog computer

Philon of Byzantium (Turkey) [~280–220 B.C.]
Automatic control theory

Phillips, Charles A. (U.S.) [1906–1985]
Cobol

Piaget, Jean (Switzerland) [1896–1980]
LOGO

Pierce, John Robinson / J. J. Coupling (U.S.) [1910–2002]
Transistor

Piggott, W(illiam) Roy (England) [1914–]
EDSAC

Pike, William H., Jr. (U.S.)
Burroughs Corporation

Piner, Stephen (U.S.)
Word processing

Pinkerton, Brian (U.S.)
Search engine

Pinkerton, John M(aurice) M(cClean) (England) [1919–1997]
LEO

Pittman, Thomas (U.S.)
Basic

Pitts, Walter H. (U.S.) [1923–1969]
Artificial Intelligence (AI)
Neural net

Plato (Greece) [427–349 B.C.]
Computer-Assisted Instruction (CAI)

Pliny the Elder / Gaius Plinius Secundus (Italy) [A.D. 23–79]
Information retrieval

Plouffe, Simon M. (Canada) [1956–]
Pi (π)

Plücker, Julius (Germany) [1801–1868]
Cathode Ray Tube (CRT)

Poage, James F. (U.S.)
Snobol

Poe, Edgar Allan (U.S.) [1809–1849]
Computer chess

Poincaré, (Jules) Henri (France) [1854–1912]
Chaos theory

Polansky, Ivan P. (U.S.)
Snobol

Poley, Stanley (U.S.) [1928–1975]
Assembler

Pomerene, James Herbert (U.S.) [1920–]
IAS computer
Stretch

Pontryagin, Lev Semyonevich (Russia) [1908–1988]
Automatic control theory

Pople, Harry E., Jr. (U.S.)
Expert system

Popov, Aleksandr Stepanovitch (Russia) [1859–1906]
Radio

Post, Emil Leon (Poland / U.S.) [1897–1954]
Church-Turing thesis
Recursion

Postel, Jonathan B. "Jon" (U.S.) [1943–1998]
TCP / IP

Poulsen, Valdemar (Denmark) [1869–1942]
Magnetic tape

Powers, James (U.S.) [1871–1935]
Punched card
Remington Rand

Pratt, John Jonathan (U.S.) [1831–1905]
Typewriter

Pratt, Vaughan Ronald (Australia / U.S.) [1944–]
Pattern matching

Press, William Henry (U.S.) [1948–]
Numerical analysis

Prigonine, Ilya (Russia / Germany / Belgium) [1917–2003]
Cybernetics

Prinz, Dietrich G. (Germany / England) [1903–]
Computer chess
Ferranti, Ltd.

Prokhorov, Alexandr Mikhailovich (Australia / Russia) [1916–2002]
Laser

Prosper, Charles (U.S.)
GE 200 series

Ptolemy I Soter (Greece / Egypt) [~365–283 B.C.]
Library

Ptolemy II Philadelphius (Greece / Egypt) [308–246 B.C.]
Library

Ptolemy III Euergetes (Greece / Egypt) [~265–210 B.C.]
Library

Pugh, Alexander L., III (U.S.)
Dynamo

Walker, John (U.S.)
CAD / CAM

Wall, Larry (U.S.) [1955–]
Perl

Wallace, David Lee (U.S.) [1928–]
Natural language

Wallace, Robert "Bob" (U.S.) [1949–2002]
Shareware

Wang, An (China / U.S.) [1920–1990]
Magnetic core
Word processing

Wang, Charles B. (China / U.S.)
Computer Associates

Wang, Hao (China / U.S.) [1921–1995]
Automatic theorem proving

Wang, Li-Chen (China / U.S.)
Basic

Ware, Willis Howard (U.S.) [1920–]
IAS computer
JOHNNIAC

Warnier, Jean-Dominique (France) [1920–1990]
Flowchart

Warnock, John E. (U.S.) [1941–]
Adobe Systems, Inc.
Desktop publishing
PostScript

Warren, David H. D. (U.S.)
Prolog

Warren, James (U.S.) [1940–]
Apple II
Personal Computer (PC)

Watson, Arthur Kittredge "Dick" (U.S.) [1919–1974]
IBM

Watson, John (U.S.)
Time sharing

Watson, Thomas Augustus (U.S.) [1854–1934]
Telephone

Watson, Thomas J(ohn) (U.S.) [1874–1956]
Card Programmed Calculator
    (CPC)

Harvard Mark I
IBM
National Cash Register (NCR)
SSEC

Watson, Thomas J(ohn), Jr. (U.S.) [1914–1993]
Card Programmed Calculator
    (CPC)
IBM
SABRE

Watt, James (England) [1736–1819]
Automatic control theory
Slide rule

Weaver, Warren (U.S.) [1894–1978]
Information theory
Machine translation

Weeg, Gerard P. "Jerry" (U.S.) [1927–1977]
MISTIC

Wegbreit, Ben (U.S.) [1944–]
Extensible language

Wegstein, Joseph Henry (U.S.) [1922–]
Spreadsheet

Wei, Pei (Taiwan / U.S.)
Browser

Weierstrass, Karl Theodor Wilhem (Germany) [1815–1897]
Difference calculation

Weintraub, Joseph (U.S.) [1965–]
Turing test

Weir, Sir Cecil (England) [1885–1960]
International Computers Limited
    (ICL)

Weiser, Mark (U.S.) [1952–1999]
Ubiquitous computing

Weiss, David M. (U.S.) [1943–]
Information hiding

Weiss, J. (Germany)
Semiconductor

Weizenbaum, Joseph (U.S.) [1923–]
Eliza
Linked list

Welch, John F. "Jack" (U.S.) [1936–]
General Electric (GE)

Welker, Heinrich (Germany / France)
Transistor

Wells, Donald (U.S.)
Image processing

Wells, Mark Brimhall (U.S.)
Computer chess
MANIAC

Wenner-Gren, Axel Leonard (Sweden) [1881–1961]
Alwac III-E

Werbos, Paul John (U.S.) [1947–]
Perceptron

Westervelt, Franklin Herbert (U.S.) [1930–]
Time sharing

Westinghouse, George (U.S.) [1846–1914]
Radio

Wetli, Kaspar (Germany) [1822–1889]
Planimeter

Wetzel, Donald (U.S.) [1930–]
Automatic Teller Machine (ATM)

Wexelblat, David (U.S.) [1967–]
X Window System

Wexelblat, Richard L. (U.S.) [1938–]
Computer science

Wheatstone, Sir Charles (England) [1802–1875]
Telegraph

Wheeler, David John (England) [1927–]
Assembler
EDVAC
Initial orders
ORDVAC
Sorting

Whetten, Nathan Rey (U.S.) [1928–]
Computerized Tomography (CT)

Whipple, Richard (U.S.)
Basic

Chronological Index

This is an index of significant dates (year only) that occur in the main entries. All dates before about A.D. 1000 are necessarily approximate. By turning to the cited article the reader can ascertain the event associated with the subject that took place in the year given.

20,000 B.C.
Natural language

3500 B.C.
Abacus

2500 B.C.
Pi (π)
Positional number system

1650 B.C.
Pi (π)

870 B.C.
Zero

375 B.C.
Automaton

350 B.C.
Binary number system

300 B.C.
Euclidean algorithm

290 B.C.
Library

270 B.C.
Automatic control theory

230 B.C.
Number theory
Sieve of Eratosthenes

80 B.C.
Antikythera mechanism

850
Algorithm

1202
Recurrence relation

1275
Lull logic machine

1500
Quipi

1580
Variable

1595
Sector

1614
Logarithm

1617
Napier's bones

1623
Schickard calculator

1629
Slide rule

1637
Analytic geometry

1642
Pascal calculator

1646
Computer

1654
Discrete mathematics

1665
Calculus (Newton)

1673
Calculus (Leibniz)
Leibniz calculator

1701
Binary number system

1714
Taylor series

1725
Jacquard loom

1730
Stirling's approximation

1736
Graph theory

1739
Automaton

1750
Simpson's rule

1752
Prime number

1775
Stanhope calculator

Geographical Index

Geographical Index

Geographical Index

General Index

Note: Entries and page numbers in **bold** denote articles.

General Index

Anagram, 207

Analog computer, **9**, 11, 40, 77, 81, 83, 111, 124, 147, 163, 208, 229, 236

Analog-to-digital converter, 43, 192

Analysis of algorithms, **10**, 50, 81, 113, 123, 158, 185, 221, 239, 245, 260

Analytic Differentiator, 51

Analytic geometry, **10**, 71

Analytical Engine, **11**, 5, 10, 15, 51, 77, 102, 145, 169, 208, 209, 211, 250

AND operation, 33, 135, 159, 186

AN/FSQ series, 228, 257

Annals of the History of Computing, 132

ANSI standard, 38

Antikythera mechanism, **11**, 10

Antilogarithm, 39, 158, 159, 236

AOL. *See* **America Online (AOL)**.

Apache, **12**, 156, 190, 277

APL, **12**, 48, 95, 106, 118, 126, 191, 244

Apollo, Inc., 120, 248

Apple, **12**, 13, 14, 27, 37, 52, 78, 101, 102, 117, 120, 122, 124, 175, 176, 177, 189, 191, 206, 222, 239, 276

Apple I, 12, 13

Apple II, **13**, 9, 27, 48, 81, 137, 172, 175, 176, 189, 191, 200, 217, 242, 275

Apple DOS, 191

Apple G3 / G4, 13

Apple iMac, 150

Apple Lisa, **14**, 8, 13, 176, 275, 278

Apple Macintosh, **14**, 8, 9, 13, 48, 69, 75, 94, 114, 124, 171, 176, 189, 191, 197, 275, 278

Apple Newton, 13, 117, 201

Applet, **15**, 113, 186

Application generator, 23

Applicative language, 106

Applix 1616, 176

APT, **15**, 208

Aptiva, 131

Arcade game, 57

ArcInfo, 112

Argonne National Laboratory, 23, 25, 125

Arithmetic-Logic Unit (ALU), **15**, 28, 41, 42, 73, 119, 132, 137, 176, 184, 193, 273, 281

Arithmetic shift, 232

Arithmometer, **15**, 35, 198

Around the World in 80 Days, 86

ARPA, 16

ARPAnet, **16**, 31, 35, 87, 99, 140, 142, 182, 196, 210, 251, 278

Array, **16**, 4, 12, 71, 72, 88, 94, 99, 108, 133, 134, 167, 176, 208, 221, 232, 233, 250

Array processor, 133, 196

Ars Combinatoria, 160

Art of Computer Programming, The, 148

Artificial Intelligence (AI), **16**, 23, 51, 52, 55, 57, 60, 88, 94, 111, 117, 120, 143, 155, 156, 161, 162, 182, 183, 198, 199, 224, 240, 241, 260, 271

Artificial Life (AL), **17**, 42, 108, 111, 183, 237

ASCC. *See* **Harvard Mark I (ASCC)**.

ASCII, **18**, 33, 37, 43, 71, 84, 206, 225, 246, 265, 275

Ashton-Tate, 275

Assembler, **18**, 64, 103, 128, 141, 155, 161, 163, 210, 266

Assembly language, **19**, 74, 112, 120, 137, 141, 163, 176, 191, 210, 232, 242, 244

Assignment statement, 106

Association for Computing Machinery (ACM), **19**, 6, 59, 136, 148, 186, 229, 237, 267

Association for Shareware Professionals, 231

Associative memory, **19**, 39, 122, 135

Astounding Science Fiction, 224

Asynchronous Transfer Mode (ATM), 195

AT&T, 38, 181, 189

Atanasoff-Berry Computer (ABC). *See* **ABC**.

Atari, **20**, 12, 48, 57, 175

Atlantic Monthly, 168

Atlas, **20**, 167, 179, 191, 270

ATM. *See* **Automatic Teller Machine (ATM)**.

Atomic Energy Commission (AEC), 245

Augmented reality, **20**

AutoCAD, 39

Autocode, **21**, 49

Autodesk, 39

Automata theory, **21**, 43, 44

Automatic control theory, **21**, 66, 106, 147

Automatic Language Processing Advisory Committee (ALPAC), 162

Automatic programming, **22**

Automatic Teller Machine (ATM), **23**, 87, 239

Automatic theorem proving, **23**, 17, 42, 159, 210

Automation, **24**, 60, 154, 224

Automaton, **24**, 55, 199, 224

Auxiliary storage, **25**, 41, 80, 90, 99, 102, 118, 165, 166, 167, 169, 179, 193, 200, 217, 239, 248, 270

AVIDAC, **25**, 125

AVL tree, 260

Axiom, 112, 117, 210

B-0 compiler, 49

B-box, 167

Baby machine, 166, 167

Back up, 169

Backgammon, 57

Backus-Naur Form (BNF), **26**, 7, 46, 95, 204

About the Author

EDWIN D. REILLY is Emeritus Professor of Computer Science at the State University of New York at Albany. He is coeditor of the *Encyclopedia of Computer Science*, 4th ed. (2000) and coauthor of *VAX Assembly Language Programming* (1991) and *Pascalgorithms* (1988).